Transforming Enterprise

Transforming Enterprise

The Economic and Social Implications of Information Technology

edited by William H. Dutton, Brian Kahin, Ramon O'Callaghan, and Andrew W. Wyckoff

The MIT Press
Cambridge, Massachusetts
London, England

This book was set in Sabon by SNP Best-set Typesetter Ltd., Hong Kong.

Printed and bound in the United States of America.

Library of Congress Cataloging-in-Publication Data

Transforming enterprise : the economic and social implications of information technology / edited by William H. Dutton . . . [et al.].
    p.   cm.
Includes bibliographical references and index.
ISBN 0-262-04221-5 (alk. paper) — ISBN 0-262-54177-7 (pbk. : alk. paper)
1. Information technology. 2. Information technology—Social aspects.
I. Dutton, William H., 1947–.
T58.5.T75 2005
303.48′33—dc22

                                                                2004055178

10 9 8 7 6 5 4 3 2 1

Contents

Preface

Brian Kahin

Over the years, research on IT-enabled change has followed changes in technology and the market. In the 1980s researchers looked at the phenomenon of personal computing in work, education, and recreation. In the 1990s, the growth of the Internet raised a much broader set of questions, with different institutional and policy implications. The use of the World Wide Web for advertising, promotion, and, finally, consumer transactions drove household use to mass medium levels, while the emergence of the Web technology as infrastructure for business reframed the problem of linking computers and productivity.

Quickly internationalized but colored by American entrepreneurship, the Web triggered vigorous discourse around principles and policies for "global electronic commerce." Measuring this phenomenon while understanding the social and economic implications became a priority for policymakers as well as businesses. The focus on electronic commerce was overshadowed by a more general research-based case that there was discernible "digital economy." This digital economy embraced the IT sector and extended into every other sector. It served as the principal driver for a "new economy" characterized by sustained productivity gains that defied the traditional boom and bust of the business cycle.

But there was a bust, and it took the wind out of this ever-expanding phenomenon. Yet with the bust comes the possibility of a new set of insights about the essential and ephemeral aspects of all that has gone before.

The exuberance that characterized information technology in the late 1990s is in short supply today. Spam, viruses, worms, and denial of service attacks dominate the news, echoing anxiety about terrorism in the tangible world. Change, such as it is, is led not by start-ups but by the leaders of the old economy. New applications and opportunities are few. Absorption and consolidation are the order of the day. Investment in underlying infrastructure has collapsed. Litigation rises as blame is sought for failure. Moore's law seems to matter less and less, except to game players hungry for realism.

Yet deep processes unleashed by information technology continue to transform human activity. Markets, value chains, firms, transactions, business models, institutions, innovation, collaboration, standardization, trust, community—all have been and are being reshaped by information technology. These changes ripple across the economic fabric within and across sectors. Instead of the category-killer dot-coms in business-to-consumer electronic commerce, the manifestations of change are widely dispersed and often subtle. They twist and turn deep within the firm, in the relationship between firms, or in social interactions.

Transforming Enterprise looks at change from five perspectives that correspond to the sections of the book:

1. What is the impact on the economy as a whole? In particular, how does IT affect productivity? Familiar questions, but we keep learning more about the answers.

2. What are the implications of IT for the creation and use of knowledge, especially new knowledge that leads to real innovation in products and services? Knowledge is driving the economy in many ways, while information technology and the Internet are changing the environment in which knowledge is developed and managed. Yet both knowledge and the changing environment for knowledge are very difficult to define and measure.

3. What are the implications for how enterprise is organized? We see firms working together with new facility and in new ways, thanks to the Internet. Not just one on one, but through interlocking and overlapping networking. Knowledge is shared and marshaled with varying degrees of immediacy and formality within and across boundaries.

4. To get a practical perspective on how these developments and other phenomena play out, we take a close look at how business practice and industry structure have been within particular sectors of the economy. This line of investigation is now well established and benefits from years of results and insight.

5. Finally, we look beyond the sphere of business to the implications for home, community, and society. Here, too, there is now a substantial body of research but also continual change as the capabilities of the home user continue to expand.

There is long history behind the *Transforming Enterprise* project. In 1997, the National Science Foundation and the National Academy of Sciences developed a broad research agenda on the economic and social implications of information technology.[1] OECD did important work bringing the economic and social impact of electronic commerce to the attention of the policy community in 1997–1998.[2] The U.S. Department of Commerce published its defining work on the "digital economy" in 1998.[3] In May 1999, a number of agencies collaborated on "Understanding the Digital Economy," a conference that served as a useful model for mixing industry, government, and academic perspectives and making the best current research available to a large audience.[4]

The sudden rise of electronic commerce and a distinct digital economy demanded attention from a larger policy community concerned with trade, economic growth, and the influence of the Internet on a wide spectrum of communications and information policy issues. The speed with which these phenomena were developing and perhaps transforming the whole of the economy seemed to defy, and in fact discouraged, any well-reasoned public response. After the bust, it became easier to gain a balanced view of the process of transformation, as well as a better understanding of how it recalibrates policy frameworks.

We appreciate the many who contributed to making *Transforming Enterprise* possible. Principal funding was provided by the Digital Society and Technologies Program of the National Science Foundation. The conference was hosted by the Technology Administration's Office of Technology Policy in the Department of Commerce. The Center for Information Policy in the College of Information Studies, University of Maryland, served as the project manager. The other cosponsors, each of which contributed in special ways to the project, include the Information Society Technologies Program of the European Commission, the University of Michigan School of Information, the Center for Research on Information Technology and Organizations at the University of California, Irvine, the Berkeley Roundtable on the International Economy, the Progress and Freedom Foundation, and the Interagency Working Group on Information Technology Research and Development. For further information, the reader is directed to http://transformingenterprise.com.

There is still much to learn, but we have a foundation here for empirically grounded public debate about the future of human enterprise. By strengthening the connections between research and policy, we expect useful research, sound policy, and a better future.

Brian Kahin
Ann Arbor, Michigan

Notes

1. *Fostering Research on the Economic and Social Impacts of Information Technology, Report of a Workshop.* Washington, D.C.: National Academy Press, 1998. http://www.nap.edu/readingroom/books/esi/. See also National Science Foundation, Division of Science Resources Studies, *Social and Economic Implications of Information Technologies: A Bibliographic Database Pilot Project,* http://srsweb.nsf.gov/it_site/index.htm.

2. OECD, *The Economic and Social Impact of Electronic Commerce: Preliminary Findings and Agenda,* 1998, http://www.oecd.org/dataoecd/3/12/1944883.pdf.

3. This includes the series of internally produced reports beginning with Lynn Margherio et al., *The Emerging Digital Economy,* Departments of Commerce, Economics and Statistics Administration, 1998, http://www.esa.doc.gov/TheEmergingDigitalEconomy.cfm.

4. Archive at http://www.technology.gov/digeconomy/; Erik Brynjolfsson and Brian Kahin, *Understanding the Digital Economy,* MIT Press, 2000.

Introductory Essays

Technological Innovation in Organizations and Their Ecosystems

Ramon O'Callaghan

With the massive adoption of ICT by organizations in the past decade, business managers, academics, and policy makers are addressing questions such as "Is ICT adoption really transforming the enterprise?" "If so, how should 'transformation' be effectively managed?" "What are the implications for future research and management practice?"

In an increasingly networked environment, ICT requires a different governance approach since ICT infrastructures need to be conceived, deployed, and managed across organizational boundaries. ICT stakeholders are part of an ecosystem that involves suppliers, partners, competitors, customers, systems providers, institutions, and government. Today the locus of competence and innovation is moving away from the company to the ecosystem, and value is being cocreated by the different actors in the network. The implications for management and research are many, and they are significant.

Adoption of ICT Innovations and Organizational Change

Many organizations adopt ICT not just to do the same old things more efficiently but also to do things differently (process innovation) or do new things (product innovation). Thus, ICT adoption often implies innovation and change. But what drives organizations to adopt an innovation in the first place? Innovations are adopted primarily on the basis of some expected organizational benefits. Research has shown, however, that compatibility and complexity also influence the likelihood of adoption of an innovation (Tornatzky and Klein, 1982; Rogers, 1983). This suggests that the specific organizational context in which a new ICT is introduced determines the extent to which it is compatible with existing practices. Likewise, the level of ICT literacy and experience of the potential adopters will determine the perceived complexity of the ICT application. Thus, the factors influencing adoption are not only the features of the ICT being adopted (the "attributes of the innovation") but also the characteristics of the organizational context (such as people, processes, systems, structure, or culture).

An innovation's adoption is also affected by peer pressure attributed to the "diffusion effect," that is, the cumulatively increasing degree of influence upon an individual or an organization to adopt or reject an innovation, resulting from the activation of peer networks in a social system (Rogers, 1983). In a business environment, the diffusion effect translates into competitive pressure. Firms may end up adopting because of perceived competitive necessity (or even sheer imitation) rather than as a result of a cost/benefit assessment. That was certainly the case with many e-commerce projects before the "dot-com" bubble burst. ICT adopted on this basis yielded little or no return on investment.

Technology diffusion is also a knowledge management process. As the cumulative number of adopters increases, the collective knowledge about the technology also increases, and this influences the adoption decision. The key factor is knowledge rather than peer pressure. This is particularly true of complex technologies where adopters face "learning by using" (Rosenberg, 1982). This perspective views technology diffusion in terms of organizational learning, skill development, and knowledge barriers (Attewell, 1992). Firms may delay adoption of complex technologies until they obtain sufficient technical know-how to implement and operate successfully. Know-how and organizational learning are thus potential barriers to adoption. When knowledge barriers are overcome, diffusion speeds up.

Successful technology implementations, however, require both behavioral changes and organizational changes. The organizational change literature views implementation of ICT as a dynamic process involving the introduction of a technical change into an existing social system. From this perspective, organizations need to manage the introduction of technology as a change strategy (Beer, 1980; Kotter et al., 1979), and this requires management commitment, political support, and resources. This perspective suggests that effective IT implementation requires the alignment of the technology and the organization that operates it.

Alignment, in practice, can be achieved essentially in three ways (McKersie and Walton, 1991). Some of the enabling conditions may already exist or can be developed in anticipation of the introduction of the ICT system, and the organization can "pull" the technology into place by the users rather than push it into place by managers. Alternatively, managers can design the new technology and the operating organization at the same time. This pattern has the advantage of allowing mutual adaptation of the technical and social subsystems of an ICT implementation (Leonard-Barton and Kraus, 1988). In a third option, management focuses exclusively on implementing the technology, letting the technology drive subsequent organizational adaptation. In this case, a firm moves ahead with the introduction of ICT, leaving existing organizational arrangements in place, and subsequently attends to organizational changes on a responsive or adaptive basis. In conclusion, ICT adoption is intrinsically associated with organizational change, and therefore an effective implementation requires necessarily good change management.

From Organizational Change to Business Performance: Productivity Paradox or Management Paradox?

Research on the link between organization design and business performance has a long tradition (e.g., Chandler, 1962; Thompson, 1967; Galbraith, 1977; Caves, 1980; Quinn, 1980; Porter, 1985). In the past decade, the role of ICT as enabler of organizational design and organizational transformation became a topic of interest in both the information systems literature as well as the general management literature (Hammer, 1990; Scott-Morton, 1991; Davenport, 1993; Hammer and Champy, 1993). By redesigning the way existing business processes were performed and using ICT to enable new ones, some organizations were able to achieve significant improvements in key business drivers, such as cost, quality, service levels, or lead times.

Yet these successes did not seem to make an impact on productivity figures at the macroeconomic level. Robert Solow's famous quip that "You can see the computer age everywhere but in the productivity statistics" provoked a great deal of debate. If IT investments do not yield any clear advantages, why do so many organizations continue to invest heavily in IT? The suggestion that IT does not bring benefits to organizations seems to go against intuition and common sense. Subsequent research has tried to explain away the "IT productivity paradox" (e.g., Brynjolfsson and Hitt, 1998; Willcocks and Lester, 1999; Triplett, 1999).

A closer look at the IT productivity paradox reveals that it has several facets, depending on the level of analysis (Pilat and Wyckoff, in this book). Traditionally, the statistics at the macroeconomic level have been inconclusive. Increases in IT investment spending levels in some developed countries have coincided with a decrease in the productivity growth. Industry-level productivity figures suggest that the apparent decrease in productivity growth is largely due to a limited growth in productivity of office work in the service industry, which in turn has had the highest IT spending levels.

At the organizational level things look different. Case studies show that some organizations have been able to derive large benefits through IT (e.g., Wal-Mart, Dell Computer, Charles Schwab). Pilat and Wyckoff, as well as Brynjolfsson and Hitt (also in this book), show that the use of ICT is positively linked to firm performance. Other studies reveal substantial differences between organizations that utilize IT in a successful versus an unsuccessful way (Brynjolfsson and Hitt, 1998). This variation may be due to differences in organizational conditions such as ICT experience, top management commitment, and organizational politics (Weill, 1990; Strassmann, 1990).

The IT productivity paradox should not be a disquieting problem for managers. After all, there seem to be many opportunities for individual organizations to use IT in innovative and profitable ways. The question for managers is not whether IT pays off in general but what IT applications should be deployed in their respective organizations. In the end, the difference between IT success or failure may well be the ability to evaluate the benefits and strategic potential versus the cost and risks of proposed IT investments, and having the right management processes in place to plan and execute IT projects.

Value Creation and IT Governance

Over the years the focus of IT management has been shifting from efficiency-related issues to the question of how to deliver business value with

IT. In the early days, IT managers were only asked to complete IT projects within time and budget constraints. This may have worked in a period when IT merely had automating effects, which could easily be justified on the basis of cost savings, but it has become unacceptable today, when investments have "transforming" effects, e.g., improving quality, flexibility, and the innovation ability of organizations.

Research findings suggest that good evaluation and decision-making practices might well contribute to the ultimate value to be gained from investments (Weill and Olsen, 1989; Willcocks, 1994). Firms that better assess what they expect from their ICT projects and also manage ICT investments from this perspective seem to be more successful than others that do not have formal evaluation procedures in place.

The organizational structures and processes to ensure that ICT delivers value and is aligned with the strategic goals of the firm are known as "ICT governance." ICT governance brings together the different stakeholders to assess and take responsibility for ICT projects. The explicit attention to investment evaluation and stakeholder involvement helps create a shared vision and generate commitment to the business outcome.

Stakeholders, however, have personal views on the desirability of each project. Conflicting interests may complicate cooperation, and communication problems may arise from differences in background and expertise. ICT decision making will always have a political nature to some extent (Markus, 1983). Research on ICT governance should target the difficulties that such a process faces and the ways to overcome them. These lie in the areas of benefits assessment and management, cost analysis, risk management, and, also, stakeholder communications and organizational politics (Renkema, 1999).

Transcending Organizational Boundaries

Although effective management of organizational change is imperative for achieving IT benefits, the cases of leading companies (e.g., Wal-Mart, Dell, and others) also indicate that the locus of change and innovation is no longer confined within the boundaries of the firm. Some of the most dramatic changes, in fact, have taken place at the level of supply chains or business networks or on an industry level.

Kraemer and Dedrick (in this book) analyze the transformation of the PC industry, which since the mid-1990s has been using direct sales channels, demand-driven production, and modular production networks. Within these networks, firms are flexible in designing value chains for different products and markets, with each firm selecting a different mix that takes into account its own capabilities and strategies. The structure of the industry's global production network changed, making it possible to coordinate design, production, and logistics on a regional or global basis. As a result, PC makers have been able to locate these activities where costs are low and key skills are available, or close to major markets. Also, the use of IT, the Internet, and e-commerce have enabled and supported the shift from supply-driven to demand-driven production and the creation of more flexible, information-intensive value chains to support this complex process. This change has led to dramatic reductions in inventory, better use of assets, and leaner operations throughout the industry.

Kraemer and Dedrick conclude that the sources of competitive advantage in the new IT-enabled organization are the substitution of information for inventory, better matching of supply and demand, and the ability to tap into external economies in the global production network. External economies can be accessed by any firm, but demand-driven organizations are best positioned to take advantage of these economies because they can use real-time information to drive the production network in response to demand changes.

In another chapter of this book, Boy Lüthje examines new models of outsourced manufacturing (contract manufacturing and electronics manufacturing services) in globalized production networks in the electronics industry. He analyzes the interaction of new information networks with the restructuring of production, work, and the global division of labor. He concludes that information technology is not the driver of organizational change per se but part of a complex shift in the social division of labor that ultimately is related to the demise of vertically integrated mass manufacturing. In this context, information technology and Internet-based models of supply-chain management do facilitate vertical specialization.

Lüthje also raises the question of network governance. He discusses the trend in centralization of supply-chain management in electronic components. The issue is how to orchestrate complex networks of corporate actors and their interaction in global marketplaces. This suggests that further research should address coordination and regulation issues in networks, as well as the role of standards and institutions.

Toward Customer-Centric Ecosystems

As discussed earlier, one of the most significant industry-level transformations has been the shift from supply-driven to demand-driven value chains. This "reversal" of the value chain together with the Internet empowers consumers in ways that were unimaginable just some years ago. Consumers can create virtual communities and engage in an active dialogue with manufacturers of products and services. At the same time, consumers constitute a source of knowledge that companies can exploit. This transforms the traditional notion of "core competence" (Prahalad and Hamel, 1990). Competence now becomes a function of the collective knowledge available in the ecosystem, i.e., an enhanced network comprising the company, its suppliers, its distributors, its customers, its partners, and its partners' suppliers and customers.

From a research viewpoint, this implies that the unit of analysis needs to shift from the (extended) enterprise to the larger ecosystem. The firm becomes a node in the enhanced network of organizations in the ecosystem. The implication for business managers is that this ecosystem provides opportunities to address customer needs in a unique way and allows value chains to be redesigned around the consumer. In this customer-centric approach, firms are no longer producers of products or services but (co)developers of customer experiences. And customers can play an active role in this.

In order to harness customer competence, companies have to engage customers in an active, explicit, and ongoing dialogue, mobilize consumer communities, manage customer diversity, and cocreate personalized experiences with customers (Prahalad and Ramaswamy, 2000). Organizations that can "sense and respond" rapidly by moving information to mobilize resources and knowledge in the network will emerge as the "winners" in the network economy (Bradley and Nolan, 1998; Kraemer and Dedrick, in this book).

Managing customer experiences is about managing the interface between a company and its customers. Products and services will have to be more "intelligent" and adapt themselves to changing users' needs, not the other way around (Prahalad and Ramaswamy, 2000). The method by which customers and companies communicate is also an integral part of creating an experience. Companies will have to manage—and integrate—several distribution channels and ensure that the quality of fulfillment and the personalized experience are consistent across channels. The challenge will be to develop the infrastructures needed to support such a multichannel, multipartner network.

The Enabling ICT Infrastructure

Extended enterprises, supply chains, business ecosystems, and industrial clusters are complex systems based on the networking of organizations, the cooperation of the players, and flexible access to resources. They are communities that share business, knowledge, and infrastructure in a highly dynamic way. In order to enable the network of partners to collaborate, sense consumers' needs, and deploy resources rapidly, a new technology infrastructure is required.

Such infrastructure calls for a dynamic aggregation of network and software services to facilitate interorganizational interactions. The key elements of the infrastructure are software components and agents that show evolutionary and self-organizing behavior, i.e., they are subject to evolution and to self-selection based on their ability to adapt to the local business requirements (Nachira, 2002). Future research should focus on network architectures that are pervasive, adaptive, self-configuring, and self-healing.

Additional approaches may include modeling. In another chapter in this book, Nagurney introduces the concept of "supernetworks," that is, networks that are "above and beyond" existing networks. Her supernetwork framework captures decision making by economic agents (e.g., consumers, producers, and intermediaries) in the context of today's networked economy. The decisions often entail trade-offs between the use of physical versus communication networks. Such a framework can be used to model the behavior of individual decision makers as well as their interactions in the complex network system.

The Death of Distance?

Much has been made of the potential of ICT to enable a despatialization of economic activity. Cairncross (1997 and 2001) posits that, with the introduction of the Internet and new communications technologies, distance as a relevant factor in the conduct of business is becoming irrelevant. She contends that the "death of distance" will be the single most important economic force shaping all of society over the next half century.

Despite the bold predictions, however, geography and location still matter. Porter's identification of local agglomerations, based on a large-scale empirical analysis of the internationally competitive industries for several countries, has been especially influential, and his term "industrial cluster" has become the standard concept in this field (Porter, 1998, 2001). Also, the work of Krugman (1991, 1996) has been concerned with the economic theory of the spatial localization of industry. Both authors have argued that the economic geography of a nation is key to understanding its growth and international competitiveness.

Clusters facilitate the transmission of knowledge—particularly tacit knowledge, which cannot move as freely or easily from place to place as codified knowledge. Research on ecosystems should thus include a knowledge management perspective. In another chapter in this book, Mason and Apte provide a model for how knowledge transforms enterprises. Further research could focus on the potential contribution of ICTs that gives priority to socially mediated tacit skill sets and learning processes as prerequisites for the effective use of these technologies within a complex adaptive system.

The use of diverse combinations of ICTs within and between clusters is likely to have implications for the meaning of proximity. In traditional clusters, the need for physical proximity has led to regional agglomerations. But how will new ICTs affect traditionally perceived needs for physical proximity and introduce "virtual" proximity as a complement to physical proximity? Can "virtual" clusters be expected to emerge and/or develop, in part, as a result of the widespread application of ICTs? What combinations of physically proximate and "virtual" arrangements best augment the social and economic performance of networked clusters? Future research should look at clusters as geographically proximate complex organizational systems of learning and economic and social activity that are globally networked and enabled by the effective use of ICTs (see O'Callaghan, in this book).

Conclusions and Research Directions

ICT is the tool that helps design, change, innovate, and learn in the emerging models of the network economy. By examining ICT-enabled transformation from various perspectives, the authors in this book illuminate the processes of adoption and diffusion of ICT and the role that ICT plays in organizational design, process change, knowledge management, and value creation. It has implications for management, policy, and research.

Future research on ICT-enabled transformation should start by changing the unit of analysis and focus on understanding the new (inter)organizational forms, the new drivers of value, the new ICT infrastructures, and the new governance approach. The traditional business paradigm revolved around the firm. The new paradigm regards the firm as a node in an ecosystem—a network of partners that collaborate to create customer experiences and intelligent products that can adapt themselves to evolving customers needs.

Research should address emerging issues such as, How is value created and apportioned among the players in the ecosystem? How do companies in the ecosystem establish a dialogue with consumers? How can ICT be used to integrate the different partners seamlessly and provide unique customer experiences? How can resources be deployed rapidly to respond in real time? How does the network deal with complexity? What governance approach should be put in place to deploy and manage interorganizational ICTs when the stakeholders are a constellation of partners interacting in the ecosystem? What institutional arrangements will foster or impede the development and effectiveness of the new systemic models? What are the policy implications?

References

Attewell, P. (1992) "Technology Diffusion and Organizational Learning: The Case of Business Computing." *Organization Science* 3(1), February.

Beer, M. (1980) *Organizational Development and Change: A Systems View.* Santa Monica: Goodyear Press, pp. 45–69.

Bradley, S. P., and R. L. Nolan, eds. (1998) *Sense and Respond: Capturing Value in the Network Era.* Boston: Harvard Business School Press.

Brynjolfsson, Erik, and Lorin M. Hitt (1998) "Beyond the Productivity Paradox: Computers Are the Catalyst for Bigger Changes." *Communications of the ACM* 41(8): 49–55.

Cairncross, Frances (1997) *Death of Distance: How the Communications Revolution Will Change Our Lives and Our Work.* Boston: Harvard Business School Press.

Cairncross, Frances (2001) *The Death of Distance 2.0: How the Communications Revolution Will Change Our Lives.* London: Texere Publishing.

Caves, R. (1980) "Industrial Organization, Corporate Strategy and Structure." *Journal of Economic Literature* 18 (March): 64–92.

Chandler, A. D. (1962) *Strategy and Structure: Chapters in the History of the American Industrial Enterprise.* Cambridge, Mass.: MIT Press.

Davenport, T. H. (1993) *Process Innovation: Reengineering Work through Information Technology.* Boston: Harvard Business School Press.

Galbraith, J. R. (1977) *Organizational Design.* Addison Wesley Publishing Co.

Hammer, M. (1990) "Reengineering Work: Don't Automate, Obliterate." *Harvard Business Review,* July–August.

Hammer, M., and J. Champy (1993) *Reengineering the Corporation: A Manifesto for Business Revolution.* Harper Business.

Kotter, J., L. A. Schlesinger, and V. Gathe (1979) *Managing the Human Organization.* Boston: Harvard University.

Krugman, P. (1991) *Geography and Trade.* Cambridge, Mass.: MIT Press.

Krugman, P. (1996) "The Localization of the Global Economy." In *Pop Internationalism,* ed. P. Krugman. Cambridge, Mass.: MIT Press.

Leonard-Barton, D., and W. A. Kraus (1988) "Implementation as Mutual Adaptation of Technology and Organization." *Research Policy* 17.

Markus, L. M. (1983) "Power, Politics and MIS Implementation." *Communications of the ACM* 26.

McKersie, Robert B., and Richard E. Walton (1991) "Organizational Change." In *The Corporation of the 1990's: Information Technology and Organizational Transformation,* ed. Michael S. Scott-Morton. Oxford University Press, pp. 245–277.

Nachira, F. (2002) "Towards a Network of Digital Business Ecosystems Fostering Local Development." European Commission, IST Programme, Brussels. http://www.nachira.net/de/docs/discussionpaper.pdf

Porter, Michael E. (1985) *Competitive Advantage: Creating and Sustaining Superior Performance.* New York: The Free Press.

Porter, M. (1998) "Clusters and the New Economics of Competition." *Harvard Business Review* 76(6): 77–90.

Porter, M. (2001) *Clusters of Innovation: Regional Foundations of U.S. Competitiveness.* Washington, D.C.: Council on Competitiveness.

Prahalad, C. K., and G. Hamel (1990) "The Core Competence in the Corporation." *Harvard Business Review*, November–December, pp. 79–91.

Prahalad, C. K., and V. Ramaswamy, (2000) "Co-opting Customer Competence," *Harvard Business Review*, January–February, p. 79.

Quinn, J. B. (1980) *Strategies for Change: Logical Incrementalism.* Homewood, Ill.: Irwin.

Renkema, Theo J. W. (1999) *The IT Value Quest: How to Capture the Business Value of IT-Based Infrastructure.* Chichester, England: John Wiley & Sons.

Rogers, E. M. (1983) *Diffusion of Innovations*, Third Edition. New York: The Free Press.

Rosenberg, N. (1982) *Inside the Black Box: Technology and Economics.* Cambridge University Press.

Scott-Morton, M. S., ed. (1991) *The Corporation of the 1990's: Information Technology and Organizational Transformation.* Oxford University Press.

Strassmann, Paul A. (1990) *The Business Value of Computers.* New Canaan, CT: Information Economics Press.

Thompson, J.D. (1967) *Organizations in Action.* McGraw-Hill.

Tornatzky, L.G., and R.J. Klein (1982) "Innovation Characteristics and Innovation Adoption-Implementation: A Meta-Analysis of Findings." *IEEE Transactions on Engineering Management* EM-29.

Triplett, Jack E. (1999) "The Solow Productivity Paradox: What Do Computers Do to Productivity?" *Canadian Journal of Economics* 32(2): 309–334.

Weill, P. (1990) *Do Computers Pay Off? A Study of Information Technology Investment and Manufacturing Performance.* Washington, D.C.: ICIT Press.

Weill, P., and M. Olsen (1989) "Managing Investment in Information Technology: Mini Case Examples and Implications." *MIS Quarterly*, March, 2–17.

Willcocks, L. (1994) "Introduction: Of Capital Importance." In *Information Management: The Evaluation of Information Systems Investments*, ed. L. Willcocks. London: Chapman & Hall.

Willcocks, L., and S. Lester, eds. (1999) *Beyond the IT Productivity Paradox.* Chichester, England: John Wiley and Sons.

Continuity or Transformation? Social and Technical Perspectives on Information and Communication Technologies

William H. Dutton

The Importance of Social Research on Technology

New and continuing trajectories of rapid advances in information and communication technologies (ICTs) have created an array of new digital products, services, and media, from the cell phone to the Internet and World Wide Web. The significance of these innovations across all sectors of society made the Internet and other ICTs key defining technologies of the late 20th century.

This book is anchored in an assumption that the Internet and related ICTs are transforming the way the world communicates, works, and learns. Nevertheless, the implications of these technical innovations are the subject of tremendous contention. There are many who argue that the Internet has had little or no impact on major areas of business, community life, governance, education, and other sectors (Woolgar 2002; Robins and Webster 2002). They see continuity rather than change. And even among those who maintain that the Internet is a force for change, there are countervailing interpretations of the nature of that change.

For example, some observers emphasize how the technology will support centralized control and coordination of global enterprises (Beniger 1986), while others point to its potential for devolved empowerment (Castells 2000; Lessig 2002). Similarly, there have been claims that the Internet has increased social isolation and undermined civic life, while others find it has strengthened existing communities centered around locality and created new forms of virtual community built on shared interests (Katz and Rice 2002; Wellman and Haythornthwaite 2002; and Wellman and Chen this volume).

The book examines the transformational potential of digital technologies from a variety of economic, technological, organizational, and community perspectives. All these are rooted in social contexts that influence the people making the multitude of decisions that interact to determine outcomes of the design, production, dissemination, use, and consumption of ICTs (Dutton 1996).

Perspectives on current and future societal implications of 21st century ICTs are grounded in studies of computers and society that started in the 1950s. These have subsequently expanded to take account from a growing number of disciplinary vantage points of the rapid advances and growth in ICT capabilities and use. This broadening of use, particularly around the Internet and Web, has made ICTs an intrinsic aspect of everyday life in many societies, spawning increased studies of the technology's implications for communities and society.

As a background and introduction to the contributions of this book, this chapter outlines four major research traditions focused on the transformational potential of ICTs that have been developed since the 1950s. It highlights their insights and limitations and suggests that there is a need for fresh approaches, as undertaken elsewhere in the book.

Influence and the Mass Media

An early perspective of social research on ICTs evolved after the second World War from studies of propaganda and the political implications of newspapers, radio, and other mass media. Anchored in sociology, political science, and later in communication studies, this influence perspective created many useful models of media effects, such as "agenda-setting," which highlights the critical role in shaping access to news of journalists, publishers, and other media "gatekeepers." The focus of such research was generally on the content of messages delivered through the media and their influence on those exposed to them, directly or indirectly (Katz and Lazersfeld 1955).

As awareness grew that all communication media are based on the technologies that have converged into ICTs, researchers increasingly looked at ICTs within this media "influence" tradition (Rogers 1986; Dutton et al. 1987). Some have asked whether the interactive character of emerging media, such as digital and Web-TV, would make them more engaging and, therefore, more powerful in shaping attitudes, beliefs, and values. Others have focused on how the profusion of media and channels can segment audiences in ways that might erode their quality and integrative effect in providing the common experiences or "shared text" of a community.

However, the ICT-based technical change that is reconfiguring access to audiences has challenged the very idea of "mass media" that dominated communication industries and media research throughout the 20th century. The influence perspective becomes even less applicable outside media studies, such as in assessing the implications of ICTs for the economy.

Marshall McLuhan anticipated this problem in the early 1960s, when he argued that there had been too much emphasis on the content of the message. In claiming "it is the medium that shapes and controls the scale and form of human association and action," McLuhan (1964: 9) indicated his belief that television's ability to reconfigure access to messages was more significant than whatever message was conveyed. By the 1980s, widespread recognition by mass media scholars of the emergence of so-called "new media" led many communication researchers to revisit McLuhan's work, since he provided a way of thinking beyond the messages conveyed by mass media to the societal implications of the transformation under way in the very media of communication, which was being ushered in by cable and satellite as well as computer-based communications. Just as McLuhan argued that satellite communication enabled global reach and that this shift in access to audiences was more important than the messages conveyed, I and others have tried to focus attention on the ways ICTs can reconfigure access to people, information, services, and technologies (Dutton 1999 and this volume).

"Social Impact" versus "Social Shaping" Perspectives

In parallel with the rise of a media effects tradition, but in the fields of engineering and computer sciences, initially, a critical literature emerged around the social impacts of technology. Widespread fascination with the technical ingenuity and growing capabilities of ICTs has been a consistent focus of attention within this literature: from the emergence of the "computer age" in the 1950s and '60s, through the "microchip revolution" in the 1970s, to the "superhighway" and "information society," on to future "e-everything" digitized globalization. This focus led to a strong technologically deterministic viewpoint that suggests changes tied to such innovation follow a technological logic that is, to some extent, independent of human will. It also points to technological change as leading to "social impacts" that are on a predetermined trajectory propelled by the logic of the technology and the industries producing it (Winner 1986; MacKenzie 1999: 39).

With ICTs, these impacts have been viewed mainly in terms of conflicting utopian versus dystopian perspectives. For example, the classic dystopian novel of the 20th century, George Orwell's *1984*, pivoted around the emergence of two-way television as an instrument of propaganda and electronic surveillance. The growing centrality of the Internet and Web has generated updated Orwellian visions of multimedia surveillance, together with new fears that ICT networks and services are isolating individuals, undermining democratic processes, and destroying jobs. On the other hand, many influential protagonists have argued that the Internet and Web promote global community, a more participatory and open "electronic democracy," and an information economy bringing fulfillment, empowerment, and new kinds of work and leisure (Negroponte 1995; Gates 1995). Still others dismiss the importance of ICTs, focusing instead on underlying social and economic forces (Woolgar 2002).

In practice, however, technological change usually exhibits a two-edged thrust, which is an underlying reason for the continuing social tensions as well as benefits surrounding technical innovations. For instance, the use of ICTs as part of wider social and economic changes has led to major shifts in patterns of employment, with jobs disappearing in some traditional industries, particularly manufacturing, while others are created, especially in services (Freeman 1996).

The industry-wide reverberations of such innovations are fueled by the technological convergence at the heart of ICTs. For instance, the growing cost effectiveness, versatility, and ubiquitousness of ICTs has changed the cost structures of telephone, film, television, and other industries in ways that make their sustainable outcomes uncertain (Gilder 1994; Dutton 1999).

Dystopian and utopian scenarios appear highly plausible because they are often based on the technical features and capabilities of ICTs that are seen to have "social impacts" as drivers of change. However, critics of the social impacts perspective argue that the outcomes of technical-based innovation are shaped by complex social and political choices, not just technology, which means the actual implications of the development and use of ICTs are unpredictable, for example, in relation to issues such as privacy, freedom of speech, and employment (MacKenzie and Wajcman 1985; Dutton 1996). Yet, much of the earliest research into social aspects of ICTs was led by specialists involved directly in developing the technology and so adopted the view that social impacts followed technologically driven trajectories.

The social-impact approach was usually based on rational forecasts of the social opportunities and risks created by particular features of the technology, such as the way a computer's storage capacity enables organizations to create huge databanks of information about individuals. Yet, empirical research usually found that rational expectations based on such capabilities were seldom realized (Kling 2000). For instance, computers did enable organizations to create huge databanks containing personal information, but managers and professionals tended to use them within prevailing data processing paradigms that aimed simply to improve the efficiency of existing practices (Westin and Baker 1972). It took decades before the emergence of new ICT paradigms changed information collection in ways that opened potentially major new threats to privacy and surveillance (Lyon 2001).

The perils of predicting impacts of new ICT capabilities when they are being shaped in the unpredictable cauldron of social and organizational life is illustrated by the bursting of the dot-com bubble at the turn of the century. The key lesson from this is that technical and engineering feats do not in themselves translate into successful innovations in real social settings. The crash was the result of poor business and financial decisions, not a failure of the technology, as indicated by the continuing growth in the use of the Internet for e-business and other "e-everything" activities. It also indicates that the technological-determinist "social impact" model of ICT innovation needs to be rethought, since it is an oversimplified depiction of how outcomes are determined.

The implications of the way factors related to the geography of space and place can be facilitated or constrained by access to ICTs has also led to predictions that haven't been fulfilled in practice. For instance, there is evidence that the use of ICTs can actually make location more rather than less important, as had been thought originally (Goddard and Richardson 1996).

The poor track record in forecasting the implications of technical change led to fundamental criticisms of the simplified, linear models of cause and effect used by the technological determinist. Alternative "social shaping" approaches to research on the design and implications of technologies have therefore emphasized ways of exploring the broader individual and social processes and choices that determine the designs and outcomes of innovation (Williams and Edge 1996; Kling 2000). In examining the Internet and society, this broader perspective is vital because the Internet touches so many everyday social, organizational, and individual activities. Such approaches also recognize the centrality of users and consumers as well as developers and producers as active participants in shaping the designs and outcomes of technical innovations.

That said, the design and development of ICTs are not neutral. Advances in ICTs reshape access, but they do not determine access. Technology, like policy, tells us how we are supposed to do things and makes some ways of doing things more rational and practical than others. Also, the biases designed into technological artifacts and systems can be even more enduring than legislation (Winner 1986: 19–39; Lessig 1999), as well as creating a social and technical momentum that is difficult to reverse (Hughes 2000). For instance, the rise of e-mail is biased toward speeding up interpersonal communication. Individuals can choose to slow it down, but this often takes a conscious personal effort and strong organizational will to overcome expectations created by Internet speed.

Since technology also encompasses the knowledge essential to its use, its control is bound up with issues of who has access to the skills, equipment, and know-how essential to design, implement, and use it. Changes in technology can restrict access to all these resources but can equally expand access, for example, by the way simple Web user interfaces have given easy and quick access to worldwide stores of information. Likewise, a social choice such as the decision to learn a new human or computer language affects access to technology, jobs, and people.

These perceptions have led to a vision of centralized, draconian social control, perhaps through an unelected high-tech elite, as a significant theme of social research on ICTs (and mass media). However, some social researchers observing more recent advances argue the opposite, particularly in relation to the Internet's ability to place more control in the hands of users (de Sola Pool 1983; Lessig 1999).

The "Information Society" Thesis

By focusing on the technology, social-impact research naturally gave high priority to the principal functions for which ICTs were designed: processing and communicating information. For example, Bell (1999 [1973]) wrote a seminal work on the "postindustrial" information society. He saw information as the key economic resource in this new era, not raw materials or financial capital as in earlier agricultural and industrial societies.

The most significant trend toward an information society was the shift in the majority of the labor force from agriculture and manufacturing to services, largely through the growth of "information work." This includes a broad array of jobs related to the creation, transmission, and processing of information, ranging from programmers and Web designers to teachers, researchers, consultants, journalists, game designers, and call-center staff. Increasing importance has also been attached to knowledge created from information and to the power shifts involved in the growth of a knowledge elite who understand how to work with data, knowledge, information systems, simulation, and related analytical techniques.

These conceptions of the information society have increasingly defined the public's understanding of social and economic change tied to ICTs. However, the changes in work, technology, and power posited by this thesis have all been challenged (Miles 1990; Freeman 1996; Castells and Himanen 2002), as has the general identification of information as the key resource of the economy (Robins 1992). The emphasis given to the information sector has also risked shifting attention away from other still-important parts of the economy, such as agriculture, manufacturing, and healthcare.

Information is a highly variable currency, but it is not always wanted or valued. It can lose its value, as for a day-old newspaper, or retain or enhance its value over time, as with a literary or film classic. If defined narrowly as being about "facts," the term "information" becomes far too limiting as a depiction of the social role of ICTs. But if defined very broadly, for instance, as anything that "reduces uncertainty," then information seems to be so all-encompassing that it becomes virtually meaningless. Moreover, information can create, rather than reduce, uncertainty. For example, those more informed about a topic are often less certain about its properties than many who are less informed (MacKenzie 1999). And information is not a new resource, as has frequently been suggested. It has been important in every sector of the economy throughout history (Castells 2000 [1996]). What is new is how choices in the design and use of emerging ICTs are reconfiguring access to information, as well as to people, services, and technologies (Dutton 1999 and this volume).

Another problem with the information society thesis is a bias toward the long-term analyses of its underpinning notion of a staged progression of an economy: from agricultural to industrial, information, and whatever comes next, such as a "knowledge society." An alternative forecast of social and economic evolution is the idea of long economic development cycles caused by successive waves of technological revolutions (Freeman 1996). This argues that the invention of a new technology, like the steam engine or computer, can have applications across the economy that affect many facets of our lives, well beyond what might be labeled as information work.

Both the staged and cyclical theories are attractive because they promise some level of predictability. However, there is no consensus on the identification and validation of these stages or cycles. A long lag of one to two decades exists between the invention of a radically new technology and its impact on the way things are done, since it takes time for people to change habits and beliefs before accepting a new paradigm for how they do things. But such lags do not always eventually yield to progress or a new stage of development.

The "e-society" and other popular conceptions of an information society or age might therefore capture the social significance of the increasing centrality of ICTs such as the Internet but fail to provide adequate insights into its role in social change.

Management and Policy Strategies

A focus on the goals and strategies of actors, instead of the capabilities of the technology, has been most fully developed within the management field. In the early decades of its use from the 1950s, the computer was seen by management theorists and many practitioners as an extension of, and means for realizing, the prevailing management paradigms of the times. And empirical research indeed discovered that those who control decision making tend to adopt and use ICTs in ways that follow and reinforce existing centralized or decentralized patterns of control within organizations and society (Danziger et al. 1982).

From this management strategy perspective (Danziger et al. 1982), the major implications of ICTs grow from the goals and strategies of managers, enabled rather than determined or shaped by technological characteristics. That view remains central to contemporary management research, but often within a management paradigm placing less emphasis on the value of top management control and more on the virtues of innovation and networking (Castells 2000 [1996]). This is illustrated by a growing literature on the value of networks in reconfiguring organizational structures and processes (see Damaskopoulos, O'Callaghan, Nagurney, all in this volume).

All the perspectives highlighted in this chapter demonstrate the potential strategic significance of ICTs across all sectors of society, particularly given the technology's malleability in meeting human, social, and organizational needs in different contexts. Also a strategic perspective moves beyond the constraints of technologically determinist viewpoints by recognizing that technologies are inherently social, in that they are designed, produced, and used by people. Moreover, their design and use involves know-how, which is itself a social attribute (MacKenzie and Wajcman 1985: 3). Technologies are also social in that they define, but do not determine, how people do things, thereby making some paths more economically, culturally, or socially rational than others.

The outcomes of strategic "digital choices" can influence social behavior long after the choices are made. Too frequently, however, choices about the design and use of ICTs are delegated to others as if they had no social significance. Yet such choices about the Internet and other ICTs reconfigure access in ways that have consequences for the communicative power through which enterprises and society at large can be transformed, and transform for better or worse (Dutton, this volume). This reconfiguration process is shaped by the actions and interactions of many actors in many arenas involving ICT-based innovation. For instance, a specialist might be pursuing a technically elegant network design, while a top manager is primarily seeking cost reductions. This places major constraints on the predictability of outcomes based on assessments of strategic aims, unless the varied goals of different actors in the broader "ecology of games" is well understood and orchestrated (Dutton 1999).

Conclusion: Time to Think Again about Transformation through ICTs

There are other perspectives on social transformation, such as philosophy of technology and cultural studies of communication, but these can be viewed as themes within these broad approaches to thinking about technology and society. Ideas such as the information society and management strategies are important because they shape views about the way the world works and, thereby, influence the decisions of individuals, firms, and governments. That is a reason why alternative perspectives on the role of the Internet in society are more than competing theories. They are also ideas that can shape decisions in everyday life, and in once-in-a-lifetime choices. In looking at the transforming of enterprise in a variety of settings, this book seeks to help clarify many of the key forces shaping the design, implementation, and use of ICTs in ways that can transform our lives and the social and institutional structure in which they take place.

References

Bell, D. (1999 [1973]) *The Coming of Post-Industrial Society: A Venture in Social Forecasting*. New York: Basic Books.

Beniger, J. N. (1986) *The Control Revolution*. Cambridge, Mass.: Harvard University Press.

Castells, M. (2000 [1996]) *The Rise of the Network Society: The Information Age: Economy, Society and Culture*, volume 1. Oxford: Blackwell Publishers.

Castells, M., and P. Himanen (2002) *The Information Society and the Welfare State: The Finnish Model*. Oxford: Oxford University Press.

Danziger, J. N., W. H. Dutton, R. Kling, and K. L. Kraemer (1982) *Computers and Politics.* New York: Columbia University Press.

de Sola Pool, I. (1983) *Technologies of Freedom.* Cambridge, Mass.: The Belknap Press of Harvard University Press.

Dutton, W. H., ed. (1996) *Information and Communication Technologies—Visions and Realities.* Oxford: Oxford University Press.

Dutton, W. H., ed. (1999) *Society on the Line: Information Politics in the Digital Age.* Oxford: Oxford University Press.

Dutton, W. H., J. G. Blumler, and K. L. Kraemer, eds. (1987) *Wired Cities.* Boston: G. K. Hall.

Freeman, C. (1996) "The Two-Edged Nature of Technological Change: Employment and Unemployment." In *Information and Communication Technologies—Visions and Realities,* ed. W. H. Dutton. Oxford: Oxford University Press, pp. 19–36.

Gates, B. (1995) *The Road Ahead.* London: Viking.

Gilder, G. (1994) *Life After Television: The Coming Transformation of Media and American Life.* New York: W. W. Norton.

Goddard, J., and R. Richardson (1996) "Why Geography Will Still Matter." *Information and Communication Technologies—Visions and Realities,* ed. W. H. Dutton. Oxford: Oxford University Press, pp. 197–214.

Hughes, T. (2000) "Technological Momentum." In *Technology and the Future,* 8th edition, ed. A. H. Teich. New York: Wadsworth.

Katz, E., and P. F. Lazersfeld (1955) *Personal Influence: The Part Played by People in the Flow of Mass Communication.* New York: The Free Press.

Katz, J. E., and R. E. Rice (2002) *Social Consequences of the Internet.* Cambridge, Mass.: MIT Press.

Kling, R. (2000) "Social Informatics: A New Perspective on Social Research about Information and Communication Technologies." *Prometheus* 18(3): 245–264.

Lessig, L. (1999) *Code and Other Laws of Cyberspace.* New York: Basic Books.

Lessig, L. (2002) *The Future of Ideas.* New York: Vintage Books.

Lyon, D. (2001) *Surveillance Society: Monitoring Everyday Life.* Milton Keynes: Open University Press.

MacKenzie, D. (1999) "Technological Determinism." In *Society on the Line: Information Politics in the Digital Age,* ed. W. H. Dutton. Oxford: Oxford University Press, pp. 29–41.

MacKenzie, D., and J. Wajcman, eds. (1985) *The Social Shaping of Technology: How a Refrigerator Got Its Hum.* Milton Keynes and Philadelphia: Open University Press.

McLuhan, M. (1964) *Understanding Media: The Extension of Man.* London: Routledge.

Miles, I., with T. Brady, A. Davies, L. Haddon, M. Matthews, H. Rush, and S. Wyatt (1990) *Mapping and Measuring the Information Economy*. Library and Information Research Report, 77. Boston Spa, U.K.: British Library.

Negroponte, N. (1995) *Being Digital*. London: Hodder & Stoughton.

Robins, K., ed. (1992) *Understanding Information*. London: Belhaven.

Robins, K., and F. Webster, eds. (2002) *The Virtual University? Knowledge, Markets, and Management*. Oxford: Oxford University Press.

Rogers, E. M. (1986) *Communication Technology: The New Media in Society*. New York: The Free Press.

Wellman, B., and C. Haythornthwaite (2002) *The Internet in Everyday Life*. Oxford: Blackwell Publishing.

Westin, A. F., and M. A. Baker (1972) *Databanks in a Free Society*. New York: Quadrangle Books.

Williams, R., and D. Edge (1996) "The Social Shaping of Technology." In *Information and Communication Technologies—Visions and Realities*, ed. W. H. Dutton. Oxford: Oxford University Press, pp. 53–67.

Winner, L. (1986) *The Whale and the Reactor: A Search for Limits in an Age of High Technology*. Chicago: The University of Chicago Press.

Woolgar, S., ed. (2002) *Virtual Society? Technology, Cyberbole, Reality*. Oxford: Oxford University Press.

I

Transforming the Economy

1

Intangible Assets and the Economic Impact of Computers

Erik Brynjolfsson and Lorin M. Hitt

Productivity is arguably the single most important statistic for measuring the performance of an economy. Since 1995, productivity in the United States has increased by an average of over 3.1 percent per year. This is more than double the 1.4 percent productivity growth rate that prevailed from 1973 through 1995. An increasing body of evidence suggests that computers have made a significant contribution to this productivity revival. In part, this is a by-product of the tremendous technological progress in the computer industry.

There has been a long-term trend, dating back to the beginning of the century, of rapid technological progress in information processing functions. This is evident in sustained annual improvements in the price–performance ratio of computers of more than 20 percent overall. However, this trend accelerated around 1995, with faster improvements in semiconductors. Furthermore, the performance of components such as disk drives and data communications have also advanced more rapidly.

The technology improvements have led to falling real prices. Nonetheless, corporations actually increased nominal spending on computers and communications equipment (collectively referred to as information technology, or IT) in the late 1990s, leading to a dramatically growing stock of real IT capital and capabilities. This sustained growth combined with the mathematics of growth accounting—the contribution of capital asset to output growth is equal to the product of its growth rate and factor share (input quantity per unit of output)—suggests that it was only a matter of time before sufficient IT capital accumulation would have a substantial impact on output and output growth. Indeed, this logic is confirmed since the estimated growth contribution for computers in the

1990s is on the order of 1 percent per year or more (Jorgenson and Stiroh, 1999; Oliner and Sichel, 2000). While this simple story is useful for understanding the productivity revival, it is incomplete for at least two reasons. First, it is predicated on the *assumption* that computers earn a "normal" rate of return, essentially equal to the capital cost of computer assets. The essence of the "productivity paradox" alluded to by Solow (1987), Roach (1987), and others centered on the possibility that computer investment may have been mismanaged and thus had not earned the requisite returns. If computers are not earning at least "normal" returns, then the growth accounting estimates of computer contribution will be inaccurate.

The "productivity paradox" question was a knotty one to resolve in part because many of the costs and benefits of computer investment are hard to observe. For example, benefits such as product variety, timeliness, and quality change are often underestimated or ignored entirely in measures of overall output (see, e.g., Boskin et al., 1997). These benefits may be especially important for computer users (Brynjolfsson and Hitt, 1996). Nonetheless, a combination of new data, especially at the firm level, and modern econometric methods has led to a considerable body of evidence that computers earn at least their expected returns and perhaps significantly more (Lichtenberg, 1995; Brynjolfsson and Hitt, 1995, 1996, 2000; see Brynjolfsson and Yang, 1996, or Stiroh, 2002, for reviews). The presence of returns on computer capital at least equal to their user cost is important because it suggests that computer investment has contributed not only to output growth but perhaps to productivity growth as well. Indeed, a potential puzzle raised by these later studies is that the returns on computers actually appear to be substantially *above* normal, which is difficult to explain in equilibrium.

A second, and perhaps more important, limitation of this simple story of computers and growth, is that the real power of computers is the ability to engender complementary innovations. Perhaps even more so than historical examples such as the steam engine or the electric motor, computers are a "general purpose technology" (Bresnahan and Trajtenberg, 1995) whose principal contribution to economic performance arises because they facilitate complementary innovations such as new products and processes. Clearly, software and other types of technology infrastructure are important complements to computers. However, there is considerable evidence that other "organizational complements," such as business processes, decision making structures, incentive systems, human capital, and corporate culture, to name a few, play an important role in the ability of a firm to realize value from its IT investments. Acquiring and maintaining these organizational complements is a real cost to the firm but also a potential source of significant value when combined with appropriate technology investments. In other words, they represent complementary "organizational capital." However, they are not typically treated as capital assets in firm or national accounts and have therefore not been fully incorporated into economic analyses of the impact of computers.

In this review, we survey and synthesize the growing body of literature on computers and organizational complements, including theoretical models of how computers relate to specific complements, direct measures of organizational complements through case studies or econometric analyses, and indirect analyses (principally econometric studies) that reveal properties of these organization complements even if they cannot be measured directly.

Thus, while the recent macroeconomic evidence about computer contributions is encouraging, our views are more strongly influenced by the microeconomic data. The micro data suggest that the surge in productivity that we now see in the macro statistics has its roots in over a decade of computer-enabled organizational investments. The recent productivity boom can in part be explained as a return on this intangible and largely ignored form of capital.

Qualitative Case Examples

Companies using IT to change the way they conduct business often say that their investment in IT complements changes in other aspects of the organization. These complementarities have a number of implications for understanding the value of computer investment. To be successful, firms typically need to adopt computers as part of a "system" or "cluster" of mutually reinforcing organizational changes (Milgrom and Roberts, 1990). Changing incrementally, either by making computer investments without organizational change or by only partially implementing some organizational changes, can create significant productivity losses since any benefits of computerization are more than outweighed by negative interactions with existing organizational practices (Brynjolfsson, Renshaw, and Van Alstyne, 1997). The need for "all or nothing" changes between complementary systems was part of the logic behind the organizational reengineering wave of the 1990s and the slogan "don't automate, obliterate" (Hammer, 1990). It may also explain why many large-scale IT projects fail (Kemerer and Sosa, 1991), while successful firms earn significant rents.

Many of the past century's most successful and popular organizational practices reflect the historically high cost of information processing. For example, hierarchical organizational structures can reduce communications costs because they minimize the number of communications links required to connect multiple economic actors, as compared with more decentralized structures (Malone, Yates, and Benjamin, 1987; Radner, 1993). Similarly, producing simple, standardized products is an efficient way to utilize inflexible, scale-intensive manufacturing technology. However, since the cost of automated information processing has fallen by over 99.9 percent since the 1960s, it is unlikely that the work practices of the previous era will also be the same ones that best leverage the value of cheap information and flexible production. In this spirit, Milgrom and Roberts (1990) constructed a model in which firms' transition from "mass production" to flexible, computer-enabled, "modern manufacturing" is driven by exogenous changes in the price of IT. Similarly, Bresnahan, Brynjolfsson, and Hitt (2002) showed theoretically and empirically how changes in IT costs and capabilities lead to a cluster of changes in work organization and firm strategy that increase the demand for skilled labor.

In this section we discuss case evidence on three aspects of how firms have transformed themselves by combining IT with changes in work practices, strategy, and products and services; they have transformed the firm, supplier relations, and the customer relationship. These examples provide qualitative insights into the nature of the changes, making it easier to interpret the more quantitative econometric evidence that follows.

Transforming the Firm

The need to match organizational structure to technology capabilities and the challenges of making the transition to an IT-intensive production process is concisely illustrated by a case study of "MacroMed" (a pseudonym), a large medical products manufacturer (Brynjolfsson, Renshaw, and Van Alstyne, 1997). In a desire to provide greater product customization and variety, MacroMed made a large investment in computer-integrated manufacturing, coupled with major organizational changes such as the elimination of piece rates, greater worker autonomy and decision rights, and process and workflow redesign. However, the new system initially fell well short of management expectations for greater flexibility and responsiveness. Investigation revealed that line workers still retained many elements of the now-obsolete old work practices, such as maintaining long production runs and minimizing production changeovers, not from any conscious effort to undermine the change effort but simply as an inherited pattern. While these practices were valuable with the old equipment, they negated the flexibility of the new machines. Ironically, the new equipment was sufficiently flexible that the workers were able to get it to work much like the old machines! Eventually, management concluded that the best approach was to introduce the new equipment in a "greenfield" site with a handpicked set of young employees who were relatively unencumbered by knowledge of the old practices. The resulting productivity improvements were significant enough that management ordered all the factory windows painted black to prevent potential competitors from seeing the new system in action. While other firms could readily buy similar computer-controlled equipment, they would still have to identify and implement new organizational processes, an effort entailing significant cost, time, and risk.

This example is representative of a wide variety of case evidence linking the value of IT to organizational changes. Firms such as McKinsey and KPMG in order to benefit from internal "knowledge management" systems had to first redesign their organizational structure, incentives, and corporate culture to encourage knowledge sharing and reuse (Hansen, Nohria, and Tierney, 2000). Early adopters of these systems discovered that without these changes, supposedly collaborative work tools were principally used for personal productivity rather than collaboration (Orlikowski, 1992). Firms such as Cisco Systems have invested heavily to create a distinctive corporate culture (a.k.a. "I-culture") focused around information sharing, customer focus, and collaboration that they believe is critical for information-intensive businesses. Even more interestingly, they have also made deliberate efforts to communicate this culture to other firms in the belief that it can enhance the value firms receive from data communications and networking technologies, Cisco's primary products.

Changing Interactions with Suppliers

Owing to problems coordinating with external suppliers, large firms often produce many of their required inputs in-house. General Motors is the classic example of a company whose success was facilitated by high levels of vertical integration. However, technologies such as electronic data interchange (EDI), Internet-based procurement systems, and other interorganizational information systems have significantly reduced the cost, time, and other difficulties of interacting with suppliers. They have also created opportunities for the redesign of the buyer–supplier relationship.

An early successful interorganizational system is the Baxter ASAP system, which lets hospitals electronically order supplies directly from wholesalers (Vitale and Konsynski, 1988; Short and Venkatraman, 1992). The system was originally designed to reduce the costs of data entry—a large hospital could generate 50,000 purchase orders annually, which had to be written out by hand by Baxter's field sales representatives at an estimated cost of $25–$35 each. However, once Baxter computerized its ordering and had data available on levels of hospital stock, it was able to assume responsibility for managing a hospital's entire supply operation. Later versions of the ASAP system let users order from other suppliers, creating an electronic marketplace in hospital supplies.

Computer-based supply chain integration has been especially sophisticated in consumer packaged goods. For instance, Procter and Gamble (P&G) pioneered a program called "efficient consumer response" (McKenney and Clark, 1995) in which checkout scanner data passes directly to the manufacturer, triggering a process of automatic ordering, replenishment, invoicing, and payment. Manufacturers also involved themselves more in inventory decisions and moved toward "category management," where a lead manufacturer would take responsibility for an entire retail category (say, laundry products), determining stocking levels for their own and other manufacturers' products, as well as complementary items. These changes not only improved efficiency of both manufacturing and retailing but also made it possible to offer increased product variety and convenience to consumers. Without the direct computer–computer links to scanner data and the electronic transfer of payments and invoices, they could not have attained the levels of speed and accuracy needed to implement such a system.

Changing Customer Relationships

The Internet has opened up a new range of possibilities for enriching interactions with customers. Dell Computer has succeeded in attracting customer orders and improving service by placing configuration, ordering, and technical support capabilities on the web (Rangan and Bell, 1999). It coupled this change with systems and work practice changes that emphasize just-in-time inventory management, build-to-order production systems, and tight integration between sales and production planning. Dell has implemented a consumer-driven build-to-order business model, rather than using the traditional build-to-stock model of selling computers through retail stores. This not only gives Dell a 10 percent price advantage over its competitors, but also enables it to incorporate new computing technologies (e.g., microprocessors) into its computers within seven days, compared to eight weeks or more for some less Internet-enabled competitors. This is a significant advantage in an industry where adoption of new technology and obsolescence of old technology is rapid, margins are thin, and many component prices drop by 3–4 percent each month.

Other firms have also built closer relations with their customers via the web and related technologies. For instance, web retailers such as Amazon.com provide personalized recommendations to visitors and allow them to customize numerous aspects of their shopping experience and increase product selection, which generates enormous consumer surplus (Brynjolfsson, Hu, and Smith, 2003). Merely providing Internet access to a traditional bookstore would have had a relatively minimal impact without the cluster of other changes implemented by firms like Amazon. Moreover, because of this close relationship with customers, they can rely on acquiring new customers by word of mouth. The Internet has also significantly transformed customer service, a critical but costly activity. UPS handles more than 700,000 package tracking requests via the Internet every day. It costs UPS 10¢ per piece to serve that information via the web versus $2 to provide it over the phone. In addition, this process is more efficient for customers as well. UPS estimates that two-thirds of the web users would not have bothered to check on their packages if they had not had web access. Moreover, with inexpensive package tracking via the web, firms that utilize UPS can now offer their customers tracking information and other value added services (e.g., delivery notification), enhancing their customer relationships.

Large-Sample Empirical Evidence on IT, Organization, and Productivity

The case study literature offers many examples of strong links between IT and investments in complementary organizational practices. However, to reveal general trends and to quantify the overall impact, we must examine these effects across a wide range of firms and industries. In this section we explore the results from large-sample statistical analyses. First, we examine studies on the direct relationship between IT investment and business value that can provide indirect evidence on the nature of organizational complements. We then consider studies that measured organizational factors and their correlation with IT use, as well as the few initial studies that have linked this relationship to productivity increases.

IT and Productivity

While there has been considerable debate over the measured contribution of computers to productivity, there is a consistent body of evidence linking information technology to greater output and productivity, especially in studies at the firm level or below (see Brynjolfsson and Hitt, 2003). A recent meta-analysis of these studies suggests that while there is variance in the results due to different samples, time periods, and methods, the output elasticity of computers (the marginal increase in output per unit of investment) is positive, significant, and at least sufficient to offset the cost of computer investment (Stiroh, 2003). In addition, recent industry estimates imply a similar figure for analyses of industry-level data (Stiroh, 2002), contrasting with earlier industry-level results, which were often inconclusive.

While we do not review this work in detail, there are three types of findings in this literature that are consistent with the organizational complements story. First, in many recent firm-level studies, the output elasticity of computers appears to be too high, especially when comparing levels of computer capital with levels of output. One-year rates of returns to computers in these studies have been computed to be as high as 80 percent or more per year, while the capital cost (as computed by the Jorgensonian rental price) is around 42 percent (see, e.g., Brynjolfsson and Hitt, 1996, or Lichtenberg, 1995). If computers earned "normal" returns these two numbers should be equal. When fixed effects are introduced in an effort to control for otherwise unobserved differences in firms, such as their organization, management, or culture, the returns to IT drop in half. Second, recent evidence on computers and productivity growth finds that the apparent growth contribution of computers increases as longer time periods are considered. When growth calculations are done over one-year differences, computers appear to earn a return commensurate with their costs. When differences are taken over two to seven years, the rate of return rises steadily as the length of differences increases (Brynjolfsson and Hitt, 2003). Third, research using stock market valuations to determine the value of a firm's capital stock finds that computer capital appears to be valued at $5 to $20 per dollar of stock, while a dollar of ordinary capital has an apparent value close to $1 (see Brynjolfsson and Yang, 1997). One possible explanation of some of these results is that computers are simply unusually productive and valuable. However, this would not explain the rise in the computer coefficients as difference lengths are increased and is generally inconsistent with economic equilibrium—if computers really are that valuable, why are firms not investing in more projects, driving the marginal returns down to costs?

A more plausible, and fully consistent, explanation is that we are not observing the returns from computers alone but rather the returns to a system of computers and organizational complements. This is exactly what we observe in the case studies: Computerization involves much more than computers. While the total value of this computers-plus-complements system may be substantially larger than the apparent capital stock of computers alone, it may simply represent equilibrium returns on a much larger stock of capital, including both computers and organizational assets. Excluding the organizational complements from the denominator makes the output elasticities and the stock market valuations seem unduly high. Under this conjecture, the capital stock of complementary organizational assets may be several times larger than the stock of computer assets. Moreover, if these organizational assets cannot be adapted to the levels of computer assets immediately but only after a period of several years, this can also explain why the contributions rise when changes over longer time periods are considered. While the short-run returns of computers (before organizational assets can adjust) may be close to cost, the long-term returns appear to be significantly larger.

Direct Measurement of the Interrelationship between IT and Organization

Some studies have attempted to measure organizational complements directly and to show either that they are correlated with IT investment or that firms that combine complementary factors have better economic performance. Finding correlations between IT and organizational change, or between these factors and measures of economic performance, is not sufficient to prove that these practices are complements without a full structural model that specifies the production relationships and demand drivers for each factor (see Athey and Stern, 1998). However, after empirically evaluating possible alternative explanations and combining correlations with performance analyses, complementarities are often the most plausible explanation for observed relationships between IT, organizational factors, and economic performance.

The first set of studies in this area focuses on correlations between use of IT and extent of organizational change. An important finding is that IT investment is greater in organizations that are decentralized and have a greater level of demand for human capital. For example, Bresnahan, Brynjolfsson, and Hitt (2002) surveyed approximately 400 large firms to obtain information on aspects of organizational structure such as the allocation of decision rights, workforce composition, and investments in human capital. They found that greater levels of IT are associated with increased delegation of authority to individuals and teams, greater levels of skill and education in the workforce, and greater emphasis on pre-employment screening for education and training. In addition, they found that these work practices are correlated with each other, suggesting that they are part of a complementary work system. Black and Lynch (2000) reported similar findings—not only was the widespread use of computers linked to higher productivity, but they also found that "high performance work practices" and practices that encourage workers to think about ways to improve the production process are correlated with increased productivity.

Research on jobs within specific industries has begun to explore the mechanisms within organizations that create these complementarities. Drawing on a case study on the automobile repair industry, Levy, Beamish, Murnane, and Autor (2000) argued that computers are most likely to substitute for jobs that rely on rule-based decision making while complementing nonprocedural cognitive tasks. In banking, researchers have found that many of the skill, wage, and other organizational effects of computers depend on the extent to which firms couple computer investment with organizational redesign and other managerial decisions (Hunter, Bernhardt, Hughes, and Skuratowitz, 2000; Murnane, Levy, and Autor, 1999). Researchers focusing at the establishment level have also found complementarities between existing technology infrastructure and firm work practices to be a key determinant of the firm's ability to incorporate new technologies (Bresnahan and Greenstein, 1997); this also suggests a pattern of mutual causation between computer investment and organization.

A variety of industry-level studies also show a strong connection between investment in high technology equipment and the demand for skilled, educated workers (Berndt, Morrison, and Rosenblum, 1992; Berman, Bound, and Griliches, 1994; Autor, Katz, and Krueger, 1998). Again, these findings are consistent with the idea that increasing use of computers is associated with a greater demand for human capital.

Several researchers have also considered the effect of IT on macro-organizational structures. They have typically found that greater levels of investment in IT are associated with smaller firms and less vertical integration. Brynjolfsson, Malone, Gurbaxani, and Kambil (1994) found that increases in the level of IT capital in an economic sector were associated with a decline in average firm size in that sector, consistent with IT leading to a reduction in vertical integration. Hitt (1999), examining the relationship between a firm's IT capital stock and direct measures of its vertical integration, arrived at similar conclusions. These results corroborate earlier case analyses and theoretical arguments that suggested that IT would be associated with a decrease in vertical integration because it lowers the costs of coordinating externally with suppliers (Malone, Yates, and Benjamin, 1987). At a finer level of detail, Baker and Hubbard (2003) developed and tested a model that links computer-based monitoring and vehicle routing technologies to the decision on whether trucks should be owned by the driver ("owner–operators") or by a centralized fleet manager.

One difficulty in interpreting the literature on correlations between IT and organizational change is that some managers may be predisposed to try every new idea and some managers may be averse to trying anything new at all. In such a world, IT and a "modern" work organization might be correlated in firms because of the temperament of management, not because they are economic complements. To rule out this sort of spurious correlation, it is useful to bring measures of productivity and economic performance into the analysis. If combining IT and organizational restructuring is economically justified, then firms that adopt these practices as a system should outperform those that fail to combine IT investment with appropriate organizational structures.

In fact, firms that adopt decentralized organizational structures and work structures do appear to have a higher contribution of IT to productivity (Bresnahan, Brynjolfsson, and Hitt, 2002). For example, firms that are more decentralized than the median firm (as measured by individual organizational practices and by an index of such practices) have, on average, a 13 percent greater IT elasticity and a 10 percent greater investment in IT than the median firm. Firms that are in the top half of *both* IT investment and decentralization are on average 5 percent more productive than firms that are above average only in IT investment or only in decentralized organization.

Similar results also appear when economic performance is measured as stock market valuation. Firms in the top third of decentralization have a 6 percent higher market value after controlling for all other measured assets; this is consistent with the theory that organizational decentralization behaves like an intangible asset. Moreover, the stock market value of a dollar of IT capital is between $2 and $5 greater in decentralized firms than in centralized firms (per standard deviation of the decentralization measure), and this relationship is particularly striking for firms that are simultaneously extensive users of IT and highly decentralized, as shown in figure 1.1 (Brynjolfsson, Hitt, and Yang, 2002). In the figure, each firm in the sample is represented in a three-dimensional space based on its relative levels of IT (IT-intensive firms are to the right side of the figure), relative levels of organizational capital (firms that have adopted more of the "new" cluster are toward the back of the figure), and

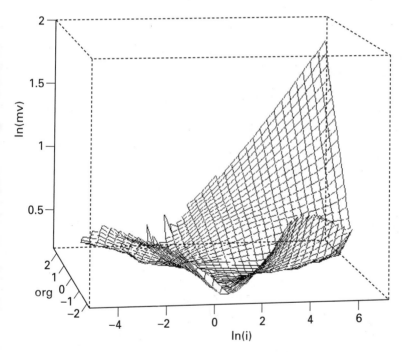

Figure 1.1
Market value as a function of IT and work organization. This graph was produced by nonparametric local regression models using data from Brynjolfsson, Hitt, and Yang 2002. Note: *i* represents computer capital, *org* represents a measure of decentralization, and *mv* is market value after controlling for all tangible assets. In each case, firms in the sample are compared to other firms in the same industry, with the median defined as zero.

relative market value (firms with more "excess" market value are toward the top of the figure). Clearly, the vast majority of the high performers are firms that simultaneously invest in both IT and organizational capital. Doing one type of investment without the other is not correlated with any significant market value. Interestingly, Bresnahan, Brynjolfsson, and Hitt (2003) found essentially the same pattern of results when they looked at productivity instead of market value.

The weight of the firm-level evidence shows that a combination of investment in technology and changes in organization and work practices facilitated by these technologies contributes to firms' productivity growth and market value. However, much work remains to be done in categorizing and measuring the relevant changes in organization and work practices and relating them to IT and productivity.

Implications for Macroeconomic Performance

Studies of the contribution of IT concluded that technical progress in computers contributed roughly 0.3 percentage points per year to real output growth when data from the 1970s and 1980s were used (Jorgenson and Stiroh, 1995; Oliner and Sichel, 1994; Brynjolfsson, 1996). Using similar methods, the estimates for the 1990s appear to be on the order of 1 percent (Oliner and Sichel, 2000; Jorgenson and Stiroh, 1999; Gordon 1998). Much of the estimated growth contribution comes directly from the large quality-adjusted price declines in the computer-producing industries. The nominal value of purchases of IT hardware in the United States in 1997 was about 1.4 percent of gross domestic product (GDP). Since the quality-adjusted prices of computers decline by about 25 percent per year, simply spending the same nominal share of GDP as in previous years represents an annual productivity increase for the real GDP of 0.3 percentage points (that is, $1.4 \times 0.25 = 0.35$).

A related approach is to look at the effect of IT on the GDP deflator. Reductions in inflation, for a given amount of growth in output, imply proportionately higher real growth and, when divided by a measure of inputs, higher productivity growth as well. Gordon (1998, p. 4) calculated that "computer hardware is currently contributing to a reduction of U.S. inflation at an annual rate of almost 0.5 percent per year, and this number would climb toward 1 percent per year if a broader definition of IT, including telecommunications equipment, were used."

These studies show that a substantial part of the upturn in measured productivity of the economy as a whole can be linked to increased real investments in computer hardware and declines in their quality-adjusted prices. However, there are several key assumptions implicit in economy- or industry-wide growth accounting approaches that can have a substantial influence on their results, especially if one seeks to know whether investment in computers is increasing productivity as much as other investment alternatives. The standard growth accounting approach begins by assuming that all inputs earn "normal" rates of return. Unexpected windfalls, whether the discovery of a single new oil field or the invention of a new process that makes oil fields obsolete, show up not in the growth contribution of inputs but as changes in the multifactor productivity residual. By construction, an input can contribute more to output in these analyses only by growing rapidly, not by having an unusually high net rate of return.

Changes in multifactor productivity growth, in turn, depend on accurate measures of final output. However, nominal output is affected by whether firm expenditures are expensed, and therefore deducted from value added, or capitalized and treated as investment. As emphasized throughout this chapter, IT is only a small fraction of a much larger complementary system of tangible and intangible assets. Current statistics typically treat the accumulation of intangible capital assets, however, such as new business processes, new production systems, and new skills, as expenses rather than as investments. This leads to a lower level of measured output and input: The organizational capital is not counted as an output when it is created and is not counted as an input in subsequent years when it is used.

The magnitude of investment in intangible assets associated with computerization may be large. Analyses of 800 large firms by Brynjolfsson and Yang (1997) suggest that the ratio of intangible assets to IT assets may be 10 to 1. Thus, the $167 billion in computer capital recorded in the U.S. national accounts in 1996 may have actually been only the tip of an iceberg of $1.67 trillion of IT-related complementary assets in the United States.

Examination of individual IT projects indicates that the 10:1 ratio is plausible. For example, a survey of enterprise resource planning projects found that the average spending on computer hardware accounted for less than 4 percent of the typical start-up cost of $20.5 million, while software licenses and development were another 16 percent of total costs (Gormley et al., 1998). The remaining costs included hiring outside and internal consultants to help design new business processes and to train workers in the use of the system. The time of existing employees, including top managers, that went into the overall implementation were not included, although these costs can also be quite substantial.

The up-front costs were almost all expensed by the companies undertaking the implementation projects. However, insofar as the managers who made these expenditures expected them to pay for themselves only over several years, the nonrecurring costs are properly thought of as investments, not expenses, when considering the impact on economic growth. In essence, the managers were adding to the nation's capital stock not only of easily visible computers but also of less visible business processes and worker skills.

How might these measurement problems affect economic growth and productivity calculations? In a steady state, it makes little difference, because the amount of new organizational investment in any given year is offset by the "depreciation" of organizational investments in previous years. The net change in capital stock is zero. Thus, in a steady state, classifying organizational investments as expenses does not bias overall output growth as long as it is done consistently from year to year. However, the economy has hardly been in a steady state with respect to computers and their complements. Instead, in the 1990s, the U.S. economy was likely adding rapidly to its stock of both types of capital faster than it was using capital built up in past years. To the extent that this net capital accumulation has not been counted as part of output, output and output growth were probably underestimated to a greater extent than input growth. Conversely, in 2001 and 2002, this pattern may well have been reversed, as firms cut back on investments in computers and their complements but continued to harvest the benefits of past years' large investments.

The software industry offers a useful example of the impact of classifying a category of spending as expense or investment. Historically, efforts on software development have been treated as expenses, but recently the government has begun recognizing that software is an intangible capital asset. Software investment by U.S. businesses and governments grew from $10 billion in 1979 to $159 billion in 1998 (Parker and Grimm, 2000). Properly accounting for this investment has added 0.15 to 0.20 percentage points to the average annual growth rate of real GDP in the 1990s. While capitalizing software is an important improvement in our national accounts, software is far from being the only, or even most important, complement to computers.

If the wide array of intangible capital costs associated with computers were treated as investments rather than expenses, the results would be striking. According to some preliminary estimates from Yang and Brynjolfsson (2003), building on estimates of the intangible asset stock derived from studies of IT project expenses as well as the estimated stock market valuations of computers, the true growth rate of U.S. GDP, after accounting for the intangible complements to IT hardware, may have been underestimated by over 1 percent per year in the 1990s. This reflects the large net increase in intangible assets of the U.S. economy associated with the computerization efforts of the 1990s. Thus, the true performance of the U.S. economy in the 1990s may have been even greater than the already impressive measured performance.

Interestingly, the same model can also help explain the even greater surge in measured productivity in recent years, just as investment in IT and intangibles appears to have slowed down. Since 2001, the economy has been earning returns on the intangible investments it made in the 1990s, converting it back into consumption. This has had the effect of *raising* productivity growth as conventionally measured, since the capital services from these previously created intangible assets have not been counted as current inputs. Because firms have been investing less in IT and the associated intangibles in the past two years or so, they have been able to devote more resources to producing current output. At the same time, they are "harvesting" the benefits of process improvements that accompanied the earlier wave of IT investment. In some cases, this involves laying off workers who are now redundant. In other cases, it simply means expanding production without commensurate increases in employment. Either way, it shows up as higher productivity in the current year.

While the quantity of intangible assets associated with IT is difficult to estimate precisely, the central lesson is that these complementary changes are economically significant and cannot be ignored in any realistic attempt to estimate the overall economic contributions of IT.

Just as the Bureau of Economic Analysis successfully reclassified many software expenses as investments and is making quality adjustments, perhaps we will also find ways to measure the investment component of spending on intangible organizational capital and to make appropriate adjustments for the value of all gains attributable to improved quality, variety, convenience, and service. Unfortunately, addressing these problems can be difficult even for single firms and products, and the complexity and number of judgments required to address them at the macroeconomic level is extremely high. Moreover, because of the increasing service component of all industries (even basic manufacturing), which entails product and service innovation and intangible investments, these problem cannot be easily solved by focusing on a limited number of "hard to measure" industries—they are pervasive throughout the economy.

Both case studies and econometric work point to organizational complements such as new business processes, new skills, and new organizational and industry structures as major drivers of the contribution of IT. These complementary investments, and the resulting assets, are likely larger than the investments in the computer technology itself. However, they go largely uncounted in our national accounts, suggesting that computerization has made a much larger real contribution to the economy than previously believed. A task for researchers and managers alike, then, is to find ways to better measure the intangibles that are increasingly important to economic growth and firm performance.

Acknowledgments

This chapter is substantially based on our paper "Beyond Computation: Information Technology, Organizational Transformation and Business Performance." *Journal of Economic Perspectives* 14, no. 4. We thank David Autor, Brad DeLong, David Fitoussi, Robert Gordon, Shane Greenstein, Dale Jorgenson, Brian Kahin, Alan Krueger, Jacques Mairesse, Dan Sichel, Robert Solow, Kevin Stiroh, Timothy Taylor, and Andrew Wyckoff for valuable comments on (portions of) earlier drafts. This work was funded in part by The MIT Center for eBusiness and by NSF Grants IIS-0085725 and IIS-9733877.

References

Athey, S., and S. Stern (1998) "An Empirical Framework for Testing Theories about Complementarities in Organizational Design." MIT Sloan School of Management Working Paper WP 4022–98.

Autor, D., L. F. Katz, and A. B. Krueger (1998) "Computing Inequality: Have Computers Changed the Labor Market?" *Quarterly Journal of Economics* (November).

Baker, George P., and Thomas Hubbard (2003) "Make v. Buy in Trucking: Asset Ownership, Job Design and Information." *American Economic Review* 93(3) (June).

Berman, E., J. Bound, and Z. Griliches (1994) "Changes in the Demand for Skilled Labor within U.S. Manufacturing Industries." *Quarterly Journal of Economics* 109 (May): 367–398.

Berndt, E. R., C. J. Morrison, and L. S. Rosenblum (1992) "High-Tech Capital, Economic Performance and Labor Composition in U.S. Manufacturing Industries: An Exploratory Analysis." MIT Working Paper 3414EFA.

Black, Sandra E., and Lisa M. Lynch (2001) "How to Compete: The Impact of Workplace Practices and Information Technology on Productivity." *Review of Economics and Statistics* (August).

Boskin, Michael J., Ellen R. Dulberger, Robert J. Gordon, Zvi Griliches, and Dale Jorgenson (1997) "The CPI Commission: Findings and Recommendations." *American Economic Review* 87(2): 78–83.

Bresnahan, T. F., and S. Greenstein (1997) "Technical Progress and Co-Invention in Computing and in the Use of Computers." *Brookings Papers on Economic Activity: Microeconomics* (January): 1–78.

Bresnahan, T. F., and M. Trajtenberg (1995) "General Purpose Technologies: 'Engines of Growth'?" *Journal of Econometrics* 65: 83–108.

Bresnahan, T., E. Brynjolfsson, and L. Hitt (2002) "Information Technology, Workplace Organization, and the Demand for Skilled Labor: Firm-Level Evidence." *Quarterly Journal of Economics* 117: 339–376.

Brynjolfsson, E. (1996) "The Contribution of Information Technology to Consumer Welfare." *Information Systems Research* 7(3): 281–300.

Brynjolfsson, E., and L. Hitt (1995) "Information Technology as a Factor of Production: The Role of Differences Among Firms." *Economics of Innovation and New Technology* 3(4): 183–200.

Brynjolfsson, E., and L. Hitt (1996) "Paradox Lost? Firm-Level Evidence on the Returns to Information Systems Spending." *Management Science* 42(4): 541–558.

Brynjolfsson, E., and L. M. Hitt (2000) "Beyond Computation: Information Technology, Organizational Transformation and Business Performance." *Journal of Economic Perspectives,* 14(4): 23–48.

Brynjolfsson, E., and L. Hitt (2003) "Computing Productivity: Firm Level Evidence." *Review of Economics and Statistics* (November).

Brynjolfsson, E., and Yang, S. (1996) "Information Technology and Productivity: A Review of the Literature." In *Advances in Computers*, ed. M. Zelkowitz, vol. 43.

Brynjolfsson, E., and Yang, S. (1997) "The Intangible Benefits and Costs of Computer Investments: Evidence from Financial Markets." *Proceedings of the International Conference on Information Systems*, Atlanta, Ga. Revised (2000).

Brynjolfsson, Erik, Lorin M. Hitt, and Shinkyu Yang (2002) "Intangible Assets: Computers and Organizational Capital." *Brookings Papers on Economic Activity: Macroeconomics* 1: 137–199.

Brynjolfsson, E., T. Malone, V. Gurbaxani, and A. Kambil (1994) "Does Information Technology Lead to Smaller Firms?" *Management Science* 40(12): 1628–1644.

Brynjolfsson, E., A. Renshaw, and M. Van Alstyne (1997) "The Matrix of Change." *Sloan Management Review* (Winter).

Brynjolfsson, Erik, Yu (Jeffrey) Hu, and Michael D. Smith (2003) "Consumer Surplus in the Digital Economy: Estimating the Value of Increased Product Variety at Online Booksellers." *Management Science* 49(11) (November).

Gordon, Robert J. (1998) "Monetary Policy in the Age of Information Technology: Computers and the Solow Paradox." Working Paper, Northwestern University.

Gormley, J., W. Bluestein, J. Gatoff, and H. Chun (1998) "The Runaway Costs of Packaged Applications." *Forrester Report*, vol. 3, no. 5, Cambridge, Mass.

Hammer, M. (1990) "Reengineering Work: Don't Automate, Obliterate." *Harvard Business Review* (July–August): 104–112.

Hansen, Morten T., Nitin Nohria, and Thomas Tierney (2000) "What's Your Strategy for Managing Knowledge?" *Harvard Business Review* (February 1).

Hitt, Lorin M. (1999) "Information Technology and Firm Boundaries: Evidence from Panel Data." *Information Systems Research* 10(9) (June): 134–149.

Hunter, Larry W., Annette Bernhardt, Katherine L. Hughes, and Eva Skuratowicz (2000) "Its Not Just the ATMs: Firm Strategies, Work Restructuring and Workers' Earnings in Retail Banking." Mimeo, Wharton School of Business.

Jorgenson, Dale W., and Kevin Stiroh (1995) "Computers and Growth." *Journal of Economics of Innovation and New Technology* 3: 295–316.

Jorgenson, Dale W., and Kevin Stiroh (1999) "Information Technology and Growth." *American Economic Review, Papers and Proceedings* (May).

Kemerer, C. F., and G. L. Sosa (1991) "Systems Development Risks in Strategic Information Systems." *Information and Software Technology* 33(3): 212–223.

Levy, Frank, Anne Beamish, Richard J. Murnane, and David Autor (2000) "Computerization and Skills: Examples from a Car Dealership." Working Paper, MIT and Harvard. Retrieved from http://web.mit.edu/flevy/www/car-paper.pdf.

Lichtenberg, F. R. (1995) "The Output Contributions of Computer Equipment and Personnel: A Firm-Level Analysis." *Economics of Innovation and New Technology* 3: 201–217.

Malone, T. W., J. Yates, and R. I. Benjamin (1987) "Electronic Markets and Electronic Hierarchies." *Communications of the ACM* 30(6): 484–497.

McKenney, J. L., and T. H. Clark (1995) "Procter and Gamble: Improving Consumer Value through Process Redesign." Harvard Business School Case Study 9-195-126.

Milgrom, P., and J. Roberts (1990) "The Economics of Modern Manufacturing: Technology, Strategy, and Organization." *American Economic Review* 80(3): 511–528.

Murnane, Richard J., Frank Levy, and David Autor (1999) "Technological Change, Computers and Skill Demands: Evidence from the Back Office Operations of a Large Bank." NBER Economic Research Labor Workshop (June).

Oliner, Stephen D., and Daniel E. Sichel (1994) "Computers and Output Growth Revisited: How Big is the Puzzle?" *Brookings Papers on Economic Activity: Microeconomics* 2: 273–334.

Oliner, Stephen D., and Daniel E. Sichel (2000) "The Resurgence of Growth in the Late 1990s: Is Information Technology the Story?" *Journal of Economic Perspectives* 14(4): 3–22.

Orlikowski, W. J. (1992) "Learning from Notes: Organizational Issues in Groupware Implementation." In *Conference on Computer Supported Cooperative Work*, ed. J. Turner and R. Kraut. Toronto, Association for Computing Machinery: 362–369.

Parker, R., and B. Grimm (2000) "Software and Real Output: Recent Developments at the Bureau of Economic Analysis." Bureau of Economic Analysis (April 7).

Radner, R. (1993) "The Organization of Decentralized Information Processing." *Econometrica* 62: 1109–1146.

Rangan, V., and M. Bell (1999) "Dell Online." Harvard Business School Case Study 9-598-116.

Roach, Stephen S. (1987) "America's Technology Dilemma: A Profile of the Information Economy." Morgan Stanley Special Economic Study (April).

Short, James E., and N. Venkatraman (1992) "Beyond Business Process Redesign: Redefining Baxter's Business Network." *Sloan Management Review* 34(1): 7–20.

Solow, R. M. (1987) "We'd Better Watch Out." *New York Times Book Review* (July 12): 36.

Stiroh, Kevin (2002) "Information Technology and the U.S. Productivity Revival: What Do the Industry Data Say?" *American Economic Review* 92(5) (December): 1559–1576.

Stiroh, Kevin (2003) "Reassessing the Role of IT in the Production Function: A Meta-Analysis." NBER/CRIW/CREST Conference on R&D, Education, and Productivity in honor of Zvi Griliches (August).

Vitale, M., and Konsynski, B. (1988) "Baxter Healthcare Corp.: ASAP Express." Harvard Business School Case 9-188-080.

Yang, Shinkyu, and Erik Brynjolfsson (2003) "Intangible Assets and Growth Accounting: Evidence from Computer Investments." MIT working paper.

2

Projecting Productivity Growth: Lessons from the U.S. Growth Resurgence

Dale W. Jorgenson, Mun S. Ho, and Kevin J. Stiroh

The unusual combination of more rapid growth and lower inflation in the United States from 1995 to 2000 touched off a strenuous debate among economists about whether improvements in U.S. economic performance could be sustained. This debate has now given way to a broad consensus that the role of information technology is the key to understanding the American growth resurgence. Questions persist about whether similar trends have characterized the leading industrialized economies. The answers to these questions are essential for resolving uncertainties about future growth that currently face decision makers in both public and private sectors.

In this chapter we review the most recent evidence on growth in the United States and G7 countries and quantify the role of information technology (IT). Despite downward revisions to the gross domestic product (GDP) and investment in the annual revisions of the U.S. National Income and Product Accounts by the Bureau of Economic Analysis (BEA) in July of 2002, we conclude that the U.S. productivity revival remains largely intact and that IT investment is the predominant source of this revival. The capital deepening contribution from computer hardware, software, and telecommunications equipment greatly exceeded the contribution from all other forms of investment to labor productivity growth after 1995. An increase in total factor productivity (TFP) growth in the IT-producing sectors also contributed to the resurgence of labor productivity, modestly augmented by a smaller increase in TFP growth elsewhere in the economy.

Jorgenson (2003) compiled detailed information on investment in information technology and economic growth in the G7 countries in the 1990s. This has two important advantages over previous international comparisons. First, the estimates of IT investment are based on national accounting data. Second, prices of information technology equipment and software are comparable among the seven countries. There is clear evidence of a surge of IT investment in seven countries, even in Japan, which experienced slowdowns in economic growth during the period 1995–2000. We next turn to the future of U.S. productivity growth for the U.S. economy, defined broadly to include business, households, and the government. Our overall conclusion is that the projections of Jorgenson and Stiroh (2000), prepared more than three years ago, are largely on target. Our base-case projection of trend labor productivity growth for the next decade is 1.64 percent per year, below the average of 2.11 percent per year for the period 1995–2000. Our projection of output growth for the next decade is only 2.74 percent per year, compared with 4.10 percent per year for 1995–2000.[1] The difference is largely due to a projected slowdown in the growth in hours worked owing to changing demographics. We conclude that the American growth resurgence of the late 1990s was not sustainable because it depended in large part on a rate of work force expansion that will not be maintained.

We emphasize that projecting growth for periods as long as a decade is fraught with uncertainty. Our pessimistic projection of labor productivity growth is only 1.02 percent per year, while our optimistic projection is 2.38 percent. The range for output growth is from 2.12 percent in the pessimistic case to 3.48 percent in the optimistic case. These ranges result from fundamental uncertainties about future patterns of investment and changes in technology in the production of IT equipment and software. Jorgenson (2001) has traced these uncertainties to variations in the product cycle for semiconductors, the most important component of computers and telecommunications equipment.

The starting point for projecting U.S. output growth is a projection of future growth of the labor force. The growth of hours worked of 1.99 percent per year from 1995 through 2000 is not sustainable because labor force growth for the next decade will average only 1.1 percent. The slowdown in the growth of hours worked would have reduced output growth by 0.89 percent, even if labor productivity growth had continued unabated. We estimate that labor productivity growth from 1995 through 2000 also exceeded its sustainable rate, however, due to exceptionally high rates of investment in information technology equipment and software. We project output growth at 2.74 percent for the next decade, below the rate of growth of output during the period 1980–1995, prior to the growth resurgence of the late 1990s. The decompositions are quite different, with stronger productivity growth offset by slower projected hours growth.

In the next section we review the historical record, extend the estimates of Jorgenson (2001) to include data for 2000 and 2001, and revise estimates of economic growth for earlier years to incorporate new information. We employ the same methodology and summarize it briefly. We compare IT investment and economic growth for the G7 countries analyzed by Jorgenson (2003). In the subsequent section we present our projections of the trend growth of output and labor productivity in the United States for the next decade. Our methodology and data sources are described in an Appendix.

Reviewing the Historical Record

Our methodology for analyzing the sources of growth is based on the production possibility frontier introduced by Jorgenson (1996, pp. 27–28). This framework encompasses substitution between investment and consumption goods on the output side and between capital and labor inputs on the input side. Total factor productivity is defined as output per unit of both capital and labor inputs. Jorgenson and Stiroh (2000), Jorgenson (2001), and Jorgenson, Ho, and Stiroh (2002) have used this methodology to measure the contributions of information technology to U.S. economic growth and the growth of labor productivity.

Our output measure is broader than the GDP concept in the U.S. National Income and Product Accounts, the nonfarm business sector that is the focus of many productivity studies (BLS, 2000; Oliner and Sichel, 2000, 2002), or the private sector measure used in Jorgenson, Ho, and Stiroh (2002). In particular, we include service flows from residential housing and consumer durables, as well as the rate of return to government capital, in output, as in Jorgenson (2001). Our output estimates reflect the most recent revisions to the U.S. National Income and Product Accounts (NIPA), released in July 2002.

Table 2.1 and figure 2.1 report our estimates of the sources of economic growth. For the period 1980–2000, output grew 3.24 percent per year. Capital input contributed 47.2 percent of this growth, or 1.53 percent per year. Labor input followed in importance with 38.6 of growth, or 1.25 percent per year. Less than 15.0 percent of output growth, 0.47 percentage points, reflects growth in TFP or output per unit of input. These results are consistent with the other recent growth accounting estimates, including CEA (2001), Jorgenson and Stiroh (2000), Jorgenson (2001), Jorgenson, Ho, and Stiroh (2002), and Oliner and Sichel (2000, 2002).

Our data also reveal substantial acceleration in output growth after 1995. The growth rate of output increased from 2.36 percent per year for 1989–1995 to 4.10 percent for 1995–2000, reflecting a substantial acceleration in IT investment and a modest deceleration in non-IT investment. For the period 1995–2001, which includes the U.S. recession that began in March 2001, output growth was 3.58 percent. This is considerably slower, and we focus our attention on the period 1995–2000 to avoid cyclical effects of the 2001 recession.

On the input side, more rapid capital accumulation contributed 0.93 percentage points to the post-1995 acceleration through 2000, while faster growth of labor input contributed 0.40 percentage points and accelerated TFP growth the remaining 0.41 percentage points. These estimates are all smaller when 2001 is included. Finally, the contribution of capital input from IT increased from 0.47 percentage points per year for 1989–1995 to 0.97 for 1995–2000, exceeding the increased contributions of all other forms of capital.

The last rows in table 2.1 present an alternative decomposition of the contribution of capital and labor inputs, described in greater detail in the appendix. The contribution of capital reflects the contributions of capital quality and capital stock, while the contribution of labor includes labor quality and hours worked. Intuitively, capital quality growth represents the shift in the composition of capital toward investment goods, such as information technology equipment and software, with higher marginal products. Similarly, labor quality growth embodies the shift in the composition of the labor force toward more highly productive workers.

Table 2.1 shows that the revival of output growth after 1995 can be attributed to two forces. First, the rising contribution of capital quality

	1980–2000	1980–1989	1989–1995	1995–2000	1995–2001	1995–2000 less 1989–1995	1995–2001 less 1989–1995
Growth in GDP (Y)	3.24	3.34	2.36	4.10	3.58	1.74	1.22
Contribution of selected output components							
Other output (Y_n)	2.57	2.81	1.83	3.04	2.68	1.21	0.85
Computer output (Y_c)	0.22	0.20	0.15	0.32	0.27	0.17	0.12
Software output (Y_s)	0.17	0.11	0.15	0.30	0.24	0.15	0.09
Communications output (Y_m)	0.12	0.10	0.08	0.19	0.14	0.11	0.06
Information technology services (Y_{it})	0.16	0.12	0.14	0.25	0.24	0.11	0.10
Contribution of capital and CD Services (K)	1.53	1.46	1.15	2.08	2.04	0.93	0.89
Other (K_n)	0.93	1.00	0.68	1.11	1.11	0.43	0.43
Computers (K_c)	0.31	0.25	0.21	0.53	0.49	0.32	0.28
Software (K_s)	0.16	0.09	0.16	0.28	0.27	0.12	0.11
Communications (K_m)	0.13	0.12	0.10	0.16	0.17	0.06	0.07
Contribution of labor (L)	1.25	1.35	0.98	1.38	1.12	0.40	0.14
Aggregate total factor productivity (TFP)	0.47	0.53	0.23	0.64	0.42	0.41	0.19
Contribution of capital and CD quality	0.56	0.45	0.39	0.96	0.93	0.57	0.54
Contribution of capital and CD stock	0.97	1.01	0.77	1.12	1.10	0.35	0.33
Contribution of labor quality	0.30	0.30	0.36	0.21	0.23	−0.15	−0.13
Contribution of labor hours	0.95	1.05	0.62	1.17	0.90	0.55	0.28

Note: A contribution of an output or input is defined as the share-weighted, real growth rate. Source: Author's calculation based on BEA, BLS, Census Bureau, and other data.

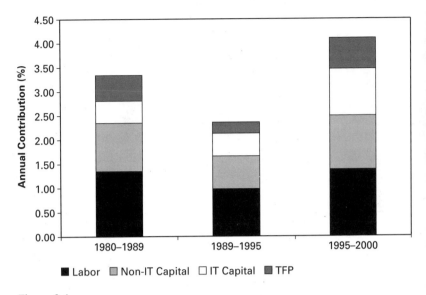

Figure 2.1
Sources of economic growth.

reflects a massive substitution toward IT capital in response to accelerating IT price declines; the growth of capital stock lagged considerably behind the growth of output. Second, the growth of hours worked surged, while labor quality growth stagnated. A fall in the unemployment rate and an increase in labor force participation drew more workers with relatively low marginal products into the work force.

Table 2.2 and figure 2.2 present estimates of the sources of average labor productivity (ALP) growth. Labor productivity is defined as output per hour worked and must be carefully distinguished from total factor productivity or output per unit of all inputs. In the appendix we show that growth in ALP can be decomposed into three components: capital deepening as a consequence of investment that provides more and better capital to workers, labor quality growth due to the increase in the proportion of more productive workers, and growth in TFP.

For the period 1980–2000 as a whole, growth in ALP of 1.63 percentage points per year accounted for 50.3 percent of output growth, owing to capital deepening of 0.86 percentage points per year, improvement of labor quality of 0.30 percentage points, and TFP growth of 0.47 percentage points. Growth in hours worked of 1.61 percentage

Table 2.2
Sources of growth in average labor productivity 1980–2001

	1980–2000	1980–1989	1989–1995	1995–2000	1995–2001	1995–2000 less 1989–1995	1995–2001 less 1989–1995
Output growth (Y)	3.24	3.34	2.36	4.10	3.58	1.74	1.22
Hours growth (H)	1.61	1.79	1.02	1.99	1.53	0.97	0.51
Average labor productivity growth (ALP)	1.63	1.55	1.34	2.11	2.05	0.77	0.71
Capital deepening	0.86	0.72	0.75	1.26	1.40	0.51	0.65
IT capital deepening	0.53	0.41	0.43	0.87	0.85	0.44	0.42
Other capital deepening	0.33	0.31	0.32	0.39	0.55	0.07	0.23
Labor quality	0.30	0.30	0.36	0.21	0.23	-0.15	-0.13
TFP growth	0.47	0.53	0.23	0.64	0.42	0.41	0.19
IT-related contribution	0.29	0.22	0.25	0.44	0.41	0.19	0.16
Other contribution	0.18	0.31	-0.02	0.20	0.01	0.22	0.03

Note: A contribution of an output or input is defined as the share-weighted, real growth rate. Source: Author's calculation based on BEA, BLS, Census Bureau, and other data.

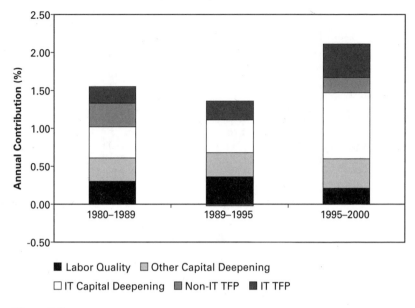

Figure 2.2
Sources of labor productivity growth.

points per year accounted for the remaining 49.7 percent of output growth.

Looking more closely at the post-1995 period, we see that labor productivity increased by 0.77 percentage points per year from 1.34 percentage points for 1989–1995 to 2.11 percentage points for 1995–2000, while hours worked increased by 0.97 percentage points per year from 1.02 percentage points for 1980–1995 to 1.99 percentage points for 1995–2000. When the recession of 2001 is included, labor productivity falls slightly, while hours growth falls considerably for 1995–2001, which underscores the remarkable strength of U.S. productivity growth during this downturn.

The labor productivity growth revival through 2000 reflects more rapid IT-capital deepening of 0.44 percentage points, partly offset by a small increase in non-IT-capital deepening of 0.07 percentage points. It also reflects accelerated productivity growth in IT production of 0.19 percentage points and in non-IT production of 0.22 percentage points. Finally, the contribution of labor quality growth fell by 0.15 percentage points.

Jorgenson (2003) compiled estimates of the contribution of IT investment to the growth of output in the G7 nations, including the four leading countries of Europe—France, Germany, Italy, and the United Kingdom—as well as Canada, Japan, and the United States. An important innovation was the introduction of "harmonized" price deflators to incorporate comparable quality adjustments for IT assets across countries. Figure 2.3 presents results for the sub-periods 1980–1989, 1989–1995 and 1995–2000. For all seven countries the contribution of IT investment accelerated after 1995, even in Japan, the only country that experienced a slowdown in growth during the late 1990s.

Jorgenson (2003) also presents growth rates of output for all seven countries before and after 1995, and we give the results in figure 2.4. The acceleration in economic growth during the last half of the 1990s was most dramatic for Canada, but both the United States and France also experienced a large increase in the rate of growth of output. A detailed analysis of the sources of economic growth for the G7 countries requires information on the impact of non-IT investment, as well as the contributions of labor input and TFP growth. Jorgenson (2003) provides an assessment of the role of information technology for all the G7 nations like that we have presented for the United States.

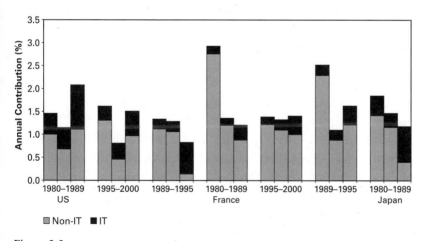

Figure 2.3
Capital input of information technology by country.

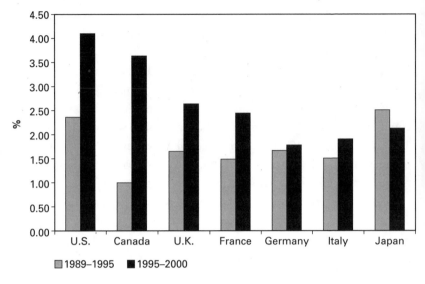

Figure 2.4
Output growth, 1989–1995 versus 1995–2000.

Projecting Productivity Growth

While there is no disagreement about the resurgence of ALP growth in the United States after 1995, there has been considerable debate about whether this is permanent or transitory. This distinction is crucial for understanding the sources of the recent productivity revival and projecting future productivity growth. Changes in the underlying trend growth rates of productivity and the work force are permanent, while cyclical factors such as strong output growth due to extraordinarily high rates of investment are transitory.

This section presents our projections of trend rates of growth for output and labor productivity over the next decade, abstracting from business cycle fluctuations. Our key assumptions are that output and the reproducible capital stock will grow at the same rate and that labor hours and the labor force will also grow at the same rate.[2] These are characteristic features of the United States and most industrialized economies over periods longer than a typical business cycle. For example, U.S. output growth averaged 3.31 percent per year for 1959–2001, while our measure of the reproducible capital stock grew 3.55 percent.[3]

We begin by decomposing the aggregate capital stock between reproducible capital stock and land, which we take to be fixed. We assume
that output and reproducible capital must grow at the same rate in the
long run. Equality between the growth rates of output and reproducible
capital is a long-run relationship that averages over cyclical and sto-
chastic elements and removes the transitional dynamics due to capital
accumulation. We construct estimates of trend output and labor productivity growth, conditional on the projected growth of the remaining
sources of economic growth.

The second part of the definition of trend growth is that the unemployment rate remains constant and hours growth matches labor force
growth. Growth in hours worked was exceptionally rapid in the
1995–2000 period, as the unemployment rate fell from 5.6 percent in
1995 to 4.0 percent in 2000, so output growth was considerably above
its trend rate.[4] We estimate hours growth over the next decade by means
of detailed demographic projections, based on Census Bureau data. We
allow hours worked to increase by 0.1 percent per year to eliminate
unemployment in excess of the "natural" rate at which inflation remains
the same.

In order to complete intermediate-term trend growth projections, we
require estimates of capital and labor shares, the IT output share, the
share of reproducible capital stock, capital quality growth, labor quality
growth, and TFP growth. Labor quality growth and the various shares
are relatively easy to project, while extrapolations of the other variables
are subject to considerable uncertainty. Accordingly, we present three sets
of projections—a base-case scenario, a pessimistic scenario, and an optimistic scenario.

We hold labor quality growth, hours growth, the capital share, the
reproducible capital stock share, and the IT output share constant across
the three scenarios. We refer to these as the "common assumptions." We
vary IT-related TFP growth, the contribution to TFP growth from non-
IT sources, and capital quality growth across these scenarios and label
them "alternative assumptions." Generally speaking for these variables,
the base-case scenario incorporates data from the long expansion of
1989–2000, the optimistic scenario assumes that the patterns of
1995–2000 will persist, and the pessimistic case assumes that the
economy reverts to 1980–1995 averages.

Common Assumptions

Hours growth and labor quality growth are relatively easy to project. The Congressional Budget Office (2002) projects growth in the potential labor force of 1.0 percent per year, a slight decrease from earlier projections. We project hours growth at 1.1 percent per year for 2002–2012 for all unemployment to decline to the natural rate. CBO (2002) does not employ the labor quality concept.

We construct our own projections of demographic trends. Ho and Jorgenson (1999) have shown that the dominant trends in labor quality growth are due to rapid improvements in educational attainment in the 1960s and 1970s and the rise in female participation rates in the 1970s. The improvement in educational attainment of new entrants into the labor force largely ceased in the 1990s, although the average educational level continued to rise as younger and better educated workers entered the labor force and older workers retired.

We project growth in the population from the demographic model of the Bureau of the Census, which breaks the population down by individual year of age, race, and sex.[5] For each group the population in each period is equal to the population in the previous period, less deaths plus net immigration. Death rates are group-specific and are projected by assuming a steady rate of improvement in health. The population of newborns in each period reflects the number of females in each age group and the age- and race-specific fertility rates. These fertility rates are projected to fall steadily.

We observe labor force participation rates in the last year of our sample period. We then project the work force by assuming constant participation rates for each sex-age group. The educational attainment of workers who are over 35 years of age in the last year of the sample is assumed to remain unchanged. For those who are younger than 35 we assume that the educational attainment of workers of a given age in the forecast period is equal to the attainment of workers of the same age in the base year.

Our index of labor quality is constructed from hours worked and compensation rates. We project hours worked by multiplying the projected population in each sex-age-education group by the annual hours per person in the last year of the sample. The relative compensation rates for each group are assumed to be equal to the observed compensation in this sample period. We project labor quality growth from our projections of hours worked and compensation per hour.

Our estimates suggest that hours growth will be about 1.1 percent per year over the next 10 years, slightly above the CBO (2002) estimate. We estimate that growth in labor quality will be 0.16 percent per year over the next decade. This is considerably lower than the 0.50 percent growth rate for the period 1980–2000, driven by rising average educational attainment and stabilizing female participation.

The capital share has not shown any obvious trend over the past 40 years. We assume it holds constant at 40.8 percent, the average for 1980–2000. Similarly, the fixed reproducible capital share has shown little change, and we assume that it remains constant at 81.1 percent, the average for 1980–2000. We assume the IT output share remains at 4.2 percent, the average for 1995–2000. This is likely to prove a conservative estimate, since IT has steadily increased in relative importance in the U.S. economy, rising from 1.7 percent of output in 1970 to 2.3 percent in 1980, 3.3 percent in 1990, and 4.7 percent in 2000.

Alternative Assumptions

Productivity growth in IT production has been extremely rapid in recent years, with a substantial acceleration after 1995. For 1989–1995 productivity growth for IT production averaged 7.29 percent per year, while for 1995–2000 growth averaged 10.16 percent. While these growth rates are high, they are consistent with industry-level productivity estimates for high-tech sectors. For example, Jorgenson, Ho, and Stiroh (forthcoming) report productivity growth of 18.00 percent per year for 1995–2000 in electronic components, including semiconductors, and 16.75 percent in computers and office equipment.

Jorgenson (2001) argues that the large increase in IT productivity growth was triggered by a much sharper acceleration in the decline of semiconductor prices. This can be traced to a shift in the product cycle for semiconductors in 1995 from three years to two years, a consequence of intensifying competition in the semiconductor market. It would be premature to extrapolate the recent acceleration in productivity growth into the indefinite future, however, because this depends on the persistence of a two-year product cycle for semiconductors.

To better gauge the future prospects of technical progress in the semi-conductor industry, we turn to *The International Technology Roadmap for Semiconductors* (2001). This *Roadmap*, constructed every two years by a consortium of industry associations and updated annually, projects a two-year product cycle through 2005 and a three-year product cycle thereafter. This is a reasonable basis for projecting the productivity growth related to IT for the U.S. economy. Moreover, continuation of a two-year cycle provides an upper bound for growth projections, while an immediate reversion to a three-year cycle gives a lower bound.

Our base-case scenario projects IT-related growth of 8.60 percent per year, the average for 1989–2000, giving equal weight to the two-year product cycle for 1995–2000 and the three-year product cycle for 1989–1995.[6] The optimistic scenario assumes that the two-year product cycle for semiconductors remains in place so that productivity growth in IT production averages 10.16 percent per year, as it did for 1995–2000. Our pessimistic projection assumes a reversion to the three-year semi-conductor product cycle of 1980–1995, when IT-related productivity growth was 7.29 percent per year. In all cases, the contribution of IT to TFP growth reflects the 1995–2000 IT share of GDP of 4.2 percent.

The non-IT TFP contribution is more difficult to project, so we present a range of alternative estimates that are consistent with the historical record. Our base-case projection uses the average contribution from the 1990s and assumes a contribution of 0.08 percentage points. This assumes that the myriad factors that drove TFP growth in the 1990s—technical progress, resource reallocations, and increased competitive pressures—will continue into the future. Our optimistic case assumes that the contribution for 1995–2000 of 0.20 percentage points per year will continue, while our pessimistic case assumes that the U.S. economy will revert back to the slow-growth period from 1980–1995 when this contribution averaged –0.02 percent per year.

The final step in our projections is to estimate the growth in capital quality. The one-sector neoclassical growth model has capital stock and output growing at the same rate in balanced growth equilibrium. We distinguish between IT and non-IT capital, and the historical record shows that substitution between these two types of capital is an important source of output and productivity growth. For the period 1980–2000 as a whole capital quality growth contributed 0.56 percentage points to output growth as firms substituted toward IT capital inputs with higher marginal products.

An important difficulty in projecting capital quality growth from recent data, however, is that investment patterns in the 1990s may partially reflect an unsustainable investment boom in response to temporary factors like Y2K investment and the NASDAQ stock market bubble, which may have skewed investment toward IT assets. Capital quality for 1995–2000 grew at 2.34 percent per year as firms invested heavily in IT, but there has been a sizable slowdown in IT investment in the second half of 2000 and 2001. Therefore, we must be cautious about relying too heavily on the most recent investment experience.

Our base-case projection uses the average rate of capital quality growth for 1989–2000, which was 1.58 percentage points; this averages the high rates of substitution of IT for non-IT capital inputs in the late 1990s with the more moderate rates of the early 1990s. Our optimistic projection ignores the possibility that capital substitution was unsustainably high in the late 1990s and assumes that capital quality growth will continue at the annual rate of 2.34 percent for the period 1995–2000. Our pessimistic scenario assumes that the growth of capital quality will revert to the 0.95 annual growth rate for 1980–1995.

Output and Productivity Projections
Table 2.3 assembles the components of our projections and presents the three scenarios. The top panel of table 2.3 shows the projected growth of output, labor productivity, and the effective capital stock. The second panel reports the five factors that are held constant across scenarios— hours growth, labor quality growth, the capital share, the IT output share, and the reproducible capital stock share. The bottom panel includes the three components that vary across scenarios—TFP growth in IT, the TFP contribution from other sources, and capital quality growth.

Our base-case scenario puts trend labor productivity growth at 1.64 percent per year and trend output growth at 2.74 percent per year. Figure 2.5 presents our projection of labor productivity growth and its decomposition, while figure 2.6 gives the corresponding projection of output growth. Projected productivity growth falls short of our estimate of 2.11 percent for 1995–2000 and even the 2.05 percent for 1995–2001.

Table 2.3
Output and labor productivity projections, total economy

	Projections		
	Pessimistic	Base-case	Optimistic
	Projections		
Output growth	2.12	2.74	3.48
ALP growth	1.02	1.64	2.38
Effective capital stock	1.72	2.22	2.82
	Common assumptions		
Hours growth	1.10	1.10	1.10
Labor quality growth	0.157	0.157	0.157
Capital share	0.408	0.408	0.408
IT output share	0.042	0.042	0.042
Reproducible capital stock share	0.811	0.811	0.811
	Alternative assumptions		
TFP growth in IT	7.29	8.60	10.16
Implied IT-related TFP contribution	0.31	0.36	0.43
Other TFP contribution	−0.02	0.08	0.20
Capital quality growth	0.95	1.58	2.34
Implied capital deepening contribution	0.64	1.10	1.66

Notes: In all projections, hours growth and labor quality growth are from internal projections, capital share and reproducible capital stock shares are 1980–2000 averages, and IT output shares are for 1995–2000. Pessimistic case uses 1980–1995 average growth of capital quality, IT-related TFP growth, and non-IT TFP contribution. Base case uses 1989–2000 averages, and optimistic case uses 1995–2000 averages.

Output growth is considerably slower due to the projected slowdown in hours growth. Hours grew at 1.99 percent per year for 1995–2000 and 1.53 percent per year for 1995–2001, compared to our projection of 1.1 percent for the next decade. Capital stock growth is projected to fall in the base-case to 2.22 percent per year.

Our base-case scenario incorporates the underlying pace of technical progress in semiconductors embedded in the *International Technology Roadmap* projection and puts the contribution of IT-related TFP below that of 1995–2000 as the semiconductor industry eventually returns to a three-year product cycle. Slower growth is partly offset by the larger

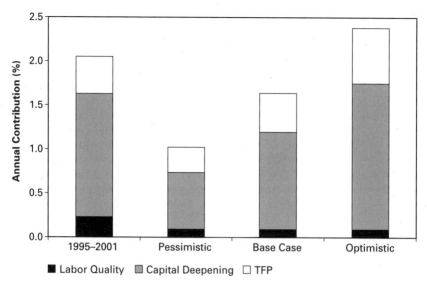

Figure 2.5
Range of labor productivity projections.

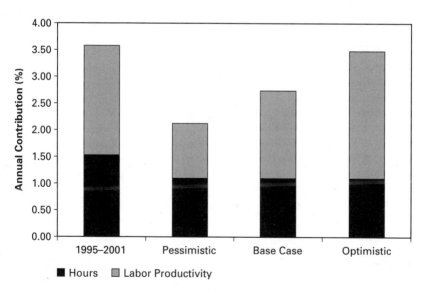

Figure 2.6
Range of output projections.

IT output share. Other TFP growth also makes a smaller contribution. Although the slower pace of capital input growth is partly offset by slower hours growth, capital deepening is insufficient to raise the projected growth rate to the observed growth rate for 1995–2001. Our optimistic scenario projects labor productivity growth at 2.38 percent per year, reflecting our assumption of continuing rapid technical progress in IT production. In particular, the two-year product cycle in semiconductors is assumed to persist for the intermediate future, driving

Table 2.4
Output and labor productivity projections, private domestic economy

		Projections		
	1995–2001	Pessimistic	Base–case	Optimistic
		Projections		
Output growth	3.93	2.43	2.92	3.84
ALP growth	2.27	1.33	1.82	2.74
Effective capital stock	2.85	1.89	2.27	2.98
		Common assumptions		
Hours growth	1.66	1.10	1.10	1.10
Labor quality growth	0.381	0.157	0.157	0.157
Capital share	0.426	0.428	0.428	0.428
IT output share	0.051	0.051	0.051	0.051
Reproducible capital stock share	0.779	0.778	0.778	0.778
		Alternative assumptions		
TFP growth in IT	9.35	7.19	8.59	10.21
Implied IT-related TFP contribution	0.48	0.27	0.39	0.51
Other TFP contribution	−0.02	0.21	0.08	0.19
Capital quality growth	2.58	0.98	1.78	2.67
Implied capital deepening contribution	1.59	0.76	1.26	1.95

Notes: In all projections, hours growth and labor quality growth are from internal projections, capital share and reproducible capital stock shares are 1980–2000 averages, and IT output shares are for 1995–2000. Pessimistic case uses 1980–1995 average growth of capital quality, IT-related TFP growth, and non-IT TFP contribution. Base case uses 1989–2000 averages, and optimistic case uses 1995–2000 averages.

Table 2.5
Projection revisions, private domestic economy

	1995–2000 current, less 1995–2000 old	Projections		
		Pessimistic	Base-case	Optimistic
		Projections		
Output growth	−0.05	0.00	−0.42	−0.25
ALP growth	−0.03	0.00	−0.42	−0.25
Effective capital stock	−0.06	−0.07	−0.41	−0.30
		Common assumptions		
Hours growth	−0.02	0.00	0.00	0.00
Labor quality growth	0.13	−0.11	−0.11	−0.11
Capital share	−0.01	0.00	0.00	0.00
IT output share	0.00	0.00	0.00	0.00
Reproducible capital stock share	−0.01	−0.03	−0.03	−0.03
		Alternative assumptions		
TFP growth in IT	−0.12	−0.19	−0.21	−0.12
Implied IT-related TFP contribution	−0.02	−0.10	−0.05	−0.01
Other TFP contribution	−0.14	0.13	−0.15	−0.14
Capital quality growth	0.22	0.14	0.03	0.22
Implied capital deepening contribution	0.05	0.03	−0.16	−0.03

Notes: Revisions result from annual revisions to BEA source data from 1999–2001, revised labor quality data and projections, and a revised projection for hours growth. In all projections, hours growth and labor quality growth are from internal projections, capital share and reproducible capital stock shares are 1980–2000 averages, and IT output shares are for 1995–2000. Pessimistic case uses 1980–1995 average growth of capital quality, IT-related TFP growth, and non-IT TFP contribution. Base case uses 1989–2000 averages, and optimistic case uses 1995–2000 averages.

Table 2.6
Projection comparison, total economy versus private domestic economy

	Total economy (1995–2001), less private economy (1995–2001)	Projections		
		Pessimistic	Base-case	Optimistic
		Projections		
Output growth	−0.38	−0.31	−0.18	−0.36
ALP growth	−0.26	−0.31	−0.18	−0.36
Effective capital stock	−0.15	−0.17	−0.05	−0.16
		Common assumptions		
Hours growth	−0.13	0.00	0.00	0.00
Labor quality growth	0.00	0.00	0.00	0.00
Capital share	−0.02	−0.02	−0.02	−0.02
IT output share	−0.01	−0.01	−0.01	−0.01
Reproducible capital stock share	0.03	0.03	0.03	0.03
		Alternative assumptions		
TFP growth in IT	−0.03	0.10	0.01	−0.05
Implied IT-related TFP contribution	−0.07	0.04	−0.03	−0.08
Other TFP contribution	0.03	−0.23	0.00	0.01
Capital quality growth	−0.29	−0.03	−0.20	−0.33
Implied capital deepening contribution	−0.22	−0.12	−0.16	−0.29

Notes: Total economy includes the government sector. In all projections, hours growth and labor quality growth are from internal projections, capital share and reproducible capital stock shares are 1980–2000 averages, and IT output shares are for 1995–2000. Pessimistic case uses 1980–1995 average growth of capital quality, IT-related TFP growth, and non-IT TFP contribution. Base case uses 1989–2000 averages, and optimistic case uses 1995–2000 averages.

rapid TFP growth in IT production, as well as continued substitution toward IT capital input and rapid growth in capital quality. In addition, non-IT TFP growth continues at the pace for 1995–2000. Finally, the pessimistic projection of 1.02 percent per year growth in labor productivity assumes that underlying trends in TFP growth and growth in capital quality revert back to the sluggish growth rates of the 1980–1995 period and that the three-year product cycle for semiconductors begins immediately. Even with the larger share of IT, labor productivity growth in this scenario will fall short of the rates seen in the 1970s and 1980s.

Conclusions

Our primary conclusion is that a consensus has emerged about trend rates of growth for output and labor productivity. Our methodology assumes that trend growth rates in output and reproducible capital are the same and that hours growth is constrained by the growth of the labor force along a balanced growth path. While productivity is projected below the pace seen in the late 1990s, we conclude that the U.S. productivity revival is likely to remain intact for the intermediate future. Similar projections for the leading industrialized economies will have to await the availability of detailed information on productivity similar to that we have employed for the United States.

Our second conclusion is that trend growth rates are subject to considerable uncertainty. For the U.S. economy this can be identified with the future product cycle for semiconductors and its impact on the production of other high-tech gear. The switch from a three-year to a two-year product cycle in 1995 produced a dramatic increase in the rate of decline of IT prices. This is reflected in the investment boom of 1995–2000 and the massive substitution of IT capital for other types of capital that took place in response to price changes. The issues that must be confronted by policymakers are whether this two-year product cycle can continue and whether firms will continue to respond to the dramatic improvements in the performance/price ratio of IT investment goods.

The lessons from the U.S. growth resurgence are, first, that investment in information technology equipment and software has increased in relative importance in all the G7 countries. This lesson has not been fully absorbed by analysts who focus on trends in the growth of output or labor productivity alone. Second, future trends in economic growth depend on future labor force growth, as well as future changes in technology. While trends in labor force growth are relatively easy to project, considerable uncertainty will continue to characterize projections of new developments in technology.

A complete understanding of the role of information technology in G7 requires a full accounting for recent economic growth, as presented in Jorgenson (2003), which incorporates quality-adjusted prices for IT assets. The U.S. National Income and Product Accounts currently employ such deflators for computer hardware, and portions of software and telecom equipment. This is critical since failure to capture these important quality improvements leads both capital input and output to be severely understated.

Appendix

Methodology

In the *production possibility frontier* output (Y) consists of consumption goods (C) and investment goods (I), while input (X) consists of capital services (K) and labor input (L). Output can be further decomposed into IT investment goods—computer hardware (I_c), computer software (I_s), communications equipment (I_m)—and non-IT output (Y_n). Capital services can be similarly decomposed into the capital service flows from computer hardware (K_c), software (K_s), communications equipment (K_m), and non-IT capital services (K_n).[7] The input function (X) is augmented by *total factor productivity* (A). The production possibility frontier can be represented as

$$Y(Y_n, I_c, I_s, I_m) = A \cdot X(K_n, K_c, K_s, K_m, L) \qquad (1)$$

Under the standard assumptions of competitive product and factor markets, and constant returns to scale, Equation (1) can be transformed into an equation that accounts for the sources of economic growth:

$$\overline{w}_{Y_n} \Delta \ln Y_n + \overline{w}_{I_c} \Delta \ln I_c + \overline{w}_{I_s} \Delta \ln I_s + \overline{w}_{I_m} \Delta \ln I_m = \overline{v}_{K_n} \Delta \ln K_n$$
$$+ \overline{v}_{K_c} \Delta \ln K_c + \overline{v}_{K_s} \Delta \ln K_s + \overline{v}_{K_m} \Delta \ln K_m + \overline{v}_L \Delta \ln L + \Delta \ln A \qquad (2)$$

where $\Delta x \equiv x_t - x_{t-1}$, $\bar{w}$ denotes the average output shares, $\bar{v}$ the average input shares, and $\bar{w}_{Y_n} + \bar{w}_{I_c} + \bar{w}_{I_s} + \bar{w}_{I_m} = \bar{v}_{K_n} + \bar{v}_{K_c} + \bar{v}_{K_s} + \bar{v}_{K_m} + \bar{v}_L = 1$. The shares are averaged over periods t and $t - 1$. We refer to the share-weighted growth rates in Equation (2) as the *contributions* of the inputs and outputs.

Average labor productivity is defined as the ratio of output to hours worked, so that $ALP = y = Y/H$, where the lower-case variable (y) denotes output (Y) per hour (H). Equation (2) can be rewritten in per hour terms as

$$\Delta \ln y = \bar{v}_{K_n} \Delta \ln K_n + \bar{v}_{K_{IT}} \Delta \ln K_{IT} + \bar{v}_L (\Delta \ln L + \Delta \ln H) + \Delta \ln A \tag{3}$$

where $\bar{v}_{K_{IT}} = \bar{v}_{K_c} + \bar{v}_{K_s} + \bar{v}_{K_m}$.

Equation (3) decomposes ALP growth among three components. The first is *capital deepening*, defined as the contribution of capital services per hour and allocated between non-IT and IT components. The interpretation of capital deepening is that increases in capital per worker enhance labor productivity in proportion to the capital share. The second component is *labor quality improvement*, defined as the contribution of increases in labor input per hour worked. This reflects changes in the composition of the work force and raises labor productivity in proportion to the labor share. The third component is *total factor productivity* growth, which raises ALP growth point for point.

In an interindustry production model like that of Jorgenson, Ho, and Stiroh (forthcoming), the growth of TFP reflects the productivity contributions of individual industries. It is difficult, however, to create the detailed industry data needed to measure industry-level productivity in a timely and accurate manner. The Council of Economic Advisors (CEA, 2001), Jorgenson and Stiroh (2000), Jorgenson (2001), Jorgenson, Ho, and Stiroh (2002), and Oliner and Sichel (2000, 2002) have employed the price dual of industry-level productivity to estimate TFP growth in the production of IT equipment and software.

Intuitively, the idea underlying the dual approach is that declines in relative prices for IT investment goods reflect productivity growth in the IT-producing industries. We weight these relative price declines by the shares in output of each of the components of IT investment in order to estimate the contribution of IT production to economy-wide TFP growth. This enables us to decompose aggregate TFP growth as

$$\Delta \ln A = \bar{u}_{IT} \Delta \ln A_{IT} + \bar{u}_n \Delta \ln A_n \qquad (4)$$

where $\bar{u}_{IT}$ represents IT's average share of output, $\Delta \ln A_{IT}$ is IT-related productivity growth, and $\bar{u}_{IT} \Delta \ln A_{IT}$ is the contribution to aggregate TFP from IT production. Non-IT productivity growth $\Delta \ln A_n$ includes productivity gains in other industries, as well as reallocations of inputs and outputs among sectors.

We estimate the contribution to aggregate TFP growth from IT production $\bar{u}_{IT} \Delta \ln A_{IT}$ by estimating output shares and growth rates of productivity for computer hardware, software, and communications equipment. Productivity growth for each component of investment is the negative of the rate of price decline, relative to the price change of capital and labor inputs. The output shares are the final expenditures on these investment goods, divided by total output.[8] Finally, the contribution of non-IT productivity growth $\bar{u}_n \Delta \ln A_n$ is derived from Equation (4) as a residual.

Data Sources

We briefly summarize the information required to implement Equations (1) to (6); more detailed descriptions are available in Jorgenson (2001) and Jorgenson, Ho, and Stiroh (2002). Our capital service estimates are based on the Tangible Wealth Study, published by the BEA and described in Lally (2002). This includes data on business, household, and government investment for the U.S. economy through 2001. We construct capital stocks from the investment data by the perpetual inventory method. We assume that the effective capital stock for each asset available for production is an average of current and lagged stocks. The data on tangible assets from BEA are augmented with inventory data to form our measure of the reproducible capital stock. The total capital stock also includes land and inventories.

Finally, we estimate the service flow for each component of capital stock by multiplying the rental price by the effective capital stock, as suggested by Jorgenson and Griliches (1996). Our estimates of rental prices incorporate the asset-specific differences in asset prices, tax rates, tax lifetimes, and depreciation rates presented by Jorgenson and Yun (2001). This is essential for understanding the productive impact of IT investment because IT capital inputs have dramatically higher rates of decline of asset prices and depreciation rates.

We refer to the ratio of capital services to capital stock as *capital quality*, so that

$$\Delta \ln KQ = \Delta \ln K - \Delta \ln Z \tag{5}$$

where KQ is capital quality, K is capital services, and Z is effective capital stock. The effective capital stock Z is a quantity index of 70 types of structures and equipment, plus land and inventories, using investment goods prices as weights. The flow of capital services K is a quantity index of the same stocks, using rental prices as weights. The difference in growth rates is the growth rate of capital quality. Capital quality increases as firms invest relatively more in assets with higher marginal products like information technology equipment and software.

Labor input is a quantity index of hours worked that takes into account the distribution of the work force by sex, employment class, age, and education. The weights used to construct the index reflect the compensation of the various types of workers. In the same way as for capital, we define *labor quality* as the ratio of labor input to hours worked, so that

$$\Delta \ln LQ = \Delta \ln L - \Delta \ln H \tag{6}$$

where LQ is labor quality, L is labor input, and H is hours worked. Labor quality rises as firms hire relatively more highly skilled and highly compensated workers.

Our labor data incorporate individual microdata on hours worked and compensation per hour from the Censuses of Population for 1970, 1980, and 1990 and the annual Current Population Surveys (CPS) for 1964–2001. We take total hours worked for employees directly from the NIPA (table 6.9c), self-employed hours worked for the non-farm business sector from the BLS, and self-employed hours worked in the farm sector from the Department of Agriculture.

Acknowledgments

The authors are grateful to Jon Samuels for excellent research assistance. The BLS and BEA have kindly provided data and advice. The views expressed in this chapter are those of the authors only and do not necessarily reflect those of the Federal Reserve System or the Federal Reserve Bank of New York.

Notes

1. We focus on the period 1995–2000 to avoid the cyclical effects of the 2001 recession. We discuss estimates for the period 1995–2001 later in the chapter. Note also that productivity growth for our broad coverage of the U.S. economy is somewhat slower than the nonfarm business sector.

2. The assumption that output and the capital stock grow at the same rate is a property of balanced growth equilibrium in the standard neoclassical growth model.

3. Reproducible assets exclude land.

4. These unemployment rates are annual averages for the civilian labor force, 16 years and older, from BLS.

5. The details of the population model are given in Bureau of the Census (2000).

6. Note that we explicitly exclude 2001 because the cyclical declines associated with the 2001 recession obscure the underlying trends.

7. Note that our output and capital service flow concepts include the services of residential structures and consumer durables, as well as government structures and equipment. See Jorgenson (2001) for details.

8. Output shares include personal consumption expenditures, gross private domestic investment, government purchases, and net exports for each type of IT equipment and software. Note that the use of the price dual to measure technological change assumes competitive markets in IT production. As pointed out by Aizcorbe (2002), the market for many IT components, notably semiconductors and software, is not perfectly competitive and part of the drop in prices may reflect changes in markups rather than technical progress. However, Aizcorbe concludes that the decline in markups accounts for only about one-tenth of the measured decline in the price of microprocessors in the 1990s.

References

Aizcorbe, Ana (2002). "Why are Semiconductor Prices Fallings So Fast? Industry Estimates and Implications for Productivity Measurement." Federal Reserve Board, Mimeo, February.

Bureau of Labor Statistics (2002). "Multifactor Productivity Trends, 2000." USDL 02-128, March 12.

Bureau of the Census (2000). "Methodology and Assumptions for the Population Projections of the United States: 1999 to 2100." www.census.gov/population/www/projections/natproj.html.

Colecchia, Alessandra, and Paul Schreyer (2002). "ICT Investment and Economic Growth in the 1990s: Is the United States a Unique Case? A Comparative Study of Nine OECD Countries." *Review of Economic Dynamics* 5(2) (April): 408–443.

Congressional Budget Office (2002). "The Budget and Economic Outlook: An Update." Washington, D.C.: Government Printing Office, August.

Council of Economic Advisors (2001). "Annual Report of the Council of Economic Advisors." In *Economic Report of the President*, January.

Ho, Mun Sing, and Dale W. Jorgenson (1999). "The Quality of the U.S. Workforce, 1948–95." Harvard University, manuscript.

International Technology Roadmap for Semiconductors 2001. Austin, Texas: Sematech Corporation, December. http://public.itrs.net.

Jorgenson, Dale W. (1996) "The Embodiment Hypothesis." In *Postwar U.S. Economic Growth*. Cambridge, Mass.: MIT Press.

Jorgenson, Dale W. (2001) "Information Technology and the U.S. Economy." *American Economic Review* 91(1) (March): 1–32.

Jorgenson, Dale W. (2003) "Information Technology and the G7 Economies." *World Economics* 4(4) (October–December): 1–32.

Jorgenson, Dale W., and Zvi Griliches (1996) "The Explanation of Productivity Change." In *Postwar U.S. Economic Growth*, Cambridge, Mass.: MIT Press.

Jorgenson, Dale W., and Kevin J. Stiroh (2000). "Raising the Speed Limit: U.S. Economic Growth in the Information Age." *Brookings Papers on Economic Activity* 2000(1): 125–211.

Jorgenson, Dale W., and Kun-Young Yun (2001) *Lifting the Burden: Tax Reform, the Cost of Capital, and U.S. Economic Growth*. Cambridge, Mass.: MIT Press.

Jorgenson, Dale W., Mun S. Ho, and Kevin J. Stiroh (2002). "Projecting Productivity Growth: Lessons from the U.S. Growth Resurgence." *Federal Reserve Bank of Atlanta Economic Review* 87(3) (third quarter): 1–14.

Jorgenson, Dale W., Mun S. Ho, and Kevin J. Stiroh (forthcoming) "Growth in U.S. Industries and Investments in Information Technology and Higher Education." In *Measurement of Capital in the New Economy*, ed. Carol Corrado, John Haltiwanger, and Charles Hulten. Chicago: University of Chicago Press.

Lally, Paul R. (2002) "Fixed Assets and Consumer Durable Goods for 1925–2001." *Survey of Current Business* 82(9) (September): 23–37.

Oliner, Stephen D., and Daniel E. Sichel (2000) "The Resurgence of Growth in the Late 1990s: Is Information Technology the Story?" *Journal of Economic Perspectives* 14(4) (Fall): 3–22.

Oliner, Stephen D., and Daniel E. Sichel (2002). "Information Technology and Productivity: Where Are We Now and Where Are We Going?" *Federal Reserve Bank of Atlanta Economic Review* 87(3) (third quarter): 15–44.

3

The Impacts of ICT on Economic Performance: An International Comparison at Three Levels of Analysis

Dirk Pilat and Andrew W. Wyckoff

The recent economic slowdown starting in 2000 has laid to rest some of the myths that surrounded information and communications technology (ICT) during the boom of the late 1990s: The business cycle is not dead, stock market valuations must be backed by sound profit expectations, and the ICT sector is not immune to downturns.[1] Nearly all assessments of the future role of ICT are more sober today than they were several years ago. In addition, new data—at the level of the economy, specific industries, and the firm—are now available, providing new empirical insights that were not possible a few years ago and allowing the exploration of questions such as the following: Why have some countries invested more in ICT than others? What factors help firms in seizing the benefits from ICT? How precisely does ICT affect firm performance? What policies should governments undertake to help firms benefit from ICT? In this sense, it is a better time to examine the role of ICT and the economy now that the hype of the new economy is over.

The State of ICT Diffusion

The economic impact of ICT is closely linked to the extent to which different ICTs have diffused. This is partly because ICT is a network technology; the more people and firms that use the network, the more benefits it generates. While ICT investment has accelerated in most OECD countries over the past decade, the pace of that investment differs widely. ICT investment rose from less than 15 percent of total nonresidential investment in the business sector in the early 1980s to between 15 percent and 30 percent in 2000. In 2000, the share of ICT investment was particularly high in the United States, Finland, and Australia (figure

3.1). These shares did not change much in 2001 in the countries for which data are available, although overall ICT investment declined somewhat in some countries, such as the United States and Canada. This suggests that ICT investment has not been affected disproportionally by the slowdown compared with other types of investment.

The rapid growth in ICT investment has been fueled by a rapid decline in the relative prices of computer equipment and the growing scope for the application of ICT (Jorgenson, 2001). The benefits of lower ICT prices have been felt as both firms investing in these technologies and consumers buying ICT have benefited from lower prices. The lower prices of ICT are only one of the drivers of investment, however; firms have also invested in ICT because it offers large potential benefits.

Another determinant of the economic impacts associated with ICT is the size of the ICT sector. Having an ICT-producing sector can be important, since ICT production has been characterized by rapid technological progress and has been faced with very strong demand. In 2000, value added in the ICT sector represented between 4 percent and 17 percent of business sector value added (figure 3.2), while about 6–7 percent of

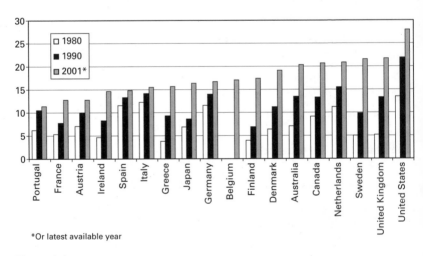

*Or latest available year

Figure 3.1
ICT investment in selected OECD countries (as a percentage of nonresidential gross fixed capital formation, business sector). Note: Estimates of ICT investment are not yet fully standardized across countries, mainly owing to differences in the capitalization of software in different countries. See Ahmad 2003. Source: OECD, Database on Capital Services.

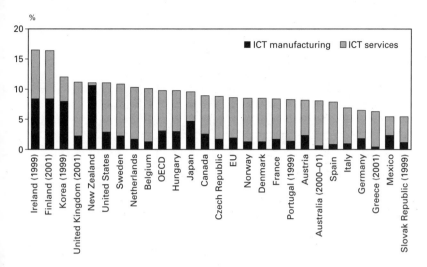

Figure 3.2
Share of the ICT sector in business sector value added, 2000 (or latest available year). There are some small differences in the definition of the ICT sector. See source for detail. Source: OECD, *Science, Technology, and Industry Scoreboard*, 2003.

total business employment in the OECD area could be attributed to ICT production.[2] While parts of the ICT sector are currently experiencing a slowdown, these shares are unlikely to change much in the short term.

A third indicator of ICT diffusion is the proportion of businesses that use the Internet to make purchases and sales (figure 3.3). This is not available for all OECD countries but shows a large number of firms using the Internet for sales or purchases in the Nordic countries (Denmark, Finland, Norway, and Sweden) as well as in Australia, the Netherlands, and New Zealand. In contrast, only a few firms in Greece, Italy, Portugal, and Spain use the Internet for sales or purchases. Monetary estimates of electronic commerce suggest that electronic commerce is growing, albeit more slowly than originally envisaged. However, it still accounts for a relatively small proportion of overall sales. For the few countries that currently measure this, Internet sales in 2000/2001 ranged between 0.2 percent and 2 percent of total sales. In the second quarter of 2002, 1.2 percent of all retail sales in the United States were carried out through computer-mediated networks, up from 1.0 percent in the second quarter of 2001.

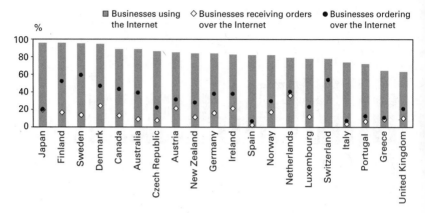

Figure 3.3
Proportion of businesses using the Internet for purchases and sales, 2002 (or latest available year), percentages of businesses with ten or more employees. The results of the Eurostat survey are based on a selection of industries, which changes slightly across countries. Estimates for Japan, Australia, New Zealand, the Netherlands, Canada, Switzerland, and the United Kingdom differ slightly from those in other countries; see source for details. Source: OECD, *Science, Technology, and Industry Scoreboard*, 2003.

There are many other indicators that point to the role of ICT in different OECD economies (OECD, 2002*a*). In practice, the different indicators are closely correlated and tend to point to the same countries as having the highest rate of diffusion. These typically are the United States, Canada, New Zealand, Australia, Northern European countries such as Denmark, Finland, and Sweden, as well as the Netherlands. It is therefore likely that the largest economic impacts of ICT should also be found in these countries.

Factors Affecting the Diffusion of ICT

Why is the diffusion of ICT so different across countries? Previous work has already noted several factors, such as a lack of relevant skills, a lack of competition, or the high costs in certain OECD countries (OECD, 2001). From a firm's perspective, high costs are important, since they affect the possible returns that a firm can extract from its investment. Firms not only incur costs in acquiring new technologies but also in making them effective in the workplace and in using them on a daily basis. Costs related to personnel, telecommunication charges, and organizational change are therefore also important. Some cross-country evidence is available on how these factors may have affected diffusion. A first factor concerns the costs of ICT hardware. Since ICT hardware is traded internationally, prices should not vary too much across countries. The available evidence suggests otherwise, however. Detailed price comparisons of ICT goods show that over much of the 1990s, firms in the United States and Canada enjoyed considerably lower costs of ICT investment goods than firms in European countries and Japan (OECD, 2001). Barriers to trade, such as nontariff barriers related to standards, import licensing, and government procurement, may partly explain the cost differentials (OECD, 2002*b*). The higher price levels in some countries may also be associated with a lack of competition within countries. International differences in the costs of telecommunication are also considerable.

Evidence on barriers to the uptake of ICT can also be drawn from firm-level surveys. These ask firms and consumers about the barriers they face in using the Internet and electronic commerce. Some interesting patterns emerge (OECD, 2002*a*). As regards Internet access, lack of security and slow or unstable communications were considered the key problems in European countries. Other problems, such as lack of know-how or personnel and high costs of equipment or Internet access, were considered less of a problem. Surveys on the barriers to Internet commerce also provide insights (figure 3.4). These suggest that legal uncertainties (uncertainty over payments, contracts, terms of delivery, and guarantees) are important in several countries. Business-to-consumer transactions are typically hampered by concerns about security of payment, the possibility of redress in the on-line environment, and privacy of personal data. For business-to-business transactions, the security and reliability of systems that can link all customers and suppliers are often considered more important. Cost considerations also remain an important issue for businesses in several countries, while logistic problems were also cited frequently.

Commercial factors were also cited by many businesses as a factor in not taking up Internet commerce, for example, because Internet commerce might threaten existing sales channels. Existing transaction models or strong links with customers and suppliers along the value chain may

% citing specific barriers

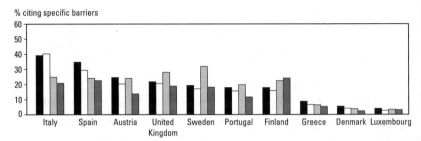

- ■ Uncertainty in payments
- ▨ Cost of developing and maintaining an e-commerce system
- ☐ Uncertainty concerning contracts, terms of delivery and guarantees
- ■ Logistic problems

% citing specific barriers

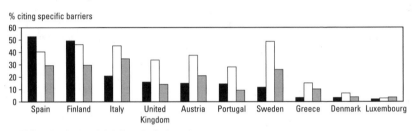

- ■ Consideration for existing channels of sales
- ☐ Goods and services available not suitable for sales by e-commerce
- ▨ Stock of (potential) customers too small

Figure 3.4
Barriers to Internet commerce faced by businesses, 2000 (percentage of
businesses using a computer with 10 or more employees). Source: OECD
2002*a*, *Measuring the Information Economy*, based on Eurostat, E-commerce
Pilot Survey.

discourage businesses from introducing new sales models. In many cases,
the goods and services on offer by a particular firm were not considered
suitable for Internet commerce, while firms in several countries considered the market too small. Some of these considerations differ by the size
and activity of firms; for example, large firms found logistical barriers
more important than small firms.

More elaborate analysis of this survey evidence provides further
insights into the factors explaining ICT uptake. Using recent data for
Switzerland, Hollenstein (2002) found that the anticipated benefits and
costs of adoption, the firm's ability to absorb knowledge from other firms
and institutions, experience with related technologies, and international

competitive pressure are among the main factors explaining ICT adoption, with sectoral differences also playing an important role.

There is also cross-country evidence that regulations in product and labor markets may affect ICT investment (figure 3.5). Product market regulations typically limit competition, which is important to spur ICT investment because it forces firms to seek ways to strengthen performance relative to competitors and also because it helps lower the costs of ICT. Moreover, product market regulations may limit firms in the ways

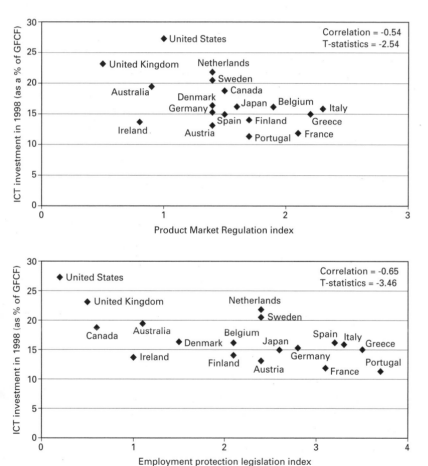

Figure 3.5
Countries with strict product and labor market regulations have lower ICT investment. Source: ICT investment from OECD Database on Capital Services; regulations from Nicoletti et al. 1999.

that they can extract benefits from their use of ICT. For example, they may not be able to extend beyond traditional sectoral boundaries (e.g., software firms offering financial services). Labor market regulations also play a role since they have an impact on the organizational changes that may be needed to make ICT work. If firms cannot adjust their workforce or organization and make ICT effective within the firm, they may decide to limit investment or relocate. These links between regulations and ICT investment have been confirmed through econometric analysis; Gust and Marquez (2002) found that regulations impeding workforce reorganizations and competition between firms hinder investment in ICT. Bartelsman et al. (2002) confirmed these findings.

These factors already point to some areas that are relevant for policy. For example, measures to increase competition can help bring down costs, labor market and education policies may help reduce skill shortages, and risk and uncertainty may be tackled by a well-designed regulatory framework.

ICT's Impact on Growth

What precisely are the impacts that ICT can have on business performance and growth? Three effects can be distinguished. First, as a capital good, investment in ICT contributes to overall capital deepening and therefore helps raise labor productivity. Second, rapid technological progress in the production of ICT goods and services may contribute to productivity growth in the ICT-producing sector. And third, greater use of ICT may help firms increase their overall efficiency in using capital equipment and labor (so-called multifactor productivity, or MFP). Moreover, greater use of ICT may contribute to network effects, such as lower transaction costs, higher productivity of knowledge workers, and more rapid innovation, which will improve the overall efficiency of the economy. This section discusses the empirical evidence for these effects on the basis of aggregate and sectoral data; the next section examines the evidence from firm-level studies.

The Impact of Investment in ICT

Evidence on the role of ICT investment across countries is primarily available from the macroeconomic level, e.g., see Colecchia and Schreyer (2001) and Van Ark et al. (2002). Both studies show that ICT has been a very dynamic area of investment due to the steep decline in ICT prices, which has encouraged investment in ICT. While ICT investment accelerated in most countries, the pace of that investment and its impact on growth differ considerably.

For the countries for which data are available, growth accounting estimates show that ICT investment typically accounted for between 0.3 and 0.8 percentage points of growth in GDP per capita over the 1995–2001 period (figure 3.6). The United States, Canada, and Australia received the largest boost, Japan and the United Kingdom a much smaller one, and France and Portugal the smallest. Software accounted for up to a third of the overall contribution of ICT investment to GDP growth. With the decline in investment in some countries over 2001–2002, the

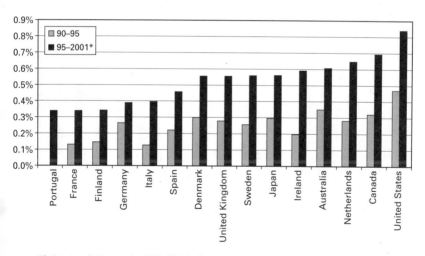

*Or latest available year, i.e.,1995–2000 for Denmark, Finland, Ireland, Japan, the Netherlands, Portugal, and Sweden.

Figure 3.6
The contribution of investment in ICT capital to GDP growth (percentage points contribution to annual average GDP growth, total economy). Source: OECD estimates based on Database on Capital Services. See Schreyer et al. 2003 for details.

contribution of ICT investment to growth has fallen somewhat, although it is likely to pick up once the recovery takes hold.

The Role of ICT-Producing and ICT-Using Sectors

Evidence of the impact of ICT can also be found from sectoral data, notably in the relative contributions of ICT-producing and ICT-using sectors to overall growth performance. The ICT-producing sector is of particular interest for several countries since it has been characterized by very high rates of productivity growth. Figure 3.7 shows that in most countries, the contribution of ICT manufacturing to overall labor productivity growth has risen over the 1990s. This can partly be attributed to more rapid technological progress in the production of certain ICT goods, such as semiconductors, which has contributed to more rapid

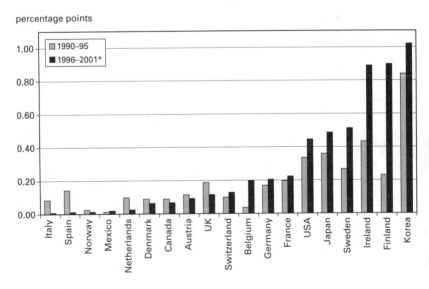

* Or latest available year.

Figure 3.7
The contribution of ICT manufacturing to aggregate labor productivity growth. Note: 1991–1995 for Germany; 1992–1995 for France and Italy; 1993–1995 for Korea; 1996–1998 for Sweden, 1996–1999 for Korea and Spain; 1996–2000 for Belgium, France, Germany, Ireland, Japan, Mexico, Norway, and Switzerland. Source: Pilat et al. 2002 and OECD STAN database.

and thus to higher growth in real volumes (Jorgenson,

ICT manufacturing made the largest contributions to aggregate productivity growth in Finland, Ireland, and Korea, where close to 1 percentage point of aggregate productivity growth in the 1995–2000 period was due to ICT manufacturing. The ICT-producing services sector (telecommunications and computer services) plays a smaller role in aggregate productivity growth but has also been characterized by rapid progress (Pilat et al., 2002). The contribution of this sector to productivity growth increased in several countries over the 1990s, notably in Finland, Germany, and the Netherlands. Some of the growth in ICT services is due to the emergence of the computer services industry. These services are important for ICT use since firms in these sectors offer key advisory and training services and also help develop appropriate software.

A much larger part of the economy uses ICT in the production process. Indeed, several studies have distinguished an ICT-using sector composed of industries that are intensive users of ICT (McGuckin and Stiroh, 2001; Pilat et al., 2002). Examining the performance of these sectors over time can help point to the role of ICT in aggregate performance. Figure 3.8 shows the contribution of the key ICT-using services (i.e., wholesale and retail trade, finance, insurance, and business services) to aggregate productivity growth over the 1990s.

The graph suggests small improvements in the contribution of ICT-using services in Sweden, Canada, and the United Kingdom and substantial increases in Australia, Ireland, Mexico, and the United States.[3] The United States has experienced the strongest improvement in productivity growth in ICT-using services over the 1990s, which is linked to more rapid productivity growth in wholesale and retail trade and in financial services (securities). This result for the United States is confirmed by several other studies (e.g., McKinsey, 2001; Triplett and Bosworth, 2002).

Stronger growth in labor productivity in ICT-producing and ICT-using industries could simply be due to greater use of capital. Estimates of MFP growth adjust for changes in the use of capital and can help show whether ICT-using sectors have indeed improved overall efficiency. Breaking aggregate MFP growth down into its sectoral contributions can

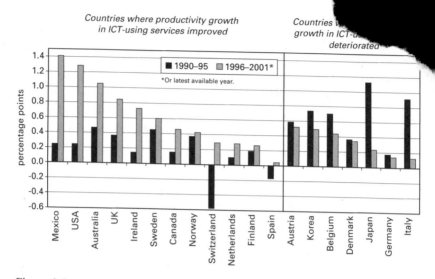

Figure 3.8
The contribution of ICT-using services to aggregate productivity growth. Note: See figure 3.7 for period coverage. Estimates for Australia refer to 1996–2001. Source: Pilat et al. 2002 and OECD STAN database.

also help show whether changes in MFP growth should be attributed to ICT-producing sectors, to ICT-using sectors, or to other sectors. MFP estimates at the sectoral level are available for only a limited number of OECD countries due to the limited availability of estimates of capital stock or capital services (Pilat et al., 2002). For the United States, several detailed industry studies suggest that MFP in certain services improved over the second half of the 1990s. For example, a recent study by Triplett and Bosworth (2002) estimated that MFP growth in wholesale trade accelerated from 1.1 percent annually to 2.4 percent annually from 1987–1995 to 1995–2000. In retail trade, the jump was from 0.4 percent annually to 3.0 percent, and in securities the acceleration was from 2.9 percent to 11.2 percent. Combined with the relatively large weight of these sectors in the economy, this translates into a considerable contribution to more rapid aggregate MFP growth of these ICT-using services.

There is therefore some evidence of strong MFP growth in the United States in certain ICT-using services. Some studies also suggest how these productivity changes due to ICT use could be interpreted. First, a considerable part of the pickup in productivity growth can be attributed to price declines where firms such as Wal-Mart used innovative practices, 2001). ...g ICT, to gain market share from competitors (McKinsey, 2001). The larger market share for Wal-Mart and other productive firms raised average productivity and also forced their competitors to improve performance. Among the other ICT-using services, securities accounts also for a large part of the pickup in productivity growth. Its strong performance has been attributed to a combination of buoyant financial markets (i.e., large trading volumes), effective use of ICT (mainly in automating trading processes), and stronger competition (McKinsey, 2001; Baily, 2002). These impacts on MFP are therefore primarily linked to more efficient use of labor and capital due to ICT.

Spillover effects may also play a role, however, since ICT investment started earlier, and was stronger, in the United States than in most OECD countries (Colecchia and Schreyer, 2001; Van Ark et al., 2002). Moreover, previous work has pointed out that the U.S. economy might be able to achieve greater benefits from ICT since it got its fundamentals right before many other countries (OECD, 2001). The combination of sound macroeconomic policies, well-functioning institutions and markets, and a competitive economic environment may be at the core of the U.S. success. A recent study by Gust and Marquez (2002) confirmed these results and attributed relatively low investment in ICT in European countries partly to restrictive labor and product market regulations that have prevented firms from getting sufficient returns from their investment.

The United States is not the only country where ICT use may already have had impacts on MFP growth. Studies for Australia (e.g., Parham et al., 2001) suggest that a range of structural reforms have been important in driving the strong uptake of ICT by firms and have enabled these investments to be used in ways that generate productivity gains. This is particularly evident in wholesale and retail trade and in financial intermediation, the main drivers of Australian productivity gains in the 1990s.

ICT and Firm-Level Performance

Does ICT Use Matter?
The previous section showed that ICT investment and ICT production have contributed to growth in a variety of countries. It also showed that

ICT-using industries in the United States and Australia experienced a strong increase in productivity growth in the second half of the 1990s. Few other countries have thus far experienced similar gains. Nevertheless, much of the current interest in the potential impacts of ICT on growth is linked to the potential benefits arising from its use in the production process. If the rise in MFP due to ICT were only a reflection of rapid technological progress in ICT production, there might not be effects of ICT use on MFP in countries that are not already producers of ICT. For ICT to have benefits on MFP in countries that do not produce ICT goods, the use of ICT would need to be beneficial too. The sectoral evidence presented above suggests that this might be the case for the United States and Australia. Moreover, some aggregate data suggest that the growth in MFP may also be associated with the productivity-enhancing benefits from the use of and investment in ICT (figure 3.9).

The macro evidence may not be convincing, however. Indeed, more convincing evidence on the impact of ICT use can be drawn from firm-level evidence. ICT use may have several impacts at this level. For

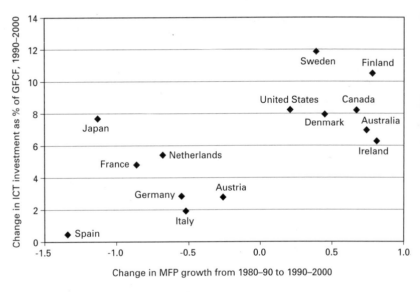

Correlation coefficient = 0.66; T-statistic = 3.03.

Figure 3.9
Change in MFP growth and increase in ICT investment. Source: ICT investment from OECD, Database on Capital Services; MFP growth from OECD 2003.

example, it may help firms gain market share at the cost of less productive firms, which could raise overall productivity. In addition, the use of ICT may help firms expand their product range, customize the services offered, or respond better to client demand, i.e., to innovate. Moreover, ICT may help reduce inefficiency in the use of capital and labor, for example, by reducing inventories. These effects might all lead to higher productivity growth. These, and related, effects have long been difficult to capture in empirical studies, contributing to the so-called "productivity paradox." However, a growing number of firm-level studies provide evidence on such impacts.

The Impacts of ICT at the Firm Level

A number of survey articles summarize the early literature on ICT, productivity, and firm performance (e.g., Brynjolfsson and Yang, 1996). Many of these studies tended to find no, or a negative, impact of ICT on productivity. Most of these early studies also primarily focus on labor productivity and the return to computer use, not on MFP or other impacts of ICT on business performance. Moreover, most of these studies used private sources, since official sources were not yet available. Recent work by statistical offices, using large official databases, has provided many new insights into the role of ICT.

The Use of ICT and Advanced Technologies Is Positively Linked to Firm Performance   There is evidence from many firm-level studies that ICT use has a positive impact on firm performance. These impacts can vary. Figure 3.10 illustrates a typical finding and shows that Canadian firms that used either one or more ICTs had a higher level of productivity than firms that did not use these technologies. Moreover, the gap between technology-using firms and other firms increased between 1988 and 1997, as technology-using firms increased relative productivity compared to nonusers. The graph also suggests that some ICTs are more important in enhancing productivity than other technologies, communication network technologies being particularly important.

The evidence shown in figure 3.10 is confirmed by many other studies, which also point to other impacts of ICT on economic performance. For example, firms using ICT typically pay higher wages. In addition, the

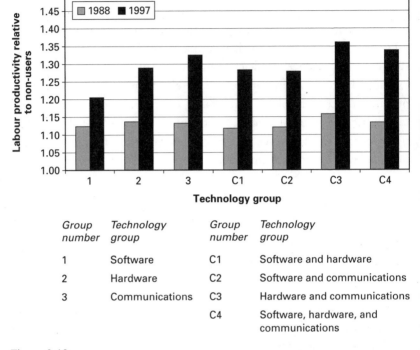

Group number	Technology group	Group number	Technology group
1	Software	C1	Software and hardware
2	Hardware	C2	Software and communications
3	Communications	C3	Hardware and communications
		C4	Software, hardware, and communications

Figure 3.10
Relative productivity of advanced technology users and nonusers (manufacturing sector in Canada, 1988 versus 1997). Source: Baldwin and Sabourin 2002.

studies show that the use of ICT does not guarantee success; many of the firms that improved performance thanks to their use of ICT were already experiencing better performance than the average firm. Moreover, the benefits of ICT appear to depend on sector-specific effects and are not found equally in all sectors.

There is also evidence that ICT can help firms in the competitive process. For the United States, Doms et al. (1995) found that increases in the capital intensity of the product mix and in the use of advanced manufacturing technologies are positively correlated with plant expansion and negatively with plant exit. For Canada, Baldwin and Sabourin (2002) found that a considerable amount of market share is transferred from declining firms to growing firms over a decade. Those technology users that were using communications technologies or that combined technologies from several different technology classes increased their relative productivity the most. In turn, gains in relative productivity were accompanied by gains in market share.

Computer Networks Play a Key Role   Some ICTs may be more important to strengthen firm performance than others. Computer networks may be particularly important, since they allow a firm to outsource certain activities, to work closer with customers and suppliers, and to better integrate activities throughout the value chain. For the United States, Atrostic and Nguyen (2002) directly linked computer network use (both EDI and Internet) to productivity. They found that average labor productivity is higher in plants with networks and that the impact of networks is positive and significant after controlling for several production factors and plant characteristics. Networks are estimated to increase labor productivity by roughly 5 percent.

Similar work has been carried out for Japan. Motohashi (2001) found that the impact of direct business operation networks on productivity is much clearer than that of back office supporting systems, such as human resource management and management planning systems. Firms with networks are also found to outsource more production activities. For Germany, Bertschek and Fryges (2002) show that the more firms in an industry that already use business-to-business (B2B) electronic links, the more likely it is that the firm will also implement B2B.

Firms in the Services Sector also Benefit from ICT   Thanks to improved data, the work with firm-level statistics is also broadening to the services sector. For example, Doms, Jarmin, and Klimek (2002) showed that growth in the U.S. retail sector involves the displacement of traditional retailers by sophisticated retailers introducing new technologies and processes, thus confirming the sectoral evidence discussed above. For Germany, Hempell (2002) showed significant productivity effects of ICT in the German service sector. Experience gained from past process innovations helps firms to make ICT investments more productive. ICT investment may thus have contributed to growing productivity differences between firms, and potentially also between countries. For the Netherlands, Broersma and McGuckin (2000) found that computer investments have a positive impact on productivity and that the impact is greater in retail than in wholesale trade.

Factors That Affect the Impact of ICT

The evidence summarized above suggests that the use of ICT does have impacts on firm performance. However, these effects occur primarily, or only, when accompanied by other changes and investments, including investment in skills and organizational change. This is also confirmed by many empirical studies that suggest that ICT primarily affects firms where skills have been improved and organizational changes have been introduced. The role of these complementary factors was raised by Bresnahan and Greenstein (1996), who argued that users help make investment in technologies, such as ICT, more valuable through their own experimentation and invention. Without this process of "coinvention," which often has a slower pace than technological invention, the economic impact of ICT may be limited. This section looks at some of the factors that affect the uptake of ICT and the main complementary factors for ICT investment.

ICT Use Is Complementary to Skills   A substantial number of firm-level studies address the interaction between technology and human capital and their joint impact on productivity performance. For the United States, Krueger (1993) found that workers using computers were better paid than those who do not use computers. Doms et al. (1997) found no correlation between technology adoption and wages, however, and concluded that technologically advanced plants pay higher wages both before and after the adoption of new technologies. A more recent study by Luque and Miranda (2000) found that technological change in U.S. manufacturing was skill biased, however.

For Germany, Falk (2001*a*) found that firms with a higher diffusion of ICT employ a larger fraction of workers with a university degree as well as ICT specialists. A greater penetration of ICT is negatively related to the share of both medium- and low-skilled workers. For France, Entorf and Kramarz (1998) found that computer-based technologies are often used by workers with higher skills. These workers become more productive when they get more experienced in using these technologies. Caroli and Van Reenen (1999) found that French plants that introduce organizational change are more likely to reduce their demand for unskilled workers than those that do not. Shortages in skilled workers may reduce the probability of organizational changes. Greenan et al.

(2001) examined the late 1980s and early 1990s and found strong positive correlations between indicators of computerization and research on the one hand and productivity, average wages, and the share of administrative managers on the other hand. They also found negative correlations between these indicators and the share of blue-collar workers.

For the United Kingdom, Haskel and Heden (1999) found that computerization reduces the demand for manual workers, even when controlling for endogeneity, human capital upgrading, and technological opportunities. Caroli and Van Reenen (1999) found evidence that human capital, technology, and organizational change are complementary and that organizational change reduces the demand for unskilled workers.

Studies for Canada also point to the complementarity between technology and skills. For example, Baldwin et al. (1995) found that use of advanced technology was associated with a higher level of skill requirements, leading to a higher incidence of training and increased expenditure on education and training. A more recent study (Sabourin, 2001) found that establishments adopting advanced technologies often reported labor shortages of scientists, engineers, and technical specialists. However, the most technologically advanced establishments were often able to solve these shortages.

Organizational Change Is Key to Making ICT Work   Closely linked to human capital is the role of organizational change. Studies typically find that the greatest benefits from ICT are realized when ICT investment is combined with other organizational changes, such as new strategies, new business processes and practices, and new organizational structures. Several U.S. studies with official statistics have addressed this link to human capital and organizational change. For example, Black and Lynch (2001) found that the implementation of human resource practices is important for productivity, e.g., giving employees greater voice in decision-making, profit-sharing mechanisms, and new industrial relations practices. In another study (2000), they found that firms that reengineer their workplaces to incorporate high-performance practices experience higher productivity and higher wages.

For Germany, Bertschek and Kaiser (2001) found that the introduction of organizational changes raises overall labor productivity. Falk (2001*b*) found that the introduction of ICT and the share of training expenditures are important drivers of organizational changes, such as the introduction of total quality management, lean administration, flatter hierarchies, and delegation of authority.

For France, Greenan and Guellec (1998) found that the use of advanced technologies and the skills of the workforce are both positively linked to organizational variables. An organization that enables communication within the firm and that innovates at the organizational level seems better able to create the conditions for a successful uptake of advanced technologies. Moreover, these changes also seemed to increase the ability of firms to adjust to changing market conditions through technological innovation and the reduction of inventories.

For the United Kingdom, Caroli and Van Reenen (1999) found that organizational change, technology, and skills were complementary. More specifically, they found that organizational change reduced the demand for unskilled workers and that it has the largest productivity impacts in establishments with larger initial skill endowments. For the Netherlands, Broersma and McGuckin (2000) also found that computer use was linked to the introduction of flexible employment practices, e.g., greater use of temporary and part-time workers.

Firm Size Affects the Impact of ICT   A substantial number of studies have looked at the relationship between ICT and firm size. Most studies find that the adoption of advanced technologies, such as ICT, increases with the size of firms and plants. Evidence for the United Kingdom (Clayton and Waldron, 2003) shows that large firms are more likely to use network technologies such as an intranet, the Internet, or EDI than small firms; they are also more likely to have their own website. However, small firms of between 10 and 49 employees are more likely to use the Internet as their only ICT network technology. Large firms are more likely to use a combination of network technologies. For example, over 38 percent of all large U.K. firms use an intranet, EDI, and the Internet and also have their own website, as opposed to less than 5 percent of small firms. Moreover, almost 45 percent of all large firms already use broadband technologies as opposed to less than 7 percent of small firms. These differences are linked to the different uses of technology. Large firms may use the technologies to redesign information and communication flows within the firm and to integrate these flows throughout the production process. Some small firms only use the Internet for marketing purposes.

Ownership, Competition, and Management Are Important   Firm-level studies also point to the importance of ownership changes and management in the uptake of technology. For example, McGuckin and Nguyen (1995) found that plants with above-average productivity are more likely to change owners and that acquiring firms tended to have above-average productivity. Plants that changed owners generally improved productivity following the change. According to the authors, ownership changes appear associated with the purchase or integration of advanced technologies and better practices into firms.

Some studies also point to the impact of competition. A study by Baldwin and Diverty (1995) found that foreign-owned plants were more likely to adopt advanced technologies than domestic plants. For Germany, Bertschek and Fryges (2002) found that international competition was an important factor driving a firm's decision to implement B2B electronic commerce.

Management also plays a role. Stolarick (1999) found that low-productivity plants may sometimes spend more on IT than high-productivity plants, in an effort to compensate for their poor productivity performance. The study suggests that management skill should therefore be taken into account as an additional factor when investigating the IT productivity paradox.

ICT Use Is Closely Linked to Innovation   Several studies point to an important link between the use of ICT and the ability of a company to adjust to changing demand and to innovate. The clearest example of this link is found in work on Germany, which draws on results from innovation surveys. For example, Licht and Moch (1999) found that information technology has important impacts on the qualitative aspects of service innovation. Hempell (2002) found that firms that have introduced process innovations in the past are particularly successful in using ICT; the output elasticity of ICT capital for these firms is estimated to be about 12 percent, about four times that of other firms. This suggests that the productive use of ICT is closely linked to innovation in general and to the reengineering of processes in particular. Studies in other countries also confirm this link. For example, Greenan and Guellec (1998) found that organizational change and the uptake of advanced technologies seemed to increase the ability of firms to adjust to changing market conditions through technological innovation.

The Impacts of ICT Use Emerge Only over Time   Given the time it takes to adapt to ICT, it should not be surprising that the benefits of ICT emerge only over time. This can be seen clearly in the relationship between the use of ICT and the year in which firms first adopted ICT. Figure 3.11 shows evidence for the United Kingdom. It shows that among the firms that had already adopted ICT in or before 1995, over 30 percent currently buy and sell using electronic commerce in 2000. For firms that adopted ICT only in 2000, less than 15 percent buy and sell using .e-commerce. The U.K. evidence also suggests that firms move towards more complex forms of electronic activity over time; out of all

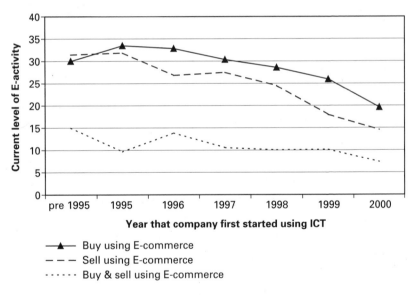

Figure 3.11
Relationship between the year of ICT adoption and the current degree of e-activity (as a percentage of all firms adopting ICT in specific year, business-weighted). Note: The graph shows the percentage of firms engaged in a specific type of e-activity in 2000, out of all the firms starting to use ICT in that year. Source: Clayton and Waldron 2003.

firms starting to use ICT prior to 1995, only 3 percent had not moved beyond the straightforward use of ICT. Most had established an Internet site or bought or sold through e-commerce. Out of the firms adopting ICT in 2000, close to 20 percent had not yet gone beyond the simple use of ICT.

Does the Impact of ICT at the Firm Level Differ across Countries?

Cross-country studies on the impact of ICT at the firm level are still relatively rare, primarily since many of the original data sources were of an ad-hoc nature and not comparable across countries. In recent years, the growing similarity of official statistics is enabling more comparative work. An example is a recent comparison between the United States and Germany (Haltiwanger et al., 2002) that examines the relationship between labor productivity and measures of the choice of technology. Figure 3.12 illustrates some of the empirical findings, distinguishing between different categories of firms according to their total level of investment and their level of investment in ICT. The first panel shows that firms in all categories of investment have much stronger productivity growth in the United States than in Germany. Moreover, firms with high ICT investment (groups 4 and 6) have stronger productivity growth than firms with low (groups 2 and 5) or zero ICT investment (groups 1 and 3).

The second panel of the graph shows that firms in the United States have much greater variation in their productivity performance than firms in Germany. This may be because U.S. firms engage in much more experimentation than their German counterparts; they take greater risks and opt for potentially higher outcomes. Other international comparisons of business performance and the impact of ICT are currently under way on the basis of micro data; these should contribute to further insights and help explain cross-country differences in the benefits that are being drawn from ICT.

Some Policy Implications

It is clear that to seize the benefits of ICT, it takes much more than simply investing in the equipment and software. Countries need policies to increase competition in telecommunications, to enhance skills and encourage labor mobility, to reduce obstacles to workplace changes, and

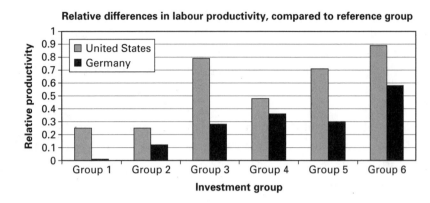

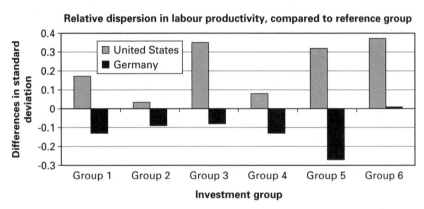

Figure 3.12
Differences in productivity outcomes between Germany and the United States.
Note: Differences are in logs and are shown relative to a reference group of zero
total investment and zero investment in ICT. The groups are distinguished on
the basis of total investment (0, low, high) and ICT investment (0, low, high).
Group 1 has low overall investment and zero ICT investment. Group 2 has low
overall investment and low ICT investment. Group 3 has high overall investment
and zero ICT investment. Group 4 has low overall investment and high ICT
investment. Group 5 has high overall investment and low ICT investment. Group
6 has high overall investment and high ICT investment. Source: Haltiwanger,
Jarmin, and Schank 2002.

to build confidence in the use of ICT. ICT is not the only factor explaining disparities in growth observed across countries, and such policies to bolster ICT will not on their own steer countries onto a higher growth path. Strengthening growth performance will thus require a comprehensive and coordinated set of actions to create the right conditions for future change and innovation, including policies to strengthen fundamentals, to foster innovation, to invest in human capital, and to stimulate firm creation. Once policies are in place that create an environment conducive to exploiting the gains from technological change, some specific measures are needed to seize the benefits from ICT.

Strengthening Competition in ICT Goods and Services   The first policy implication that can be drawn from the work concerns costs differentials and the need for sufficient competition in ICT goods and services. The available evidence suggests that differences in the costs of the technology continue to play a role in determining investment patterns. Barriers to trade, in particular nontariff barriers related to standards, import licensing, and government procurement, may partly explain these differentials. The higher price levels in other countries may also be associated with a lack of competition within countries. In time, however, international trade and competition should further erode these cross-country price differences. Policy could help to accelerate this trend, by implementing a more active competition policy and measures to promote market openness, both domestically and internationally.

Cost differentials are observed not only for ICT hardware and software but also in the associated costs of communication. For example, by August 2001, the prices of 40 hours of Internet access at peak times were lowest in the United States. The European Union average was almost three times the U.S. price level, while prices in Japan were almost double those in the United States. It is not only the liberalization of markets that is important to lower price levels but primarily the introduction of effective competition. Japan liberalized its telecommunications markets quite early but took long to reap the benefits as an effective regulatory framework took time to be established. Efforts to increase competition and continue with regulatory reform in the telecommunications industry continue to be important to enhance the uptake of ICT. Improving the conditions of access to local communication infrastructures is particularly important and will require effective policies to unbundle the local loop and establish interconnection frameworks. Such policies will also help enhance access to high-speed communication services.

Fostering an Environment for the Effective Use of ICT   Investment in ICT depends not only on the cost of the technology itself but also on the complementary investments that need to be made by firms to draw the benefits from ICT, e.g., in changing the organization of functions and tasks or in training staff. These complementary investments are often much more costly than the initial outlays for ICT investment goods. Brynjolfsson and Hitt (2000), for example, suggested that $1 USD of ICT investment may be associated with $9 USD of investment in intangible assets, such as skills and organizational practices.

Policy can help by fostering an environment that enables firms to make effective use of ICT. Adapting the organization of functions and tasks to ICT can be particularly costly to firms, since it often meets with resistance within the firm and is limited by legal constraints in several countries. Social partners and government can work together to ensure that a virtuous circle of organizational change, ICT, and productivity is set in motion. This depends on workers being given a sufficient "voice" in the firm. Institutions that allow a closer contact between management and employees can help build a high-skill, high-trust enterprise climate that facilitates change.

Adjusting the skills of workers to the new technology may also require considerable investment by firms. Having a good supply of highly qualified personnel helps, but education policies, important as they are, need to be supplemented with action in the area of adult learning. The OECD growth report (OECD, 2001) pointed to a range of policy conclusions in this area, which continue to be important for countries wishing to draw the benefits from ICT.

A third implication relates to management. Firm-level studies typically find that the firms that get the most out of their investment in ICT are those firms that were already performing well in terms of gains in both productivity and market shares. These firms improved performance by investing in ICT, by innovating, and by adapting their organization and workforce.[4] In contrast, many firms that invested much in ICT received no returns at all, since they were attempting to compensate for poor overall performance. This reinforces the view that ICT is no panacea and also points to a role for management. While governments cannot directly influence management decisions, they can help create framework conditions (e.g., disclosure rules) for good management. Policies for good corporate governance play a key role in this respect.

Rewarding the Successful Adoption of ICT   The evidence presented above shows that ICT is an enabling technology. Firms can use it to improve performance, but not all firms will succeed in making the necessary changes that are needed to make the technology work. Competition and creative destruction are key in selecting the successful firms and in making them flourish and grow. If firms that are able to make ICT work succeed and grow, the benefits for the economy as a whole are greater than if poorly performing firms survive. The exit of many dotcoms and the survival and growth of some of these firms, for example, also reflects this process of creative destruction.

Cross-country differences in experimentation may also be important for firms wishing to gain large benefits from their investments in ICT (Bartelsman et al., 2002). Firms in the United States seem to be characterized by a higher degree of experimentation than firms in other OECD countries; U.S. firms start smaller but grow much more quickly. This may be linked to less aversion to risk in the United States, linked to its market-based financial system, which provides greater opportunities for risky financing to innovative entrepreneurs. Moreover, low administrative burdens and limited restrictions in the labor market enable U.S. firms to start at a small scale, experiment and test the market, and, if successful, expand rapidly. In contrast, firms in other OECD countries are faced with higher entry costs, which offers less scope for experimentation and learning. Policies to foster new firm creation are therefore important to implement new ideas and technologies. While many new firms may not survive, start-ups may force incumbent firms to improve performance, and those that survive may contribute to improved productivity and innovation.

The state of competition also influences firms' decisions to implement electronic commerce. Many firms do not engage in e-commerce because the market is considered too small or because their products are not considered suitable for electronic delivery. In other cases, electronic commerce is seen as a rival to existing business models. These concerns can be genuine but may also reflect a conservative attitude. International competition and start-up firms can help instill greater dynamism, introduce new ideas and business models, and invigorate mature industries. Policies to enhance firm creation are key in these markets.

Building Confidence in Electronic Commerce   Businesses, governments, and consumers as well as key infrastructures such as power generation and distribution, financial markets, and transport all rely heavily on the use of information systems and networks, which are increasingly interconnected globally. This raises new issues for security since firms, governments, and consumers need to be sure that these electronic networks are stable and can be used safely under all conditions.

Much work is currently under way to address these concerns. Authentication and certification mechanisms are currently being developed to help identify users and safeguard business transactions. Guidelines that bolster and establish security protocols have been developed and need to be implemented. If e-commerce is to be an important way of doing business in the future, it will have to be reliable, secure, and safe to use under all conditions.

Unleashing Growth in the Services Sector   The growing importance of ICT also affects policies for the services sector. Service industries such as wholesale and retail trade and financial and business services are among the most important adopters of ICT. It is in these "old economy" sectors, not in the telecommunications or the dot-com sector, that the long-term impacts of ICT use should become visible. Thus far, only a few countries have clearly benefited from productivity-enhancing investment in ICT, often linked to regulatory reform and growing international competition in services delivery. This suggests that policies must take better account of the needs and characteristics of the services sector if they are to promote growth. For example, competition in many services sectors remains limited due to heavy regulatory burdens, reducing pressures to strengthen performance. Further reform of regulatory structures is needed to promote competition and innovation and to reduce barriers and administrative rules for new entrants and start-ups. International competition is also important, since services are typically less exposed to international competition. This will require the reduction of trade and foreign investment barriers to services, which can also promote the diffusion of innovative ideas and concepts across countries. Evidence from firm-level studies that foreign firms are often the first to adopt new technologies confirms that such international competition is essential. The specific characteristics of innovation in the services sector also need to be taken into account in such policies.

Harnessing the Potential of Innovation and Technology Diffusion   ICT can strengthen the ability of firms to innovate, that is, to introduce new products and services, new business processes, and new applications. For example, ICT has helped speed up the innovation process and reduce cycle times, resulting in a closer link between business strategies and performance. Moreover, ICT has fostered greater integration and networking in the economy, since it has facilitated outsourcing and improved cooperation beyond the firm, with suppliers, customers, and competitors. The roles of innovation and ICT in recent growth performance are thus closely related. Some of the recent changes in the innovation process could not have occurred without ICT. Conversely, some of the impact of ICT might not have been felt in the absence of changes in the innovation process, e.g., stronger links between scientific research and innovation (OECD, 2001). This implies that policies to harness the potential of innovation and technology diffusion, as outlined in the OECD growth study (OECD, 2001), are of great importance in seizing the benefits of ICT. Moreover, such policies help foster the kind of innovative environment in which new growth opportunities will flourish.

Making Government Programs More Efficient   Governments have put in place a range of schemes to assist firms in their uptake and use of ICT. These schemes, designed in principle to overcome market failures, have at times been counterproductive, for example, because they were not technologically neutral and subsidized specific ICT technologies that were not those considered appropriate by firms. In several cases, firms were "locked in" suboptimal technologies that were not well integrated with the technologies used by partners, often foreign, in the value chain.

Policies to extend ICT use to all firms in the economy have been an important area of government intervention over the past years, but there are doubts about whether this is still key. ICT may not be an appropriate technology for all firms, and those firms that have the largest potential to seize benefits from ICT are already intensively using the technology. Firms should move at their own pace in implementing ICT, and governments are not well placed to influence this process directly; they are better placed to create the framework conditions such as a sound macroeconomic policy and policies to enhance human capital for firms' effective use of ICT.

Notes

1. The views expressed in this chapter are those of the authors and not necessarily those of the Economic Analysis and Statistics Division, Directorate for Science, Technology and Industry, OECD, Paris, or its member countries.

2. These estimates are based on the OECD definition of the ICT sector. See OECD (2002*a*).

3. Previous studies have also shown a large contribution of ICT-using services to the pickup of productivity growth in Australia over the 1990s (Parham et al., 2001). This cannot be confirmed here, since Australia is not yet available in the OECD STAN database.

4. The management literature provides extensive discussions on how firms can make ICT work in their particular environment. These issues are not discussed here, since government policy has little role in influencing these corporate processes.

References

Ahmad, N. (2003) "Measuring Investment in Software." STI Working papers 2003/6, www.oecd.org/sti/working-papers, OECD, Paris.

Atrostic, B. K., and S. Nguyen (2002) "Computer Networks and U.S. Manufacturing Plant Productivity: New Evidence from the CNUS Data." CES Working Paper 02–01, Center for Economic Studies, Washington, D.C.

Baily, M. N. (2002) "The New Economy: Post Mortem or Second Wind." *Journal of Economic Perspectives* 16(2) (Spring): 3–22.

Baldwin, J. R., and B. Diverty (1995) "Advanced Technology Use in Canadian Manufacturing Establishments." Working Paper no. 85, Microeconomics Analysis Division, Statistics Canada, Ottawa.

Baldwin, J. R., and D. Sabourin (2002) "Impact of the Adoption of Advanced Information and Communication Technologies on Firm Performance in the Canadian Manufacturing Sector." STI Working Paper 2002/1, OECD, Paris.

Baldwin, J. R., T. Gray, and J. Johnson (1995) "Technology Use, Training and Plant-Specific Knowledge in Manufacturing Establishments." Working Paper no. 86, Microeconomics Analysis Division, Statistics Canada, Ottawa.

Bartelsman, E., A. Bassanini, J. Haltiwanger, R. Jarmin, S. Scarpetta, and T. Schank (2002) "The Spread of ICT and Productivity Growth—Is Europe Really Lagging Behind in the New Economy?" Fondazione Rodolfo DeBenedetti.

Bertschek, I., and H. Fryges (2002) "The Adoption of Business-to-Business E-Commerce: Empirical Evidence for German Companies." ZEW Discussion Paper no. 02-05, ZEW, Mannheim.

Bertschek, I., and U. Kaiser (2001) "Productivity Effects of Organizational Change: Microeconometric Evidence." ZEW Discussion Paper no. 01-32, ZEW, Mannheim.

Black, S. E., and L. M. Lynch (2000) "What's Driving the New Economy: The Benefits of Workplace Innovation." NBER Working Paper Series, no. 7479, January.

Black, S. E., and L. M. Lynch (2001) "How to Compete: The Impact of Workplace Practices and Information Technology on Productivity." *Review of Economics and Statistics* 83(3) (August): 434–445.

Bresnahan, T. F., and S. Greenstein (1996) "Technical Progress and Co-Invention in Computing and the Use of Computers." *Brookings Papers on Economic Activity: Microeconomics*, 1–77.

Broersma, L., and R. H. McGuckin (2000) "The Impact of Computers on Productivity in the Trade Sector: Explorations with Dutch Microdata." Research Memorandum GD-45, Groningen Growth and Development Centre, June.

Brynjolfsson, E., and L. M. Hitt (2000) "Beyond Computation: Information Technology, Organizational Transformation and Business Performance." *Journal of Economic Perspectives* 14: 23–48.

Brynjolfsson, E., and S. Yang (1996) "Information Technology and Productivity: A Review of the Literature." http://ecommerce.mit.edu/erik/.

Caroli, E., and J. Van Reenen (1999) "Organization, Skills and Technology: Evidence from a Panel of British and French Establishments." IFS Working Paper Series W99/23, Institute of Fiscal Studies, August.

Clayton, T., and K. Waldron (2003) "E-Commerce Adoption and Business Impact, A Progress Report." *Economic Trends*, no. 591, February.

Colecchia, A., and P. Schreyer (2001) "The Impact of Information Communications Technology on Output Growth." STI Working Paper 2001/7, OECD, Paris.

Doms, M., T. Dunne, and M. J. Roberts (1995) "The Role of Technology Use in the Survival and Growth of Manufacturing Plants." *International Journal of Industrial Organization* 13(4) (December): 523–542.

Doms, M., T. Dunne, and K.R. Troske (1997) "Workers, Wages and Technology." *Quarterly Journal of Economics* 112(1): 253–290.

Doms, M., R. Jarmin, and S. Klimek (2002) "IT Investment and Firm Performance in U.S. Retail Trade." CES Working Paper 02-14, Center for Economic Studies, Washington, D.C.

Entorf, H., and F. Kramarz (1998) "The Impact of New Technologies on Wages: Lessons from Matching Panels on Employees and on Their Firms." *Economic Innovation and New Technology* 5, 165–197.

Falk, M. (2001a) "Diffusion of Information Technology, Internet Use and the Demand for Heterogeneous Labor." ZEW Discussion Paper no. 01-48, ZEW, Mannheim.

Falk, M. (2001b) "Organizational Change, New Information and Communication Technologies and the Demand for Labor in Services." ZEW Discussion Paper no. 01-25, ZEW, Mannheim.

Greenan, N., and D. Guellec (1998) "Firm Organization, Technology and Performance: An Empirical Study." *Economics of Innovation and New Technology* 6(4): 313–347.

Greenan, N., J. Mairesse, and A. Topiol-Bensaid (2001) "Information Technology and Research and Development Impacts on Productivity and Skills: Looking for Correlations on French Firm Level Data." NBER Working Paper 8075, Cambridge, MA.

Gust, C., and J. Marquez (2002) "International Comparisons of Productivity Growth: The Role of Information Technology and Regulatory Practices." International Finance Discussion Papers, no. 727, Federal Reserve Board, May.

Haltiwanger, J., R. Jarmin, and T. Schank (2002) "Productivity, Investment in ICT and Market Experimentation: Micro Evidence from Germany and the United States." Paper presented at OECD workshop on ICT and Business Performance, December.

Haskel, J., and Y. Heden (1999) "Computers and the Demand for Skilled Labour: Industry- and Establishment-Level Panel Evidence for the UK." *Economic Journal*, 109, March, C68–C79.

Hempell, T. (2002) "Does Experience Matter? Productivity Effects of ICT in the German Service Sector." ZEW Discussion Paper no. 02-43, Centre for European Economic Research, Mannheim.

Hollenstein, H. (2002) "The Decision to Adopt Information and Communication Technologies (ICT): Explanation and Policy Conclusions." Paper presented at OECD workshop on ICT and Business Performance, Institute for Business Cycle Research (KOF), Zurich, December.

Jorgenson, D. W. (2001) "Information Technology and the U.S. Economy." *American Economic Review* 91(1): 1–32.

Krueger, A. B. (1993) "How Computers Have Changed the Wage Structure: Evidence from Microdata, 1984–1989." *Quarterly Journal of Economics*, February, 33–60.

Licht, G., and D. Moch (1999) "Innovation and Information Technology in Services." *Canadian Journal of Economics* 32(2) (April).

Luque, A., and J. Miranda (2000) "Technology Use and Worker Outcomes: Direct Evidence from Linked Employer-Employee Data." CES WP-00-13, Center for Economic Statistics, Washington, DC.

McGuckin, R. H., and S. V. Nguyen (1995) "On Productivity and Plant Ownership Change: New Evidence from the LRD." *Rand Journal of Economics* 26(2): 257–276.

McGuckin, R. H., and K. J. Stiroh (2001) "Do Computers Make Output Harder to Measure?" *Journal of Technology Transfer* 26: 295–321.

McKinsey Global Institute (2001) *U.S. Productivity Growth 1995–2000: Understanding the Contribution of Information Technology Relative to Other Factors*, McKinsey Global Institute, Washington, D.C., October.

Motohashi, K. (2001) "Economic Analysis of Information Network Use: Organisational and Productivity Impacts on Japanese Firms." Research and Statistics Department, METI.

Nicoletti, G., S. Scarpetta, and O. Boylaud (1999) "Summary Indicators of Product Market Regulation with an Extension to Employment Protection Legislation." OECD Economics Department Working Paper no. 226, Paris.

OECD (2001) *The New Economy: Beyond the Hype*, Paris.

OECD (2002a) *Measuring the Information Economy 2002*.

OECD (2002b) "Non-Tariff Barriers in the ICT Sector: A Survey." OECD, Paris, September.

OECD (2003) *The Sources of Economic Growth in OECD Countries*, Paris.

Parham, D., P. Roberts, and H. Sun (2001) "Information Technology and Australia's Productivity Surge." Staff Research Paper, Productivity Commission, AusInfo, Canberra.

Pilat, D., F. Lee, and B. Van Ark (2002) "Production and Use of ICT: A Sectoral Perspective on Productivity Growth in the OECD Area." OECD Economic Studies, no. 35, 2002/2, Paris, pp. 47–78.

Sabourin, D. (2001) "Skill Shortages and Advanced Technology Adoption." Working Paper no. 175, Microeconomics Analysis Division, Statistics Canada, Ottawa.

Schreyer, P., P. E. Bignon, and J. Dupont (2003) "OECD Capital Services Estimates: Methodology and a First Set of Results." OECD Statistics Working Papers, OECD, Paris.

Stolarick, K. M. (1999) "Are Some Firms Better at IT? Differing Relationships between Productivity and IT Spending." CES Working Paper WP-99-13, Center for Economic Studies, Washington, D.C.

Triplett, J. E., and B. B. Bosworth (2002) "'Baumol's Disease' Has Been Cured: IT and Multifactor Productivity in U.S. Services Industries." Paper prepared for Brookings Workshop on Services Industry Productivity, Brookings Institution, Washington, D.C., September.

Van Ark, B., J. Melka, N. Mulder, M. Timmer, and G. Ypma (2002) "ICT Investment and Growth Accounts for the European Union, 1980–2000." Paper prepared for DG ECFIN, European Commission, Brussels, September, http://www.eco.rug.nl/ggdc/homeggdc.html.

II
Knowledge and Innovation

4

New Models of Innovation and the Role of Information Technologies in the Knowledge Economy

Dominique Foray

Analysts of innovation, who strive to identify new forms of innovation in the contemporary context, often tend to focus on a particular form that then becomes a sort of archetype, a single form governing all developments under way. Some thinkers talk of the great contribution of scientific R&D, others of the prominent role of users and even lay people, others still of the growing importance of innovations spawned by the modularization of technologies.

However, this chapter argues that it is misguided to look for a single source driving innovation in today's knowledge economies. The experience of a range of sectors suggests that innovation depends on multiple factors involving the contribution of various sources of knowledge. Important influences include not only well established elements such as the strength of scientific R&D but also newer ones such as the active role of users or the importance of modularity as a tool for innovation.

The chapter identifies, with reference to case studies documented in various recent projects, three key factors that are becoming important drivers of innovation in the economy generally and discusses some common features such as the role of networks, information technologies, and knowledge sharing as determinants of the efficiency of any innovation process. In describing the various factors needed to fully realize innovative capacity, this chapter looks at the *sources* of innovation.

The three innovation models are:

• *Science-based innovation.* Science plays an unquestionable role in advancing knowledge. However, for a long time most major technological breakthroughs were not directly based on science. It has been the slow expansion of the model of science illuminating technology that has spawned innovation in sectors where scientific research rarely or never resulted in innovation.

• *Innovation and collaboration* among users and/or doers. New actors are becoming engaged in innovation processes, and this creates new opportunities.

• *Modular structures*, each with freedom to innovate yet joined together in a whole innovative system. This devolved character of innovation in complex technological systems creates new needs for coordination and certification.

This chapter starts by describing how each of the three models works. It then discusses how these insights might be most fruitfully used in terms of strategic analysis of "missed opportunities" for companies: If you haven't explored the scope for each model's use, you may be adopting a suboptimal strategy in terms of maximizing your potential innovative capacity.

In the last sections, some common features of the three models are discussed: the collective nature of innovation, the importance of public knowledge, and the role of information technologies. I will highlight two major contributions of information technologies to the economics of innovation in each model: minimizing incurring costs, which results from the increasing collective and decentralized nature of innovation, and reaping economic returns from the ability to organize multiple experiments.

The First Model: Science-Based Innovation

Scientific knowledge potentially contributes to creating new or improving existing products, services, processes, and organizations. A feature of the knowledge economy is that in any sector there is scope for building and expanding of a scientific knowledge base through experimental R&D. A scientific approach contributes to innovation in three different ways (David et al., 1992):

• It provides a more systematic and effective base for discovery and innovation.
• It allows for better control (quality, impact, regulation) of the new products and processes introduced.
• It may be at the origin of entirely new products or processes.

Why Does Science Contribute to Innovation?

A scientific approach is important because it makes it possible (in most cases) to conceive and carry out well defined and controlled experimental probes of possible ways to improve technological performance, and to get relatively sharp and quick feedback on the results. Well defined and controlled experimental probes require isolation of the technology from its surroundings. Experimentation is often carried out using simplified versions (models) of the object to be tested and of its environment. Using a model in experimentation is a way of controlling some aspects of reality that would affect the experiment in order to simplify analysis of the results. The ability to perform exploratory activities that would not otherwise be possible in real life is a key factor supporting rapid knowledge advances.

In an increasing number of sectors, the possibility to carry out "experimentation" generates a large scientific knowledge base. Some industrial sectors have for a long time used scientific approaches to create knowledge (electricity, chemicals). Yet most major technological breakthroughs were not directly based on science. Rather, science plays a slowly expanding role by *illuminating* technology to fuel innovation, in sectors where pure research rarely or never leads directly to innovation.

Sectoral Cases

The growing influence of the scientific approach has been particularly significant in industrial sectors. Drug discovery is a good example of a domain that has recently been characterized by a shift from a random approach through large scale screening towards a more science-guided approach relying on knowledge of the biological basis of a disease to frame a research strategy (Cockburn et al., 1999). Another recent case is innovation in the development of adhesives, which has been fueled by the use of scientific knowledge of the transition properties of certain materials, which provides a theoretical base for more effective and systematic R&D.

At the same time, this growing influence is also quite clear in people-centered professions. In the health and pharmaceutical sector, scientific methods such as randomized controlled trials are used to compare a new drug with the best existing therapy. The accepted "gold standard" of evidence in this kind of approach is "double blind" testing of a new drug, in which patients are randomly assigned to groups receiving the new treatment and an existing one or a placebo, without either themselves or the physician knowing who is in which group. In social and educational research, randomized controlled trials or randomized field trials offers great potential to generate scientific knowledge and robust evidence on a broad range of topics (Foray and Hargreaves, 2003). Such scientific approaches seem constantly to cover new ground, even in sectors that a priori appear least suited to them. Across a heterogeneous array of contexts (from drug discovery to adhesive products, therapeutic testing, and education), scientific knowledge bases of direct use to innovation are being established. The idea is not to rehabilitate the old linear, so-called "science push" innovation model but to understand and exploit all aspects of knowledge systems where there is greatest potential for knowledge advances to contribute to productivity and effectiveness. Scientific research helps speed up change in response to market signals and to the emergence of certain social demands.

Two Forms of Connection

The connection of scientific research to innovation has two distinct forms. First, scientific knowledge creation at some basic research stage allows more effective innovative research that escapes the much longer, and usually much more expensive and uncertain, process of cut and try (empiricism) (Kline and Rosenberg, 1986). Second, we note the appearance within the firm and other organizations of scientific investigation tools. Hence, the ability to organize rapidly a large number of experiments based on simulation is revolutionizing design and development work. Thomke (2001) investigated how automotive companies are currently advancing the performance of sophisticated safety systems that measure a passenger's position, weight, and height to adjust the force and speed at which airbags deploy. The availability of fast and inexpensive simulation enables massive and rapid experimentation necessary to develop such complex safety devices.

These developments all point to the idea that a wide range of research problems warrants an effort at collecting scientific data and that appropriate forms of experimentation are necessary and most often possible. One of the features of the knowledge economy is that many industries are now firmly based on complex scientific knowledge. Quite surprisingly, industries that might at first glance be considered as "low tech" are in fact "complex knowledge based"—such as the food processing industry.

The Second Model: The Role of Users—"Horizontally" Organized Innovation

This model involves the contribution of users in the innovation process not only in terms of sending market signals (which is the normal representation of what users are supposed to do to help innovators) but in terms of actively contributing to the phase of design and product development. This engagement of users in innovation is a kind of deviation from the conventional model in which commercial suppliers are the only innovators (von Hippel and von Krogh, 2003). Note that "using" here refer to the use of knowledge and technologies as distinct from the final product. In this context most workers can act as "users" of knowledge and technologies. In principle, final consumers are also users of knowledge. However, in the context of the discussion below, workers are regarded as the main users.

Why Are Users Innovators?

This is the main difference from the conventional model of innovation: Users/doers are substituting for commercial suppliers in performing innovative tasks. But do they have incentives to do so? E. Von Hippel (2001) identifies three factors that can create such incentives:

(1) Direct, tailored benefits for the user. Specific improvements in the design of a product can motivate a user to find a solution that will exactly fit with his/her specific needs and circumstances. This contrasts with the supplier's incentive to create solutions that are "good enough" for a wide range of potential users.

(2) The chance to gain from "situated" learning. Users in a very broad sense acquire a certain kind of knowledge that is particular to a specific site and/or usage. This is the case for the user of a machine tool or a medical instrument. This knowledge is itself an impetus towards innovation.

(3) The possibility of addressing a problem without the difficulty of communicating "sticky" knowledge about the problem to someone else. When knowledge is costly to transfer (for instance, knowledge about

some particular circumstances of the user), the locus of problem-solving activity can shift from supplier to user.

Institutional Forms

Users' engagement in innovation has two particular forms:

(1) The creation of technical and organizational systems through which the producer leaves it up to users to make adjustments and develop the design that suits them best. Such a partial transfer of design capabilities should occur under the following circumstances: when market segments are shrinking and customers are increasingly asking for customized products, when costs are increasing, without much possibility of passing those costs on to customers, and when producers and users need many iterations before a solution can be found (Thomke and von Hippel, 2002).

(2) The emergence and upsurge of user cooperatives that take over the function of innovation (e.g., open software, sports equipment). Users interact in a sector-defined community, designing and building innovative products for their own use, and freely reveal their design to others. Others then replicate and improve the innovation that has been revealed, and freely reveal their improvements in turn. Horizontally ordered innovation means that new ideas and methods do not necessarily flow from suppliers. Users are freed from the constraints of the willingness of their commercial suppliers to innovate (von Hippel, 2001).

Not all knowledge-sharing communities have the same combination of elements driving innovation. Some of the most favorable conditions apply in the open development of software (von Hippel, 2001). Here, success depends on (a) a critical mass of skilled users capable of finding solutions, (b) incentives to share knowledge such as rewards to reputation, and (c) very low marginal cost for writing and transmitting the information. To the extent that such factors allow the emergence and multiplication of "user-only innovation systems," this represents an important, possibly decisive development in the historical emergence of the knowledge-driven economy.

However, note that the open source example remains unusual in that user-driven innovation not only complements but potentially competes with commercial systems of manufacturing and distribution. In most cases, this is not true because the innovation at stake deals with physical improvements, in which economies of scale matter in manufacturing and distribution. In such cases (e.g., sport equipment) users are perfectly able to innovate and to share the innovation, but diffusion is still fulfilled by the commercial system.

The Third Model: Modular Structures, with Freedom to Innovate yet Joined Together as a Whole System

An important element of complexity relates to the evolution of products. New products are rarely stand-alone items; they are more often components of broader systems or structures. In modern technology, modularity is an objective that increasing numbers of firms are pursuing in order to benefit from the specialized division of labor and to create proper conditions for innovation.

Why Is Modularity a Critical Way to Manage Innovation in Industrial Organizations?

Using a modular structure means building a complex product or process from smaller subsystems that can be designed independently yet function together as a whole. As shown by C. Baldwin and K. Clark (1997), the fact that different companies or different units are working independently on modules is likely to boost the rate of innovation. Modularity can be compared to conducting multiple experiments in parallel. There is, however, a trade-off between this freedom to experiment with product design (what actually distinguishes modular suppliers from ordinary subcontractors) and the need for systemic coherence and integration. A particular class of knowledge—integrative knowledge—is, thus, required to achieve integration of modular systems. This class of knowledge covers norms and standards (for quality, for reference, and for interfaces), connective rules, certification processes, and common platforms.

The Trade-off between Freedom to Experiment and General Coordination

Imagine an organization composed of three units—two units engaged in production or design of modules and one unit called "the architect" or "the helmsman" coordinating the whole system. Such an organization must process two kinds of information to achieve its objective. One is visible information; the other is hidden information. The latter is concerned with particular needs and objectives that are specific to each of the two modules. Others do not have to know it, so it can be hidden within each unit. On the other hand, visible information is required to clarify the connective rules to be followed for achieving the integration of both modules to create a system. Modularity thus provides a mechanism in which only a fraction of processed information is shared among all agents. The literature on modularity points to several benefits of this in dealing with innovation in complex systems. One is the benefit of having more smaller modules instead of one large one. Another is specialization, since an engineer working on each module can specialize in localized design and activities. But the main benefit concerns the ability to conduct multiple experiments in parallel, with each corresponding to a different hypothesis or research option.

However, modularization also incurs costs. Human beings, unable to foresee all the uncertainties, find it impossible to enumerate and resolve all possible dependencies among modules. The more complex a system is, the more incomplete the *ex ante* design of connective rules. Thus, modularization cannot escape from a trade-off between facilitating innovation within modules and optimizing the whole system.

Coping with This Trade-off in Three Different Modular Forms

Aoki and Takizawa (2002) evaluated three generic forms of modularity in terms of their ability to cope with this trade-off:

(1) *"Hierarchical decomposition"* describes a system whose key features are predesigned by a single "architect." This architect is specialized in processing exclusively the visible information and determines the connective rules prior to the design of the modules. Even if something occurs in the environment after activities in the respective modules begin, only the architect can decide changes in the connective rules. Each module is engaged in processing idiosyncratic information required only for its activity, given the visible information transferred by the architect.

(2) *Information assimilation* is a system in which the architect leads but does not create inflexible system features. Connective rules continue to be fine tuned even after the activities in the respective modules begin. Information about changing conditions is exchanged between the architect and the modules as well as between the modules. In other words, the visible information is propagated back and forth between the architect and the modules.

(3) *Evolutionary connection* involves multiple architects and multiple agents engaged in the design of each module, with continuous assimilation of new information. Activities are carried out in parallel and duplicated. Each agent is engaged in processing not only hidden information but also processes the visible information independently in a limited way. Through information exchanges between the modules and the architects, a few connective rules may emerge in an evolutionary way. The architects will select and combine already designed modules that are most compatible with the connective rule that they select to form a product system. The market finally evaluates which system will have the highest value.

Clearly, the more complex a system is, the more efficient forms (2) and (3) will be, since they can adapt connective rules to complex and evolving conditions rather than fully predefining them.

Three Models to Fill Up with Innovative Capacity

The three sources or models of innovation (or modes of knowledge production) reviewed above are summarized in table 4.1. Yet these ideal types are rarely identifiable in a pure form. They are born at certain points in history, in specific limited domains. Their importance grows as they combine and hybrids are formed. Many "real" innovation processes are the result of combinations between the different models.

From a strategic point of view, this framework does not imply that every model will be equally relevant for each business activity but rather that each model's potential needs to be explored, in order to optimize the strength of innovation in a particular business activity.

In the following sections, I discuss some common features of the three models.

Networks and the Collective Nature of Innovation

Analysis of the three models clearly shows that knowledge that is useful for innovation is not solely generated by processes of intraorganizational conversion, combination, adaptation, and extension but also by processes that are collectively organized by industries and other larger domains of interorganization relations. The production and collective adoption of science-based/user-based/modularity-based innovations occur in different spaces:

Table 4.1
Three models of innovation for the knowledge economy

	Model 1	Model 2	Model 3
Innovative opportunities	Scientific advances	User needs and capabilities	Modular systems
Critical actors and relationships	University–start-up enterprises Large companies with R&D capacity	Horizontal communities Producer–user and design capabilities transfer	Architect and module designers Rules for their interaction (several generic forms)
Contributions to the knowledge base of the sector	Scientific knowledge of direct relevance to improving processes and products	Practical knowledge that may be widely adopted within a community	Integrative knowledge (norms, standards, certification, common infrastructure)
	Generic information	User specific (invisible) information	Visible information

- University–industry (start-up and large companies),
- Users–producers and users' communities,
- Industry consortia and architects–module designers.

A network is, thus, a useful metaphor to describe the interorganizational nature of innovation. However, a metaphor is not the same thing as a well worked out economic model or management tool, involving the provision of incentives and the design of coordination mechanisms appropriate to the particular class of network considered. University–industry, users–producers, users' communities, and architects–module designers represent different classes of network, raising each particular problems of incentive, coordination, and organizational tension.

The Importance of the Public Domain of Knowledge and Information
In each of the models described (science-based, user-based, and modular-based), the existence of a freely accessible stock of knowledge is crucial. The efficiency of innovation processes is fundamentally dependent on this domain of "public" knowledge and information. By public domain

we do not necessarily mean the public sector "controlled by the state." We are referring more generally to areas in which knowledge is shielded from mechanisms of private appropriation and in which knowledge and information are revealed and shared. "Government-controlled property" (such as national R&D laboratories) and "inherently public property" (such as collective actions giving rise to horizontal systems of innovation where knowledge is shared and reused among users) are the two pillars of the public sphere.

Public Knowledge in the Science-Based Innovation Model

The public dimension of the first source of innovation is very clear. Knowledge resulting from basic research is generic and fundamental. Accordingly, its "social returns" will be far higher if it can be used by a multiplicity of innovators. The free circulation of this knowledge facilitates cumulative research, increases opportunities for innovation, and enhances the quality of results (since everyone can examine them and try to reproduce them). This free circulation is at the heart of the organization model of science, which historically has proved its efficiency. In this model, the public sector of scientific research produces public knowledge, which can be used freely by industry. This pool of knowledge is an extremely important input for private R&D. It is generally considered that the existence of public knowledge generates (at least in the immediate vicinity) an increase in private returns to investments in R&D.

Public Knowledge in the User-Based Innovation Model

Freely revealing new solutions and ideas is a necessary condition for the functioning of communities of users. In these communities multiple potential sources of innovation are identified and each member of the community can benefit from them. If this condition were not met, each user would be obliged to make all the adjustments that user desired, which would substantially increase the overall cost of the process. It would consequently have no chance of competing with "average" solutions (more or less suited to everyone) offered at a lower cost by commercial systems. The sharing and circulation of innovation is therefore essential to ensure a minimum of efficiency.

Public Knowledge in the Modular-Based Innovation Model

The public dimension of the third source of innovation is less known but equally evident. It results from the collective creation of quasi-public goods in private markets. It is essential to preserve public access and the sharing of "essential" technological or informational elements, the visible information composing the standard of a product or an industry. As in the preceding cases, this poses thorny problems of compromise between the collective aspect of innovation and the safeguarding of private interests.

Knowledge Openness as a Key Feature in Each Model

The shared collection of basic knowledge provides the building blocks for new inventions in each model. Without a continuous flow of public knowledge, innovation is, therefore, likely to be blocked. In each model, the organization of innovation and the norm of knowledge sharing are intimately related. In other words, any alteration of the norm of knowledge sharing would put the full process at risk. This is the case for open science, open source projects as well as any users' community and consortia that are created to produce standards, connectives rules, and other kinds of integrative knowledge. In all these cases, the norm of knowledge openness is at the heart of the institutional structure. This means that private benefits are not expected to be derived by intellectual creators from the creation of exclusivity mechanisms and knowledge producers are not seeking a right to exclude (for example, a patent) as a means to extract an economic rent from their production. Private benefits do exist in systems based on knowledge openness, but they derive from a different set of mechanisms (reputation, reciprocity, private advantages generated by the free diffusion of the innovation such as the creation of a standard) (von Hippel and von Krogh, 2003). The insight provided by this argument is that free revealing of information into an "intellectual commons," much as it is done by academics in scientific fields, might be a better way to advance social welfare than intellectual property protection mechanisms such as the patent grant.

Information Technologies' Critical Role in the Economics of Innovation

The last common feature of the three models of innovation concerns information technologies. These new technologies, which first emerged in the 1950s and then really took off with the advent of the Internet, have breathtaking potential. They enable remote access to information and the means of acquiring knowledge. In addition to transmitting written texts and other digitizable items (music, pictures), they also allow users to access and work upon knowledge systems from a distance (e.g., remote experimentation), to take distance-learning courses within the framework of interactive teacher–student relations (tele-education), and to have unbelievable quantities of information—a sort of universal library—available on their desktops.

Information technologies can affect knowledge creation in a number of different ways (Foray, 2003). For a start, the mere fact that one has the capacity to create such a wealth of information is truly revolutionary. Even in the not-too-distant times of a couple of decades ago, it was a laborious task for students to produce a roundup of the "state of the art" in a particular subject or discipline and an uphill struggle to remain abreast of the latest findings in their study field.

Does the advent of increasingly high-performance access and communication network new technologies signal an end to that evolution? Clearly not, for an enormous amount of progress remains to be made in such areas as information search systems. But this might almost be said to be the culmination of what the French medievalist Georges Duby once called the "relentless pursuit of instruments of knowledge" that has preoccupied humankind since the dark ages.

Second, ITs allow more flexibility as regards the constraint of physical proximity in many cognitive activities (e.g., distance learning, remote access). Access from a distance not only to writing but also to other modes of expression of knowledge (especially gestures and voices) revolutionize possibilities for learning. It is true that many cognitive activities cannot be coordinated by virtual means alone. Direct face-to-face exchanges are important when they enable other forms of sensory perception to be stimulated apart from those used within the framework of electronic interactions (Olson and Olson, 2003). However, the influence of distance is waning now that the technological capacity is available for knowledge sharing, remote access and teamwork, and organizing and coordinating tasks over wide areas.

Third, ITs are at the base of new modes of knowledge production. (1) They enhance creative interaction not only among scholars and scientists but, equally, among product designers, suppliers, and the end customers. The creation of virtual objects that can be modified ad infinitum and are instantly accessible to one and all serves to facilitate collective work and learning. In that respect, the new possibilities that computers have opened up for numerical simulation represent another significant departure from prior experience. (2) ITs enable the exploration and analysis of the contents of gigantic databases, which is in itself a potent means of knowledge enhancement (in natural, human, and social sciences and management alike). Research stimulated by such possibilities has a strong influence in some areas of managerial work.

Fourth, the above three ways in which information technologies affect knowledge creation can be combined in the development of large-scale decentralized systems for data gathering and calculation and the sharing of findings. Such extensive systems characterize the research being done these days in the fields of astronomy, oceanography, and so on.

Finally, ITs provide powerful opportunities for collective actions—that is to say the sharing of "rich" messages among a very large number of people—and as such are the right tools for the creation and expansion of virtual communities (Mansell and Steinmueller, 2000).

The list above identifies very broad and general impacts of ITs on the production and organization of knowledge. I will now explore deeper the detailed role of ITs as a unique means to increase the efficiency of the three models of innovation.

Minimizing Incurring Costs

To better understand some common features of the application of information technologies (IT) in each model, it is useful to identify a general transformation in the division of work and organization of tasks, which is a facet of the three models. Each model is characterized by a shift in commercial supplier responsibilities to university scientific laboratories (in the first model), users capabilities (in the second model), and module designers (in the third model). This shift results in adding value, increasing flexibility, and cutting costs. However, some transaction and coordination costs are likely to rocket as a result of the increasingly collective and decentralized nature of innovation in each model. Thus, IT represents the key technology infrastructure enabling such a shift in commercial supplier responsibilities, while keeping coordination and learning costs below some tolerable level.

As a collaborative technology (from email to "co-laboratories"), IT is likely to minimize the potential increase of coordination costs due to a more complex distributed system of innovation. There is, for instance, a very interesting result produced by Lakhani and von Hippel (2000), showing that the success of any system of "free" user-to-user assistance in open source software communities is based on the very low marginal cost for writing and transmitting the information. The willingness of information providers to contribute what they know is related to the cost to them of doing so, and the intensive use of information technology is critical to minimize this cost.

As a learning technology, IT is likely to minimize the potential increase of learning costs due to the fact that new classes of agents have to deal with innovation tasks. For instance, tool kits based on high-quality computer-based simulation, rapid prototyping tools, and user-friendly interfaces are critical to support the transfer of design tasks from suppliers to users. In such a case, the tool kit must enable customers to run repeated trial-and-error experiments and tests rapidly and efficiently, and it should let customers work in a familiar design language, making it cheaper for customers to adopt the tool kit (Thomke and von Hippel, 2002).

Harnessing the Benefits of Multiple Experiments

Furthermore, ITs play a key role in each model to reap economic returns from the ability to organize multiple experiments. This is a second great common feature of the three models. Each model represents a way to organize multiple experiments. In the first model, the appearance and adoption of scientific tools in firms considerably strengthen the ability to organize rapidly a large number of experiments to generate reliable knowledge about a particular design. In the second model, the open software case illustrates that the developer communities are particularly innovative because they are designed to allow multiple experiments. Finally, it has been already strongly stressed that the ability to carry out multiple experiments is a great feature of modular structures. However, this particular class of learning processes creates information and data abundance, involving the risk of information congestion and increasing costs of storing, searching, retrieving, and combining data. In this sense, ITs provide unique opportunities to cope with such problems, enabling innovators to reap economic returns from multiple experiments.

Conclusion

The three sources of innovation or modes of knowledge production that seem to be gaining in importance are as follows:

The first major trend concerns the increasingly scientific nature of research methods. In more and more sectors, the "epistemic culture" of science for knowledge production is growing in importance.

Users' increasingly marked engagement in knowledge production represents a second trend.

Finally, the increasing complexity and modularity of industrial architecture makes it more critical than ever to produce "integrative knowledge," such as standards, norms, and common architectures and platforms.

Each of these models is characterized by some common features related to the importance of a freely accessible stock of knowledge as well as the unique role of ITs in enhancing the efficiency of innovation processes. Maintaining the public dimension of knowledge as well as promoting the development and use of ITs as knowledge instruments are, therefore, two great challenges for innovation policy in the age of the knowledge-based economy.

References

Aoki, M., and H. Takizawa (2002) "Modularity: Its relevance to industrial architecture." The Saint Gobain Centre for Economic Research.

Baldwin, C., and K. Clark (1997) "Managing in an age of modularity." *Harvard Business Review*, September–October.

Cockburn, I., R. Henderson, and S. Stern (1999) "The diffusion of science-driven drug discovery: Organizational change in pharmaceutical research." Working Paper no. 7359, NBER, Cambridge, Mass.

David, P. A., D. C. Mowery, and W. E. Steinmueller (1992) "Analyzing the economics payoffs from basic research." *Economics of Innovation and New Technology*.

Foray, D. (2003) *The Economics of Knowledge*. Cambridge: MIT Press.

Foray, D., and D. Hargreaves (2003) "The production of knowledge in different sectors: A model and some hypotheses." *London Review of Education* 1(1).

Kline, S., and N. Rosenberg (1986) "An overview of innovation." In *The Positive Sum Strategy*, ed. R. Landau and N. Rosenberg. National Academy Press.

Lakhani, K., and E. von Hippel (2000) "How open source software works: Free user-to-user assistance." MIT Sloan School of Management, Working Paper no. 4117.

Mansell, R., and W. E. Steinmueller (2000) *Mobilizing the Information Society.* New York: Oxford University Press.

Olson, G., and J. Olson (2003) "Mitigating the effects of distance on collaborative intellectual work." *Economics of Innovation and New Technology* 12(1).

Thomke, S. (2001) "Enlightened experimentation: The new imperative for innovation." *Harvard Business Review*, February.

Thomke, S., and E. von Hippel (2002) "Customers as innovators: A new way to create value." *Harvard Business Review*, April.

von Hippel, E. (2001) "Learning from open-source software." *MIT Sloan Management Review*, Summer.

von Hippel, E., and G. von Krogh (2003) "Open source software and the private-collective innovation model: Issues for organization science." *Organization Science* 14(2) (March–April).

5

Using Knowledge to Transform Enterprises

Richard O. Mason and Uday M. Apte

What does it mean to say that we live in a knowledge society, or that people do knowledge work? And, if these concepts characterize modern society, what are the processes by which enterprises are changed so that knowledge is more integral to their being?

These two questions have motivated our inquiry. The search for answers must begin with an examination of the nature of knowledge itself and the role it plays in enterprises and society. Armed with an understanding of knowledge, we have found it useful to frame our exploration in the context of a model. This general model tracks the flow of knowledge from the initial investments made in inquiry necessary to produce it, through the processes of generation, codification, transfer, and distribution, to the resulting knowledge's ultimate use in productive activities we call "business processes." The outcome of business processes results in goods and services and in organizational structures and divisions of labor that constitute the transformations people experience.

Knowledge does not accomplish these changes alone, however. It must be married with facilitating technology—information and communication technology (ICT). Knowledge and ICT are the yin and the yang of transformation. Each standing alone is impotent, but joined together they can become a powerful force for change. The history of the industrial revolution illustrates this point. So, we begin our exploration there.

The Role of Knowledge in Enterprise

Knowledge Transforms Society

A society is transformed by the goods and services it produces and the processes it uses to produce them. In previous economic eras energy produced by water wheels, steam, electricity, and then oil was the main driver—the core input—of the social economic system. During the latter half of the 20th century, however, the fundamental catalytic role of energy was superseded by information and knowledge. In recent times information and communication technology, and especially the Internet, has been used to form a network that makes knowledge and information globally available.

Sociologist Manuel Castells explains: "If information technology is the present-day equivalent of electricity in the industrial era, in our age the Internet could be likened to both the electrical grid and the electric engine because of its ability to distribute the power of information throughout the entire realm of human activity. Furthermore, just as new technologies of energy generation and distribution made possible the factory and the large corporation as the organizational foundations of industrial society, the Internet is the technical basis for the organizational form of the Information Age: the network."[1] Among the special powers of a network is that it transforms enterprises by distributing knowledge and information broadly and rapidly to places where it can be acted upon.

What role does knowledge play in socioeconomic transformation? Its primary role is to inform the conduct of business processes, which in turn produce goods and services by means of an organizational structure and division of labor.

Knowledge and Information

Knowledge—the engine of transformation—requires a more detailed examination. It goes by many names: science, innovation, know-how, creativity, and technology. In all of these uses it refers to the ability of an acting agent to do something correctly. According to Daniel Bell's classic definition, it is "a set of organized statements of facts or ideas representing a reasoned judgment or an experimental result which is transmitted to others through some communication medium in some systematic form!"[2]

And, what, then, is information? Information is elementary data that have been given meaning by being interpreted in a framework or worldview. Information that is believed and that one is willing to base his or her actions on becomes knowledge. Thus, information is data interpreted; knowledge is information authenticated.

These two epistemic units—knowledge and information—have become the driving force in our society and economy. Although the two terms are often used interchangeably, there are important differences between them. Information is objective and can be collected and processed independently of its users. Knowledge resides ultimately in the mind of the user or acting agent. Consequently, many types of knowledge are deeply embedded in the mind of the knower and are hard to extract and codify. Information can be generated, distributed, and acquired in great volumes and rapidly—think of the stacks of unread papers in a typical study and the lengthy shelves of partially read or understood books. But knowledge must be assimilated, usually by a slow reflective process frequently associated with practical use or at least some kind of an exam to be taken. Information that has become knowledge can trigger change.

The general causal pattern is straightforward. Improvements in an enterprise's knowledge base result in changes in the functioning of its business processes, which in turn produce changes in the value of the goods and services it generates. This chain of events is initiated by investments in inquiry, research, and development. Importantly, it is facilitated by the use of interactive, networked information and communication technologies. At the outset of the 21st century the Internet has become one of the most promising technologies for enabling knowledge generation, codification, and distribution processes. It is the undergirding for what Castells calls "the rise of the networked society."[3]

Knowledge is something that acting agents can reliably base their behavior on. In general, it is information that has been authenticated or validated or is thought to be true. According to philosopher Gilbert Ryle, knowledge takes two forms:

- "Knowing that"—propositional or factual knowledge—and
- "Knowing how"—prescriptive, instructional, procedural knowledge, or knowledge of technique.[4]

Ryle's distinction is the most useful for understanding the role of knowledge in transformation. Nevertheless, epistemological and social-level distinctions are also useful. The scientist qua philosopher Michael Polanyi drew a line between "tacit" knowledge (personal, context-specific knowledge) and "explicit" knowledge (knowledge packaged in formal, systematic, sharable language).[5] Subjective, tacit knowledge resides only in the mind of a person. In contrast, objective, explicit knowledge can be captured in a variety of media and communicated to others. Nonaka and Takeuchi also stress the ontological dimension of knowledge.[6] They observe that knowledge-generating processes spiral between levels: individual, group, organization, interorganization, and the societal or systemic level. These authors note that knowledge can be converted from tacit to explicit and vice versa, thus giving rise to four modes of knowledge: socialization (tacit to tacit), externalization (tacit to explicit), internalization (explicit to tacit), and combination (explicit to explicit). Polanyi's and Nonaka's frameworks are useful in identifying sources and describing the processes by which knowledge is generated. Nevertheless, the functional distinction between propositional and instrumental speaks specifically to knowledge's role in changing things and making things happen. For example, the functional distinction relates directly to the distinction between product and process knowledge (frequently made in economics, operations management, and operations research).[7]

Both forms—"knowing that" and "knowing how"—are essential. Each interacts with and enhances the other. "Knowing that" includes knowledge about natural phenomena and regularities. It is the product of methods of discovery such as scientific research. "Knowing how" includes "sets of executable instructions or recipes that tell us how to manipulate nature."[8] In a frequently referenced source, economists Richard Nelson and Sidney Winter refer to "knowing how" knowledge as "routines" and stress the importance of these routines in creating wealth.[9] "Knowing how" is the product of invention, development, and engineering. Both forms of knowledge result from processes of inquiry; that is, from quests for understanding, information, data, and truth.[10]

How Knowledge Transforms Enterprise

An enterprise is transformed when new knowledge or information is introduced and acted on in one or more of its business processes. Knowledge is the fuel that activates a business process. Knowledge is the ultimate substitute. When it is used effectively it reduces the need for raw materials, labor, time, space, equipment, and funding in business processes.

Enterprises are composed of—and do their work by means of—many different kinds of business processes. A business process may be defined as "a related group of steps or activities that use people, information, [tangible materials], and other resources to create value for internal or external customers."[11] Thus, there are three interrelating, value-adding components of an enterprise's activities or business processes: (1) material conversion, (2) customer and other stakeholder contact, and (3) knowledge and information manipulation. (See figure 5.1.)[12]

Figure 5.1 summarizes a highly complex, interrelated process that is often characterized by significant time lags. Nevertheless, the model provides a useful framework with which to address the transformational capabilities of knowledge-related activities. The process described by the model begins with investments in inquiry. These investments may run the gamut from a single individual reflecting in a lonely garret on one end to a major social undertaking such as the Manhattan Project or NASA's venture to put a man on the moon on the other. In either case time,

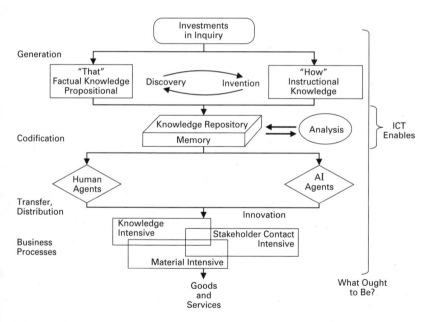

Figure 5.1
How knowledge transforms enterprises: A model. Source: Mason and Apte, *Using Knowledge to Transform*, 2003.

thought, and other resources are deployed to generate knowledge. The resulting knowledge is either "that" or "how" or a combination. It is then stored in a repository consisting of human minds and technical memory/storage devices. (In a strict interpretation, what is stored in devices is information and becomes knowledge when it is incorporated in a human mind.) Analysis techniques may be applied to the stored knowledge in order to produce more knowledge during a codification process.

The resulting knowledge is made available by means of transfer, distribution, and communication to agents—parties who will act on the knowledge and thereby convert its potential value to actual value. Human agents still are the major actors in transformations but artificial intelligent agents—ones that draw on knowledge embedded in them in the form of computer programs, routines, inference engines, or reasoning tools—increasingly play a prominent role. Going under the names of "bots" or "autonomous agents," these agents operate in the general work contexts we call a business process.[13] Business processes have three key dimensions: material, stakeholder (face-to-face) contact, and information or knowledge. Working together, elements of these three dimensions produce goods and services.

Beyond Knowledge to Social Change

Goods and services we consume affect and shape our lives. They are the grist for transforming enterprise and society. Moreover, as Marx noted, social change is created as a result of the division of labor that a business process employs, the kinds of jobs it creates and destroys, and the distribution of wealth it encourages. Today, information and communication technologies play a significant role in conducting business processes by reducing cost, speeding up handling, augmenting storage capacity, increasing the scope and range of distribution, and improving accuracy and precision.

Knowledge Is Capital and a Source of Transformation

Reaching Beyond Land, Labor, Capital, and Even Management

Increasingly, knowledge—in the form of intellectual capital—is the foundational factor of production for an enterprise. And, the network serves as its primary facilitator. Knowledge flowing through networks is creating a new form of capitalism: information and knowledge capitalism. In the economics of enterprise, properly packaged knowledge and information are now treated as assets, that is, productive items of value. Author Thomas Stewart describes: "Intelligence becomes an asset when some useful order is created out of free-floating brainpower—that is, when it is given coherent form (a mailing list, a database, an agenda for a meeting, a description of a process); when it is captured in a way that allows it to be described, shared, and exploited; and when it can be deployed to do something that could not be done if it remained scattered around like so many coins in the gutter. Intellectual capital is packaged useful knowledge."[14]

What is "Capital"? A Brief Historical Detour

The term "capital" refers to those things—including facilities and institutions such as money—that are created for the purpose of helping human beings produce goods and services. Early in the 19th century the economist David Ricardo described capital as "that part of the wealth of a country which is employed in production and consists of food, clothing, tools, raw materials, etc. necessary to give effect to labor."[15] Ricardo's model described an economy dominated by agriculture, and he wrote during the middle of the first Kondrative business cycle wave, which occurred from approximately 1780 to 1848. During this period water-powered mechanization was first used on a large scale to process agricultural products. Water wheels, cotton spinning, iron, and coal were the principal economic resources at the time. These resources were the result of innovations based on new knowledge that had been created by people like Sir Richard Arkwright (cotton spinning) and Henry Cort (puddling iron). Importantly, this new knowledge was also made mobile. It was packaged in a way that facilitated its movement from place to place and person to person. According to economic historian Joel Mokyr, leaders during the Industrial Enlightenment "sought to reduce [knowledge] access costs by surveying and cataloging artisanal practices"[16] in order to propagate best practices. During this period Diderot's *Encyclopedie* (circa 1751 to 1777), handbooks, and various periodicals were used to convey useful knowledge to a wider range of practitioners. Thus, useful, transforming knowledge was being generated, codified, transferred, and given greater value by its users and potential users. Nevertheless, the influential role this knowledge played was generally taken for granted and assumed to be just part of society. It was not formally regarded as an economic asset or as capital, at the time, even to as perceptive an observer as Ricardo.

Knowledge-based innovations created subsequent economic waves: steam-powered mechanization (circa 1848–1895), the electrification of industry, transport, and the home (circa 1895–1940), and the motorization of transportation that began with Henry Ford's mass production of the Model T (1908–1927) and took off with Alfred P. Sloan's reorganization of General Motors in 1921. The effects of this "Taylorism + Fordimus" industrial revolution are still quite evident in today's economy.[17]

There are many technological innovations that are precursors to the current age of knowledge, information, communications, and computers. In a real sense, the history of civilization has been a history of increasing improvements in knowledge, information, and communication.[18] But in this era of knowledge capitalism, knowledge is increasingly used to produce both knowledge and information and to inform the development of communication technologies (ICTs) that are in turn used to produce more knowledge and additional ICTs. Thus, knowledge production tends to unleash a reflexive, "virtuous" cycle. When did this process really get going?

The Beginnings of the Information/Knowledge Era

Business historian Alfred D. Chandler, Jr., and his associates have argued persuasively that the United States' economy has been continuously transformed by information and knowledge since Colonial times.[19] But a qualitative change occurred after World War II. Perhaps the most noteworthy tipping point was the unveiling of the ENIAC electronic computer by Eckert and Mauchly in 1946. Progress accelerated during the early 1960s with the release of IBM's popular 1401 and subsequently the company's 360 computer. Adoption of microprocessors (circa 1971) and release of the personal computer (circa 1977) extended the reach of the technology while reducing its cost and size. Telecommunication capacity has also exploded. From about 1970 to 2000, for example, the bit rate per optical fiber has increased from 280 Mbps to 160 Gbps while the cost per voice channel per year went from about $200,000 to less than $10.[20] The Internet (circa 1992) encouraged the global distribution of knowledge and information by linking these technologies together. These developments have heightened society's awareness of the value of knowledge in the economy and have made intellectual capital the most important factor in economic activity today. Because so much of the knowledge and information handled by these technologies is now encoded in 1's and 0's it is appropriate to refer to contemporary times—the outset of the 21st century—as the "digital age."

Knowledge capitalism treats "capital" in new ways. According to Stewart there are three kinds of intellectual capital, depending on where in the enterprise it primarily resides: human capital, structural capital, and customer capital.[21]

Human Capital

The Centrality of Mind

The human mind is the well-spring of knowledge. It is the source of innovation, insight, ideas, and invention. Thus, the human capital—the knowledge lodged in or available to the minds of an enterprise's members—is the essential asset. Because knowledge determines the ability of an acting agent to do something correctly, in the final analysis *it resides in the user—the agent—and not in a collection.*

Two Kinds of "Minds"

In the digital age users or acting agents may be either of two forms: human beings or artificial intelligence programs, bots, autonomous agents, and automatons.

One likely result, according to philosopher and cognitive scientist Andy Clark, is that human beings' capacity to incorporate tools, technologies, and supporting cultural practices into our lives expands our ability to think and feel. We are "natural born cyborgs," as described in his new book; we are now a species that integrates knowledge and tools, like computers and communication devices, directly into our existence. This remakes us into a new kind of person. In the preface Clark observes:

My body is an electronic virgin. I incorporate no silicon chips, no retinal or cochlear implants, no pacemaker. I don't even wear glasses (though I do wear clothes). But I am slowly becoming more and more a Cyborg. So are you. Pretty soon, and still without the need for wires, surgery, or bodily alterations, we shall be kin to the Terminator, to Eve 8, to Cable . . . just fill in your favorite fictional Cyborg. Perhaps we already are. For we shall be Cyborgs not in the merely superficial sense of combining flesh and wires, but in the more profound sense of being human–technology symbionts: thinking and reasoning systems whose minds and selves are spread across biological brain and non-biological circuitry."[22]

In the knowledge economy, people with tacit and explicit knowledge and cyborgs and machines endowed with human knowledge will work together to effect change. The 21st century will witness increased use of AI applications in business processes.[23] AI applications draw on and codify the knowledge available in human capital. Despite the fundamental nature of human capital, inanimate collections or repositories of knowledge and information are, nevertheless, crucial for creating value and wealth. These collections become part of an enterprise's structural capital.

Decoupling and Objectifying Knowledge
One key feature of the digital age is that increasing amounts of knowledge and information are separated out—disembodied—from their human carriers (or from business processes and other sources) so that they can be packaged and handled as distinct entities. In general, this is the externalization process of making tacit or embedded knowledge explicit. Information and communication technology permits organizations to acquire, package, store, process, and distribute large quantities of this separate, objectified knowledge. This knowledge can—and must—be managed. Accordingly, it should be treated as an object for investment. When knowledge and information are integrated directly into business processes and organizational designs, structural knowledge is created.

Structural Capital

Knowledge Embedded in Organizations
One chief characteristic of the digital age and of knowledge-based enterprises is that large amounts of knowledge are stored in digitized collections from which they can be retrieved rapidly and distributed broadly. In its inert form, knowledge is only of potential use, and its value is only

potential. Consequently, the value of knowledge as a capital asset is proportional to the likelihood of its use. What matters—the source of actual value—is how users, qua acting agents, react to a collection of knowledge when it is reproduced and shared with them. Structural capital is created by organization and the ways and means through which an organization's members are related to each other and communicate. It is owned by and incorporated into the enterprise. This includes its systems, organizational form, databases, patents, publications, trade secrets, copyrights, inventions, technologies, operating procedures, strategy, culture, routines, procedures, and the like. This form of capital is intellectual property. Because it is a significant source of performance and transformation, an enterprise needs to identify and protect its intellectual property, legally and otherwise.

Customer or Stakeholder Capital

Stakeholder Loyalty and Trust Are Capital

Customer capital—better termed "stakeholder" capital—is the value of the relationships an enterprise has established with its customers, suppliers, and other stakeholders. An enterprise's stakeholders include all of the parties who may affect its activities or who are affected by its activities.[24] Thus, stakeholder capital refers to the loyalty, depth of penetration, and breath of coverage of these relationships. It is measured by the likelihood that customers, suppliers, and other stakeholders will continue to do business with the enterprise and is dependent on trust and loyalty. The knowledge underlying this form of capital is the special knowledge associated with forming and maintaining relationships with people and organizations.

Goodwill

One method of measuring the value of an enterprise's intellectual capital—human, structural, and stakeholder—draws on the accounting concept of "goodwill." This measure is used to make a summary estimate of the value of several things: research and development in process, unrecorded intellectual property, the management team's capability, brand and stakeholder loyalty, public image, and, in a literal sense, the goodwill of the enterprise's customers. Seldom, however, are these values calculated directly. Rather, they are usually determined by the excess an enterprise pays over the estimated "fair market value of the net assets" of another enterprise it acquires.

Knowledge's Role in an Enterprise's Economic Recipe

Relative Intensity

All business processes function on the basis of the state of knowledge and information made available to them. Nevertheless, some depend on information more than others and in different ways. Consequently, they may be assigned to one of three basic categories, depending on the relative intensity of factors used: (1) material intensive, (2) customer or stakeholder contact intensive, and (3) knowledge intensive. Knowledge and information play a significant but somewhat different role in each of these three classes of processes. It is useful to classify a business process or enterprise according to whether its material intensity is high (H) or low (L), its stakeholder contact intensity is H or L, and its knowledge and information intensity is H or L. Using this method, material intensive enterprises may have a profile of HHL, HLH, or HLL; stakeholder contact intensive, HHL, LHL, or LHH; and knowledge and information intensive, LHH, LLH, or HLH. It is assumed that since every process has at least one crucial factor and one less significant factor, classifications of LLL and HHH are not useful. This method, however, allows for some overlays between the categories. For example, physicians work in enterprises and business processes that are both knowledge and information intensive and patient contact intensive but tend to be less material intensive. Figure 5.2 summarizes this classification approach. Note that the occupations assigned to each of the six nodes in figure 5.2 are suggestive. In practice, each actual job should be analyzed and classified. For example, a typical "Army infantry" position may be HHL, but special services personnel are more information intensive.

Table 5.1 summarizes the role that knowledge plays in each of the three categories of intensity.

The Role of Knowledge in Material Intensive Enterprises

In material intensive enterprises and business processes, knowledge and information are used to guide the manipulation of materials, to create

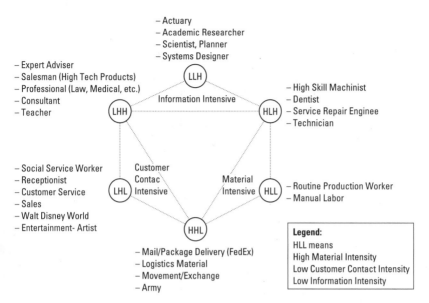

Figure 5.2
Job classification based on material, customer contact, and information intensity: Selected occupations.

new or more effective materials, to substitute for materials, or to decrease the time and cost required to process them. This is the "substitution effect" of knowledge.

Today, for example, people talk about the "smart beer can," one with less metal than its steel predecessor of several decades ago, produced with less energy consumed in its production, of higher tensile strength, stackable beyond six feet, lighter, able to withstand temperatures from 35° or lower up to 95° without damage to its contents, more manageable, and less expensive. All of these improvements are due to innovations in aluminum processing.[25] Moreover, these innovations are the result of deliberate scientific research and development that produced both factual and instrumental knowledge that could, in effect, be substituted for materials and used to improve them. This is just one illustration of how an enterprise or business process is transformed by means of applying scientific and other knowledge to physical materials. In this case the knowledge is incorporated—i.e., embedded—directly in the materials and the processes used to produce and manipulate them (see table 5.1, "Material Intensive").

Table 5.1
The role of knowledge in business processes

	Type of business process		
	Material intensive	Customer contact intensive	Information intensive
Corporate examples	Chaparral Steel, Toyota	McDonald's, Disney	AOL-Time Warner, Microsoft, Monster.com
Characteristic of knowledge used	Natural sciences, tactile knowledge, mechanical arts; empirical sciences	Humanistic, social science, and artistic knowledge; behavioral knowledge	Scientific knowledge, theoretical knowledge, enduring practical knowledge; knowledge learned in school or advanced study
Nature of jobs	Deal with materials (carry, move, transform, rearrange, test)	Highly interpersonal in nature, social context is important, location bound, deal with customer flow, manage personal experience (state of mind)	Intellectual activity; contemplation, analysis, interpretation, comparison, discussion, decision-making; deal with information (collect, transform, etc.); creativity needed; creation, organization, and use of knowledge; create knowledge about process (important since products, markets, and technologies change)
Type of intellectual capital	Structural capital; human capital	Customer capital; structural capital	Human capital; structural capital

Management challenge

Structure	Factory as a laboratory to acquire empirical knowledge; lean operations (JIT, TQM, Kaizen, cycle time reduction, etc); managing supply chains	Scripting important, manage employee–customer interaction and experience	Managing knowledge creation, incentives, and markets
Selection	Organization design closely linked to physical flow of materials; physical dexterity, manual strength, tactile skills	Location and people bound; good ambiance, pleasant surroundings, friendly, personable, appearance, outgoing personality, people person	Campus-like offices, collegial not hierarchical; intellectual abilities, level of education
Training	Drill, process rules, safety	Scripting	How to think/solve problems, continuing education, tools/techniques
Incentives	Output-based incentives	Outcome-based incentives	Output-based incentives
Measures	Output performance measures, statistical process control	Customer satisfaction measures, time/input measures	Performance difficult to measure

The Role of Knowledge in Customer or Stakeholder Contact Intensive Enterprise

Some enterprises and business processes require that—or perform best when—a human member of the enterprise interacts directly with a customer, supplier, or other stakeholder. This may be a larger set of enterprises than some technological optimists realize. Although technology provides many opportunities for people to interact with machines—i.e., distance learning—it does not replace the entire vital social needs that bring people together in community. Moreover, some interactions require in-person exchanges. Getting a haircut, for example, requires the customer to meet in person, face-to-face, with a barber or hair stylist. A physician generally wants to see a patient in the flesh before making a diagnosis and prescribing. Many types of sales and supplier negotiations are assumed to be conducted best in person. Many activities in the "experience economy"—theme parks, tourism, entertainment, and the like—require in-person interactions.

The underlying knowledge required to succeed in stakeholder contact intensive enterprises is factual knowledge about human behavior and knowledge about how to form and maintain favorable interpersonal relationships. Rather than serving primarily as a substitute for materials, this knowledge is used to enhance and enrich a relationship—the enhancing effect.

A crucial component of this knowledge is how to communicate the enterprise's message effectively to stakeholders. This is important because a loyal customer base may be an organization's most valuable asset. Sewell Cadillac of Dallas, Texas, for example, makes an exceptional effort to retain its customers for repeat business. The company estimates that the value of retaining a customer is in excess of $300,000 in lifetime sales, so Sewell engages in many activities designed to make their customer's total experience totally satisfactory from initial inquiry through ownership to repeat purchase.

Walt Disney World is an example of an organization that employs customer contact knowledge effectively. By means of a program called "guestology"—the science of the behavior of visitors to its parks—Disney studies carefully how its customers behave and what they expect when they are on the premises. Refuse cans are set out strategically, given the company's knowledge of how far a guest is willing to carry trash before dropping it on the ground. Waiting lines are designed to minimize frustration and provide some degree of favorable experience. In addition, almost every employee who comes into direct contact with guests at the park is trained to deliver "scripts." Scripting is a powerful technique for taking what an enterprise knows about how to form and maintain favorable relationships with its stakeholders and translating this knowledge into actual, semiprogrammed behaviors on the part of its employees. A script is prepared for each anticipated interaction. Like a script for a role in a play, a Disney script specifies exactly the sequence and wording to be used during each interaction. Each employee memorizes the relevant scripts for the role he or she plays and is made alert to environmental cues that trigger their use. In some cases the collection of possible scripts is substantial. McDonald's, for example, has an employee handbook of about 700 pages and by means of on-site training and its "Hamburger U" educational facility ensures that the majority of its employees know how to interact with customers in the event of any contingency (see Table 5.1, "Customer Contact Intensive").

The Role of Knowledge in a Knowledge and Information Intensive Enterprise

Many different types of media converge into a single offering as a result of digitization. Information and communication technology now permits different media forms—text, images, audio, and video—to be integrated and used to deliver various types of content such as news, information, and entertainment. The result is knowledge and information intensive enterprises such as AOL-Time Warner, Viacom, and Capital Cities/ABC. Microsoft, Netscape, and other software companies are also examples, as is Monster.com, the leading online career site on the Web. These organizations rely almost exclusively on intellectual capital to deliver their products and services. A key feature of these enterprises is that knowledge is used to produce more knowledge and to forge technologies for processing knowledge. These activities draw on the reproductive multiplier effect of knowledge and create a kind of capital with which knowledge is used to self-generate more knowledge. Software tools such as CASE tools are an example. For these firms the ratio of their market value to the replacement value of their physical assets (Tobin's Q) is quite high (Microsoft is an example). That is, the majority of the firm's total value is in intellectual capital. An extreme form of an information intensive enterprise is the virtual organization. A dictionary defines "virtual" as "of or pertaining to a device or service that is perceived to be what it is not in actuality, usually as more 'real' or concrete than it actually is."[26] Virtual organizations appear to the outside world to be complete functioning entities. But in reality they are coordinating mechanisms that link other producing entities together. They pride themselves in operating with minimal physical assets. They outsource every business process possible and coordinate these activities by means of information and communications technologies. Today many business processes, even those in traditional companies, can be performed virtually. James Martin explains: "People can be linked together with computer networks, work-group facilities, design tools, and software in general so that they cooperate closely even though they are in different locations and different organizations."[27]

How Information and Communication Technology Is Used to Handle Transformative Knowledge

Knowledge value requires movement. Inert knowledge has no value. It is actual or potential action or decision-making that confers value on knowledge. ICT, in its broadest sense, is the means used to transform knowledge into value. The "ah-ha" moment when a mind grasps a new unit of knowledge may be exciting, perhaps even an epiphany. But that knowledge does not transform enterprises until it is acted upon. Moreover, the greater the scope in which knowledge is shared, the greater the possible change that can be produced.

Any ICT system includes certain fundamental functions: input (transducer), store (memory), processor (CPU), controller (operating system, master control program), output (transducer), communication links for moving inputs and outputs, switches, and, usually, a clock for timing events. During the Colonial times in the United States, for example, paper was the predominate medium for capturing knowledge and information; quill, pencil, and occasionally the printing press were the major transducers; piles of paper constituted the store (vertical filing was not available until 1893); there were few effective aids available for calculating or processing; paper, ledgers, and index cards were the chief vehicles for output; and communication speeds were generally geared to the speed at which a horse could travel. (In 1800 it took at least six weeks for a message to travel from New York to Chicago.) This modest technological infrastructure constrained the speed and effectiveness with which knowledge could be generated, codified, transferred, and acted upon. Consequently, by today's standards, enterprises were transformed rather slowly and only occasionally radically.

Today the pace has picked up substantially and the possible order of magnitude change greatly increased. In the digital age a vast array of sensors and input devices—scanners, point-of-sale, cameras, recorders, and, of course, keyboards—serve to capture data elements over a broad expanse of input sources. Events can be captured as they happen. Electronic memory's capacity is now measured in terabytes[28] and growing. Addition, subtraction, multiplying, dividing, comparing, and a full complement of mathematical and logical functions are performed in nanoseconds[29] and, following Moore's law, are expected to approximately double in operation speed every 18 months. More sophisticated software serves to enhance calculation capability and the capacity for storage and retrieval. The rise of data warehousing and data mining techniques makes it possible to access large amounts of data cost-effectively. Outputs are effectively routed to screens, automation devices, high-speed printers, and many other multimedia information activators. Increased bandwidth[30] and connectivity make knowledge available in wireless, fiber optic, and twisted pair forms as bits move at rates up to the speed of light. Much of this new digital infrastructure is tied to the Internet (or can be) or is sent over networks and thus is available to devices like laptop computers, personal digital assistants (PDAs), smart appliances (e.g., your TiVo), and mobile phones.

This remarkable ICT infrastructure can be deployed to make knowledge available to a large number of people. Commerce Net/Nielsen estimates that as many as 175 million users are connected to the Internet. As of July 2000, 100 million people (36% of U.S. residents) owned wireless telephones. According to the Cellular Telecommunications Industry Association, new subscribers are being added at a rate of 50,000 to 60,000 a day. Observing these trends, Golobs forecasts that there will be one billion mobile phone subscribers worldwide by 2005. This "will be more than all PCs and automobiles combined."[31] The capability to make knowledge available, however, is only part of the requirement for transforming enterprises; it represents a necessary condition but is far from sufficient. Beyond the ICT infrastructure are additional social and technological requirements.

Joseph L. Badaracco, Jr., has identified four additional requirements that must be satisfied in order for knowledge to move: "For knowledge to migrate quickly, four broad conditions must hold. First, the knowledge must be clearly articulated and reside in 'packages.' Second, a person or group must be capable of opening the package, of understanding and grasping the knowledge. Third, the person must have sufficient incentives to do so, and fourth, no barriers must stop them."[32] In addition to publications, such knowledge is contained in designs, machines, and individual minds.

Knowledge that meets these four requirements may either be propositional knowledge—knowledge of "what"—or instrumental knowledge— knowledge of "how." Both types of knowledge may be applied directly or stored in repositories. An ongoing trend toward achieving lower costs per bit of storage coupled with the development of data warehousing, data mining, and more effective storage and retrieval techniques means that highly accessible repositories are being constructed and made available to users.

All of this ICT-enabled knowledge processing, however, is still not enough to transform enterprises. Enterprises, be they small family households, nonprofit organizations, government agencies, or businesses large or small, conduct their activities by means of "business processes." Any enterprise, regardless of size, type, or desired objective, operates fundamentally by transforming a collection of inputs (raw data, raw materials, etc.) into required outputs—products and services. One or more processes are employed to carry out the transformation. The factors of production used in these processes are land, labor, material and financial capital, and *knowledge*. The goods and services that are produced— the engines that transform enterprises—result from the interplay in the application of these factors. ICT-enabled knowledge processing serves to augment the knowledge component's contribution to the business process.

For these reasons, understanding and subsequently "reengineering" any business process begins by examining three things: the necessary contributions required from materials and tangible assets, the necessary contributions required from in-person contacts with customers and other stakeholders, and the necessary contributions required from knowledge and information. Next, one asks the question: How can knowledge be leveraged by improving each component's contribution to the process? With respect to materials, knowledge may be employed to replace existing materials with more effective ones—the substitution effect; with respect to in-person contacts, knowledge may be used to provide contextual information that improves interpersonal relationships—the enrichment effect. Finally, knowledge may be used to generate more knowledge and to distribute it to focal points throughout the process, so as to improve the overall speed, efficiency, or effectiveness of the business process. Analysis of a business process is facilitated if it is categorized as to its dominant component—i.e., material intensive, stakeholder contact intensive, or knowledge intensive.

This entire enterprise-transforming mechanism is mobilized by means of two types of investments:

1. Investments in inquiry to produce and acquire new knowledge, and
2. Investments in ICT targeted to deliver actionable knowledge to points in a business process where their use can make a difference.

Existing and forecasted capabilities in information and communication technology will make these investments pay off by transforming enterprises in positive ways if systems are designed to make effective use of them.

Conclusion

It is argued that knowledge and its related entity, information, serve as an engine to transform society. Since inquiry is the process by which knowledge is generated, investments in inquiry are essential for initiating and directing transformation. Knowledge is of two general types—instrumental, which describes how to do something, and propositional, which describes what something is. Both types work together to initiate a transformation.

The agents who act on knowledge—human, artificially intelligent machines, or cyborgs—may be dispersed geographically and require knowledge at future times. Consequently, repositories with widespread accessibility are needed. The context in which these agents act is usefully described as a business process. There are three generic types of business processes—material intensive, stakeholder contact intensive, and knowledge intensive. Knowledge plays an important but different role in each. It is substituted for the use of materials in material intensive business processes; it is used to enhance and enrich interpersonal contact in stakeholder contact intensive processes; and it serves to multiply and reproduce knowledge itself in knowledge intensive settings. All of these effects—substitution, enrichment, and multiplier—collectively work together to transform enterprises. These effects are the means of transformation; the outcome is twofold: (1) New goods, services, and experiences, including entertainment and education, are produced and (2) new structures of organization and divisions of labor are realized. Society is ultimately transformed by incorporating these new products and ways of working and living.

Knowledge is the active agent in enterprise transformation. Taken by itself, however, it is inert, a mere string of impotent symbols. It is energized by technology. In the twenty-first century newly evolving information and communication technologies, especially interactive, networked technologies, are becoming the motive force that is enabling more widespread knowledge-based transformations. It is knowledge processed by and flowing through technology that is changing our world.

Notes

1. Manuel Castells (2001) *The Internet Galaxy* (Oxford: Oxford University Press), p. 1.

2. Daniel Bell (1976) *The Coming of Post-industrial Society: Adventure in Social Forecasting*, 2d ed. (New York: Basic Books), p. 175.

3. Manuel Castells (2000) *The Rise of the Networked Society*, 2d ed. (Oxford: Blackwell Publishers).

4. Gilbert Ryle (1949) *The Concept of Mind* (Chicago: The University of Chicago Press).

5. See Michael Polanyi (1958) *Personal Knowledge* (Chicago: University of Chicago Press), and (1966) *The Tacit Dimension* (London: Routledge & Kegan Paul).

6. Ikyiro Nonaka and Hirofaka Takeuchi (1995) *The Knowledge-Creating Company* (New York: Oxford University Press).

7. See, for example, Kim B. Clark and Steven C. Wheelwright (1993) *Managing New Products and Process Development: Texts and Cases* (New York: The Free Press).

8. Joel Mokyr (2002) *The Gifts of Athena: Historical Origins of the Knowledge Economy* (Princeton: Princeton University Press), p. 10.

9. Richard R. Nelson and Sidney Winter (1982) *An Evolutionary Theory of Economic Change* (Cambridge, Mass.: The Belknap Press).

10. See C. West Churchman (1971) *The Design of Inquiring Systems: Basic Concepts of Systems and Organizations* (New York: Basic Books).

11. Alter, Steven (1996) *Information Systems: A Management Perspective*, 2d ed. (Menlo Park, Calif.: The Benjamin/Cummings Publishing Co).

12. Uday Apte and Richard O. Mason (1995) "Global Disaggregation of Information Intensive Services," *Management Science* 41(7): 1250–1262.

13. See http://www.wired.com/wired/archive/8.03/markets/html.

14. Thomas A. Stewart (1997) *Intellectual Capital: The New Wealth of Organizations* (New York: Doubleday), p. 67.

15. David Ricardo (1817) *Principles of Political Economy and Taxation.*

16. Mokyr, *The Gifts of Athena*, p. 12.

17. See Chris Freeman and Francisco Louca (2001) *As Time Goes By: From the Industrial Revolutions to the Information Revolutions* (Oxford: Oxford University Press), passim, esp. pp. 139–151. For a discussion of "Taylorism + Fordimus," see Thomas P. Hughes (1989) *American Genesis: A Century of Invention and Technological Enthusiasm 1870–1970* (New York: Viking Penguin).

18. See, for example, Charles Van Doren (1991) *A History of Knowledge: The Pivotal Events and Achievement of World History* (New York: Ballantine Books).

19. Alfred D. Chandler, Jr., and James W. Cortada, eds. (2000) *A Nation Transformed by Information* (Oxford: Oxford University Press).

20. R. Mansell and U. Wehn (1998) *Knowledge Societies, Information Technology for Sustainable Development* (New York: UN/OUP).

21. Stewart, Thomas A. (1997) *Intellectual Capital: The New Wealth of Organizations* (New York: Doubleday).

22. Clark, Andy (2002) *Natural Born Cyborg: Minds, Technologies and the Future of Human Intelligence,* preface (Oxford: Oxford University Press).

23. See Richard O. Mason (2003) "Ethical Issues in Artificial Intelligence," in *Encyclopedia of Information Systems*, vol. 2 (Amsterdam: Academic Press), 239–259.

24. Richard O. Mason and Mitroff, Ian I. (1981) *Challenging Strategic Planning Assumptions: Theory, Cases and Techniques,* pp. 43 and 95–100 (New York: John Wiley & Sons).

25. See, for example, Stewart, *Intellectual Capital*, pp. 3–5.

26. *Microsoft Bookshelf Computer and Internet Dictionary* (1997) Microsoft Corporation. Portions from *Microsoft Press Computer Dictionary,* 3d ed. (Redmond, Wash.: Microsoft Press, 1997.)

27. James Martin (1996) *Cybercorp: The New Business Revolution* (New York: Amacom), p. 15.

28. A terabyte is 2^{40}, or about 1000 billion bytes or a thousand gigabytes.

29. A nanosecond is one-billionth (10^{-9}) of a second.

30. Bandwidth refers to the range of frequencies that an electronic signal occupies on a given transmission medium and is measured in bits per second.

31. Thomas F. Golobs (2000) "Travel Behavior.com: Activity Approaches to Modeling the Effects of Information Technology on Personal Travel Behavior," IA TBR 2000 Conference, Gold Coast Queensland, Australia, July 2–7.

32. Joseph L. Badaracco, Jr. (1991), *The Knowledge Link* (Boston: Harvard Business School Press), p. 34.

6

Transformation through Cyberinfrastructure-Based Knowledge Environments

Daniel E. Atkins

Background

This chapter is part of a series describing the impact of information technology (IT) on transforming various economic and social dimensions of our world and in particular focuses on the impact of IT on transforming knowledge-intensive activities, especially research and learning.[1]

The term *infrastructure* emerged in the 1920s to refer collectively to the roads, bridges, rail lines, and similar public works that are required for an industrial economy to function. The term *cyberinfrastructure* has emerged recently to refer to a system of information and communication technologies together with trained human resources and supporting service organizations that are increasingly required for the creation, dissemination, and preservation of data, information, and knowledge in the "digital age." Traditional infrastructure is required for an industrial economy; cyberinfrastructure is required for a knowledge economy.

The *infrastructure* part of the word emphasizes the need to situate the acquisition and operation of information and communication technology (ICT) in a systemic context: not just buying boxes and wires, but rather as part of a system of technology, support organizations, people-trained creation, operation, and use and policies to support ongoing system enhancements in a sustainable way. Designing cyberinfrastructure also includes the identification of levels of commonality, sharing best practices, interoperability, and economies of scale. It needs to be a platform for building knowledge environments for many specialized communities/disciplines but in a way that enables communication, collaboration, and sharing of data, instruments, computational models, and other facilities both within and between communities.

Cyberinfrastructure needs to be integrating—not balkanizing. The specification of cyberinfrastructure and the alignment of many stakeholders to create and apply it is itself a complex intellectual task.

The *cyber* part of the word is useful to distinguish the properties of traditional physical infrastructure from those of infrastructure based on ICT. Sustaining and evolving it, for example, is a challenge due to the continuing exponential rate of change of ICT basic capacity and the attendant rapid depreciation curves. On the other hand cyberinfrastructure offers potential for shared use and reuse of information, facilities, and computational tools.

Science and engineering applications launched the information age and have continued to help define the leading edge of computational performance. Science and engineering research communities are now in the position to harvest the broader benefits of information technology evolution. Going beyond computation, information retrieval, or communication, these communities are starting to create integrated and comprehensive knowledge environments that can serve individuals, teams, and organizations in ways that revolutionize what they can do, how they do it, and who participates. This trend also has profound broader implications for education, commerce, and social good.

The WWW has laid a technical basis, stretched visions, and created incentive for federating diverse distributed resources to produce new organizational forms—*knowledge ecologies*—to serve a wide variety of human endeavors. The Internet, the World Wide Web, extended and augmented in ways not fully defined, can be part of a cyberinfrastructure layer on which to build and use comprehensive virtual federations of diverse resources specialized for many types of knowledge-intensive activities, including but not limited to scientific and engineering research and education.

The *resources* potentially available for federation include people, data, information, computational tools and services, and specialized instruments and facilities necessary for the functioning of a specific research team (or more generally, a specific community-of-practice.) The term *comprehensive* is included to imply that the extent and nature of the resources available through cyberinfrastructure could approach functional completeness for a specific community of practice. Astronomers, for example, can find all of the colleagues, data, literature, observational stations, and tools they need to work at the leading edge of their field. For many scientific and engineering contexts we need also to add the term *advanced*, implying both high-performance technology and leading-edge research. Science and engineering research are moving into an IT era of *advanced comprehensive virtual federation*.

Virtual Research Communities for Science and Engineering

The general idea of virtual communities or organizations mediated through information technology is not new and has been, for example, the object of design and study by the field of computer-supported cooperative work (CSCW) [1] since the 1980s. More specific to virtual research communities for science and engineering, there are three primary lines of endeavor under different names but with similar goals: *collaboratory* or *co-laboratory*, *grid*, and *e-science*. These names are not necessarily mutually exclusive and the terms are sometimes combined, as in "a grid-based collaboratory" or "a collaboratory for e-science."

The Collaboratory: A Laboratory without Walls

The concept of a *co-laboratory*, or *collaboratory*²—a laboratory without walls built upon distributed information technology—was defined at an invitational National Science Foundation (NSF) workshop at Rockefeller University [2] in 1989 and later elaborated and sanctioned in a National Research Council report published in 1993 [3]. The report includes the following assertion:

The fusion of computers and electronic communications has the potential to dramatically enhance the output and productivity of U.S. researchers. A major step toward realizing that potential can come from combining the interests of the scientific community at large with those of the computer science and engineering community to create integrated, tool-oriented computing and communication systems to support scientific collaboration. Such systems can be called collaboratories.

This report called for a major initiative by NSF to explore the design and applications of the collaboratory. Although investment at the recommended level did not occur, several collaboratory projects were funded by NSF, for example, the SPARC [4] and UARC [5] projects, and more were funded recently by the Department of Energy [6] and the National Institutes of Health [7]. The NIH National Center for Research

Resources has just issued a report on data and collaboratories in the biomedical community [8] that begins with the following statement:

Today is a time of great opportunities as biomedical research continues to harness trends in information technology, the grid, and emerging cyberinfrastructure. These trends empower teamwork and collaborative approaches that support large-scale and information-rich biomedical investigations, resulting in new insights into basic biological processes associated with health. In particular, trends in the biomedical research community include collaborative, data-driven scientific experiments, from capture through refinement to dissemination, and data discovery via integration from across spatial and temporal scales.

Overviews of collaboratory research, especially at the intersection of the technical and social dimensions, can be found in Refs. [9], [10], and [11].

Recently the NSF has funded a Science of Collaboratories (SoC) Project under the Information Technology Research (ITR) Program. The SoC Project is examining collaboratory projects over the past decade that have been funded by NSF, DOE, NIH, and other agencies—some successful and some less so. The project aims to *define, abstract, and codify the broad underlying technical and social elements* that lead to successful collaboratories. The SoC website [12] includes an inventory of about 125 collaboratory projects. This site as well as the website for the Collaboratory for Research on Electronic Work (CREW) [13] includes numerous papers and other resources about the collaboratory R&D community.

The Grid

Foster and Kesselman [14] have articulated the concept of the Grid.[3] The term *grid* was originally adopted as an analogy to the dynamic linking of electric power generators over distribution grids to balance supply and demand over a mixture of geographic regions and power companies. A computer grid is intended to provide a user extraordinary computational power by aggregating high-performance computational resources over wide areas and diverse administrative entities. The term Grid has now come to include not only linking computational power, but also data, information, instruments, and human expertise. The NSF is now funding a large Distributed Terascale (Teragrid) Project [15] to explore creating a grid of supercomputers and virtual organizations between nine high-performance computational centers and related data stores and instruments. Many of the technical activities focus on network performance and creating standards and a middleware toolkit, Globus [16], to support efficient and secure interoperability between diverse machines and their hosting domains. The Globus Project overview on the Web includes the following analogy:

The development of the World Wide Web has revolutionized the way we think about information. We take for granted our ability to access information from all over the world via the Web. The goal of the Globus project is to bring about a similar revolution with respect to computation.

A primary goal of middleware, a component of cyberinfrastructure, is to enable the seamless federation of resources across networks. It is the focus of a recent NSF Middleware Initiative (NMI) [17]. The NMI includes the following definition of middleware:

Middleware is software that connects two or more otherwise separate applications across the Internet or local area networks. More specifically, the term refers to an evolving layer of services that resides between the network and more traditional applications for managing security, access, and information exchange to (1) let scientists, engineers, and educators transparently use and share distributed resources, such as computers, data, networks, and instruments, (2) develop effective collaboration and communications tools such as Grid technologies, desktop video, and other advanced services to expedite research and education, and (3) develop a working architecture and approach that can be extended to the larger set of Internet and network users.

The NMI, among other things, is promoting synergy and commonality between major middleware grid-based activities such as Globus and the Internet2 Shibboleth Project [18] that is developing architectures, policy structures, practical technologies, and an open source implementation to support interinstitutional sharing of web resources subject to access controls in the higher education world. The emergent Web Services [19] and more ambitious Semantic Web [20] activities are also likely to be important parts of the middleware architecture of cyberinfrastructure.

Although much of the focus of the Grid is transparent interoperability between high-performance computers, the use of the Grid encompasses a broader notion of resource sharing than only computational engines and in this context a *grid community* is equivalent to a collaboratory. The term *datagrid* has also come on the scene [21]. Grids in this broader sense are about "resource sharing and coordinated problem solving in dynamic, multi-institutional virtual organizations." The international community of academic and commercial participants coordinates Grid development and application through the Global Grid Forum [22] and the related Globus World [23].

E-Science

In the U.K. research community [24] and the European Union Sixth Framework Program [25] the term *e-science* is used to describe science done through distributed global collaborations between people linked by the Internet with each other, with very large data collections, terascale computing resources, high-performance visualization, and instruments and facilities controlled and shared over the network. In describing e-science reference is made to both grid architecture and collaboratories. The United Kingdom is also calling for a 1000× increase in the computational processing power and data transport bandwidth.

In a speech to the Royal Society in May 2002, Prime Minister Tony Blair included the following references to e-science:

What is particularly impressive is the way that scientists are now undaunted by important complex phenomena. Pulling together the massive power available from modern computers, the engineering capability to design and build enormously complex automated instruments to collect new data, with the weight of scientific understanding developed over the centuries, the frontiers of science have moved into a detailed understanding of complex phenomena ranging from the genome to our global climate. Predictive climate modeling covers the period to the end of this century and beyond, with our own Hadley Centre playing the leading role internationally.

The emerging field of e-science should transform this kind of work. It's significant that the U.K. is the first country to develop a national e-science Grid, which intends to make access to computing power, scientific data repositories, and experimental facilities as easy as the Web makes access to information.

General Properties of Virtual Research Federations

The general goal of all of these efforts is to use information technology to relax barriers of time and distance in bringing together the expertise, information, tools, instruments, and facilities necessary for the discovery, dissemination, and application of knowledge. Time can be synchronous, asynchronous, or "relevant" (just in time to not slow the workflow). Distance can mean geographic, organizational, or disciplinary. Interest in virtual federations is driven in part by a growing demand in research for more global, collaborative, and multidisciplinary approaches to discovery and problem solving. Mediated by information technology, the teamwork can in theory proceed not only in physical proximity but also easily through three other variations of same and different time and place. Observational instruments, unique because of cost and/or need to be in special locations, can be shared. Arrays of smart sensors can provide unprecedented resolution or coverage in monitoring natural phenomena. Research community frameworks for multilevel, multisystem simulations can be collaboratively created and executed across grids of supercomputers. Archival data from many specialized fields can be made to interoperate in well-curated data repositories and used to drive comprehensive physical models in supercomputers or to extract new knowledge thorough data mining. Digital libraries can offer anytime and anyplace access to the complete literature of a field and software agents to help maintain current awareness.

Some fields begin a virtual community to collaborate around remote instruments and data gathering campaigns, some to collaborate in creating and using community data, and some to collaborate on creating and using complex computational models. Experience has shown, however, that if these various initial applications are successful, communities evolve virtual environments to incorporate more capabilities. They move in the direction of functional completeness—the (virtual) place to be for a specific project or field.

As mentioned earlier, the emerging use of information technology in scientific and engineering research is expanding beyond computation, email, and information retrieval. It is about the potential for new comprehensive infrastructure, environments, and organizations to empower the overall enterprise of discovery, dissemination, and use (including education, public awareness, and informed policy making). The design principles, system architecture, and analysis of impact of such IT-based environments must therefore include a merged technical and social/cultural/behavioral perspective. The past is littered in both industry and science with virtual organization projects that have failed due to lack of knowledge or consideration of the social dimensions of team activities.

Examples of What's Happening Now

Collaboratory, grid community, and e-science environments are all names for virtual federations of resources based upon information technology to create knowledge environments (or organizations) for science and engineering research. Similar work is also under way labeled with none of these terms. Recently the leadership of NSF observed a growing number of grass roots projects, many funded under the Information Technology Research (ITR) Program recommended by the Presidential Information Technology Advisory Committee, that included major activities and expenditures to create and use collaboratories, grids, or e-science environments in specific, often multidisciplinary, projects. Many also need access to the next frontier of high-performance computing and are concerned that the United States may not maintain leadership in this area. Examples of these projects include the following:

1. The National Virtual Observatory (NVO) [26], a collaboration to create standards for astronomical data collections that will be used by the wide astronomical community. It will make data easier to use, easier to find, and easier to join with other data to create a *digital sky*. A second thrust is exploring the use of high-performance computing resources for discovery in astronomy.

2. Network for Earthquake Engineering Simulation (NEES) [27], to provide an unprecedented infrastructure for research and education, consisting of networked and geographically distributed resources for experimentation, computation, model-based simulation, data management, and communication. Rather than placing all of these resources at a single location, NSF has leveraged its investment and facilitated research and education integration by distributing the shared-use equipment among nearly 20 universities throughout the United States. To ensure that the nation's researchers can effectively use this equipment, equipment sites will be operated as shared-use facilities, and NEES will be implemented as a network-enabled collaboratory. As such, members of the earthquake engineering community will be able to interact with one another, access unique, next-generation instruments and equipment, share data and computational resources, and retrieve information from digital libraries without regard to geographical location.

3. The Grid Physics Network (GriPhyN) [28] project brings together a team of experimental physicists and information technology researchers to develop Grid technologies for scientific and engineering projects that must collect and analyze distributed, petabyte-scale datasets. GriPhyN research will enable the development of petascale virtual data grids (PVDGs) through its Virtual Data Toolkit.

4. National Ecological Observatory Network (NEON) [29] will be a network of networks, a system of environmental research facilities and

state-of-the-art instrumentation for studying the environment. Each node in NEON will be a regional observatory, comprised of a core site and associated sites that are linked via cyberinfrastructure. These observatories will be geographically distributed based on the U.S. Forest Service's defined eco-regions of the United States. NEON will enable integrative research on the nature and pace of biological change at local, regional, and continental scales. Its advanced technologies and continental-scale connectivity will be used to measure all factors that affect the structure and function of ecosystems; for example, the power of genomics will be applied to predicting how the spread of invasive species will affect native biodiversity.

5. The ATLAS Experiment for the Large Hadron Collider (LHC) [30] is creating a collaboratory to support a worldwide community of physicists engaged in building and using the LHC centered at the CERN Laboratory in Switzerland. Its goal is to explore the fundamental nature of matter and the basic forces that shape the universe. ATLAS is the largest collaborative effort ever attempted in the physical sciences. There are 2,000 physicists participating from more than 150 universities and laboratories in 34 countries. The ATLAS Experiment collaboratory is also extraordinary in terms of data-handling requirements. Estimates are that by about 2012 it will need an exabyte (10^{18} bytes) archive for data from four LHC experiments. These data need to be readily available, not just in a central location but to scientists all over the world. In addition, computation rates of 0.30 petaFLOPS (0.3×10^{15} floating point arithmetic operations per second) will be required to process the data.

6. Environmental Research and Education (ERE) is a multidisciplinary field of increasing importance and is a high priority in the NSF research planning process. An NSF Environmental Research and Education Advisory Committee has just released a 10-year outlook for the NSF's ERE programs in a report entitled "Complex Environmental Systems: Synthesis for Earth, Life, and Society in the 21st Century" [31]. The report contains the following comments on building capacity:

Long-term dynamic partnerships that cross national and regional jurisdictions and international boundaries are needed to address multiscale challenges. Developing the requisite cyberinfrastructure—advanced data assimilation and curation, networking, modeling, and simulation tools for large-scale, systems-level, integrated applications—is key to making progress in the decade ahead.

Laying the Foundations for a Research Revolution

The kind of research communities highlighted above strongly support the view that new high-capacity ICT-based virtual environments are absolutely critical to their future research aspirations.

Blue Ribbon Advisory Panel on Cyberinfrastructure

In 2002 the U.S. National Science Foundation formed a Blue Ribbon Advisory Panel on Cyberinfrastructure to provide advice on the nature of this movement and the opportunities and challenges it presented to NSF, a vision of how NSF might respond, and advice on how the current major investments by NSF in cyberinfrastructure, largely in the form of supercomputing alliances, should fit into this vision. The charge included the notion that creating and using advanced cyberinfrastructure requires synergy between the computer science and engineering research communities and the greater domain science communities who want to benefit from it. Cyberinfrastructure and its use is both an object of research as well as an enabler of research. The panel conducted surveys and hearings, reviewed prior relevant recommendations, requested comments on a draft report, and has recently submitted a final report to the NSF that is now available on the Web [32].

Extrapolating in large part from prior NSF investments in cyberinfrastructure including high-performance computing, networking, middleware, and digital libraries, trends in the IT industry, and the vision and innovation coming from many research communities, the panel asserts that the capacity of information technology has crossed thresholds that now make possible a comprehensive cyberinfrastructure on which to build new types of scientific and engineering knowledge environments and organizations and to pursue research in new ways and with increased efficacy. The panel shares the belief of many who testified that advanced cyberinfrastructure could be the basis for revolutionizing the conduct of scientific and engineering research and affiliated education. It could also have broad impact in many other domains of knowledge-intensive activity.

The panel also found that such environments and organizations, enabled by cyberinfrastructure, are increasingly required to address national and global priorities such as global climate change, protecting our natural environment, applying genomics-proteomics to human health, maintaining national security, mastering the world of nanotechnology, and predicting and protecting against natural and human disasters, as well as to address some of our most fundamental intellectual questions such as the formation of the universe and the fundamental character of matter.

Although this panel was commissioned by the NSF and concentrated on the NSF research community, it also found that similar visions and needs are emerging in research communities supported by other federal agencies, most apparently by the Department of Energy (for example, Ref. [6]) and the National Institutes of Health (for example, Refs. [8] and [33]). As mentioned earlier, heavily funded e-science initiatives based on cyberinfrastructure are under way in Europe, and the Japanese recently set new records in computational modeling, one important function of cyberinfrastructure, in the Japanese Earth Simulation Center [34]. They have developed and are running a comprehensive earth systems simulator program at unprecedented levels of resolution and scale and at record-setting computation rates.

A Cyberinfrastructure for Wide Use

The NSF cyberinfrastructure panel identified the elements that should be included and how they fit within a functional stack (see the schematic in figure 6.1). Cyberinfrastructure begins with networking, operating systems, and middleware providing the generic capabilities for

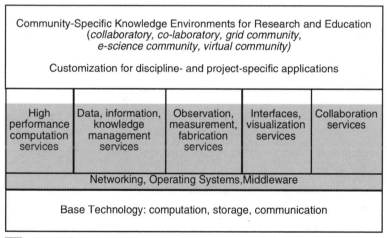

Figure 6.1
Integrated cyberinfrastructure services to enable new knowledge environments for research and education. Source: Revolutionizing Science and Engineering through Cyberinfrastructure, 2003 [32].

management, transport, and federation of systems and services (tools) described in the five columns. A community-specific, customized knowledge environment can ideally be created efficiently and effectively using facilities, tools, and toolkits provided at the cyberinfrastructure layer to federate the requisite resources.

This model assumes significant effort to capture and benefit from commonalities across science and engineering disciplines and appropriate levels of coordination and sharing of facilities and expertise to minimize duplication of effort, inefficiency, and excess cost. To achieve advanced capability it also assumes real collaboration between domain scientists and engineers and computer scientists and engineers and constructive participation by social scientists to help understand the social and cultural issues.

The panel goes on to recommend that the NSF seek significant new funding and assume the leadership of an Advanced Cyberinfrastructure Program (ACP) with close coordination with other U.S. and international R&D agencies. A central goal is to define and build cyberinfrastructure that facilitates the development of new applications, allows applications to interoperate across institutions and disciplines, ensures that data and software acquired at great expense are preserved and easily available, and empowers enhanced collaboration over distance, time, and disciplines. The individual disciplines must take the lead in defining specialized software and hardware environments for their fields based on common cyberinfrastructure, but in a way that encourages them to give back results for the general good of the research enterprise. Achieving this vision will challenge fundamental understanding of computer and information science and engineering as well as parts of social science, and it will motivate and drive basic research in these areas.

Figure 6.2 conveys complementary dimensions of an advanced cyberinfrastructure program. One dimension is technology capacity—a measure of the processing power, data volumes, and data rates that can be handled in the IT-based research environment. The other dimension is *functional comprehensiveness* of the assets needed by a research team or organization to do its work. This is a qualitative measure indicating, for example, what percentage of colleagues, relevant scientific literature, archival data, instruments, and other critical facilities are easily available

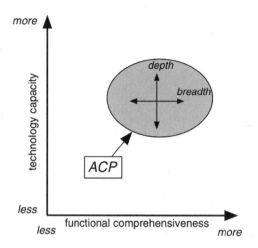

Figure 6.2
Increasing technology capacity and functional comprehensiveness of cyberinfrastructure, enabling depth and breadth approaches to discovery. Source: Revolutionizing Science and Engineering through Cyberinfrastructure, 2003 [32].

through the cyberinfrastructure-based environment, i.e., through the virtual organization—the collaboratory—the grid community.

Supercomputing initiatives have moved us higher along the technology capacity dimension. R&D sponsors have also invested in creating understanding and prototypes of important functional components of knowledge environments: digital libraries, visualization tools, and collaboration technology as well as the middleware glue to link them together. The opportunity now before the research community is to establish new working environments that are advanced, innovative, and innovating in terms of both capacity and comprehensiveness. The shaded region represents such a region and the general goal of an ACP is to move both *within* and *among* broad fields of science into these regions, and to move the regions up and to the right. As the combined region capacity and functional comprehensiveness increases and is adopted more broadly, the payoff will likely derive from enhancing both "depth" and "breadth" approaches to discovery.

In terms of depth approach, for example, atmospheric scientists could use higher-performance computation (together perhaps with denser and smarter distributed networks of sensors and with higher quality archival data) to improve the resolution and accuracy of a weather prediction model. Astronomers could use a more capable telescope to look more deeply into their favorite region of the universe.

In a breadth approach, a multidisciplinary team of earth scientists could use the availability of more computational power, more complete multidimensional data, enhanced observation capability, and more effective remote collaboration services to bring together an entire earth system simulation framework capable of supporting usefully predictive environmental simulations. Astronomers, given access to a federated "digital sky," could explore the breadth of the known universe over the entire available electromagnetic spectrum to seek, for example, rare or new objects or phenomena. We can only begin to glimpse the impact of blended depth and breadth approaches, especially as they weave together complementary expertise from multiple disciplines.

Although the NSF report stresses the vast opportunity for creating new research environments based upon cyberinfrastructure, it also notes significant risks and costs if the research community does not act quickly and at a sufficient level of investment. The dangers of delay include adoption of incompatible data formats in different fields, permanent loss of observational data due to lack of well-curated, long-term archives, increased technological balkanizations rather than interoperability among disciplines, and wasteful, redundant system-building activities among science fields or between science fields and industry.

Broader Impact and the Future

This chapter has focused most directly on IT and the conduct of scientific and engineering research. We have noted the growing needs of many research fields to work in more interdisciplinary global teams using larger, richer, and more diverse sources of expertise, information, observation, and computational tools. These needs coincide with a state of information technology that has now crossed thresholds of capacity and function that help meet these needs through cyberinfrastructure that enables virtual federation of resources to create new forms of knowledge work environments. Success at this in the science and engineering research enterprise would have enormous impact on other knowledge-intensive activities including commerce and education. The concept of advanced cyberinfrastructure and its use as a platform for virtual organizations of many types, durations, and missions obviously has application beyond scientific and engineering research. The author is aware of initiatives under way to explore research, development, and application of cyberinfrastructure in all of the areas suggested in figure 6.3. The figure is intended to imply not only that all of these areas could be innovated or even revolutionized through cyberinfrastructure but also the possibility and hope for coordinated investment and leveraging commonality in research, development, deployment, and operation of cyberinfrastructure for these complementary activities.

Impact on Education
Studies of collaboratory projects to date have shown that they have been used by researchers to help train their colleagues at other locations to use instruments, supercomputers, or other techniques for discovery; they have enabled graduate students to participate earlier in their careers in authentic experiments involving scarce or remote instruments; graduate students have benefited from mentoring through the collaboratory by leaders of their field at other institutions; and faculty at undergraduate institutions have used collaboratories to participate in first-tier research and to embellish their undergraduate teaching.

The NSF-sponsored Space Physics and Aeronomy Research Collaboratory (SPARC) [4] centered at the University of Michigan combined forces with the NASA-sponsored Windows to the Universe Project [35] to explore the concept of a dual-use collaboratory serving both a frontier research community as well as middle school earth science teachers and students. Windows to the Universe itself is a type of collaboratory

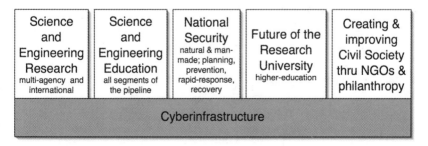

Figure 6.3
Opportunities for multiuse of cyberinfrastructure.

that provides instructional material and facilities for science teachers to build a virtual community to help each other with using the material to enrich their teaching. These teachers and students also could interact with scientists, data, and instruments in SPARC through a special portal designed for their level of understanding and participation. This dual-use experiment was not the subject of a careful longitudinal study, but anecdotal evidence suggests that collaboratories could be designed to enrich science education by making it more authentic, more fun, and more relevant. Another interesting example of a dual-use collaboratory is the University of Illinois's Bugscope Project [36].

A great hope for cyberinfrastructure-based knowledge environments is that they will democratize participation for those who have been disadvantaged because of history, location, or physical reasons. It could enable more routine and persistent diversity in the conduct of science in ways that facilitate mutual self-interest and reduce destructive competition. The original NSF Collaboratory report [2] and the new NSF report [32] treat this topic extensively.

The U.S. National Academies of Sciences in recent years has initiated a series of workshops and studies around the topic of information technology and the future of higher education. Cutting across all of these studies is the notion that information technology offers a vast array of new place- and time-independent opportunities for augmenting how the fundamental mission of teaching/learning, scholarship/research, and service/engagement is carried out. Restricted notions of "distance learning" in higher education are now broadening to that of using IT to facilitate the work of communities in knowledge creation, dissemination, and use. The cyberinfrastructure, opportunities, and challenges relevant to research is essentially the same for higher education at large—the collaboratory may be the tooling for the university of the future. More elaboration on this topic can be found in Refs. [37–40].

Impact on National Security
In February 2003 the National Science Foundation sponsored an invitational workshop on the topic of cyberinfrastructure research for homeland security [41]. The definition of *homeland security* included preparedness, response, recovery, and mitigation in the context of threats and emergencies created by man or nature. Slides from many of the

presentations are on the website and a final report from the workshop is in preparation. There was general consensus that research and development in advanced cyberinfrastructure is critical to homeland security. Special requirements in this application include mobility and rapid federation of customized, on-demand, ad hoc virtual organizations (including people, equipment, and information). Analogous to the earlier discussion of dual-use collaboratories, there is the potential for multiuse collaboratories that normally support ongoing research and education but can be mobilized and linked with other domain-specific collaboratories in response to emergencies.

An example given in Ref. [32] is the vision of a collaboratory for forestry research with computational models explicitly including fuels and chemical reactions, full two-way coupling with the atmosphere, representation of land-surface characteristics, and fuel composition and consumption. Digital forest fires could be used to test chemical agents for fighting fires at large scale in a virtual world. This research collaboratory could also be the basis for a rapid-response collaboratory that would enable a real forest fire to be modeled in real time based on data from sensors in the field. Potentially, the model running in a predictive mode could anticipate a fire blowout event in time to move a fire fighting crew to safety.

Impact on Civil Society

The nonprofit, nongovernmental, voluntary, and community-based sectors of society are generally not using the term cyberinfrastructure, but they are increasingly exploring the role of information technology in building and supporting virtual communities. One area of growing interest is the role of cyberinfrastructure in building and sustaining networks of engaged institutions, including higher education.

The engaged institution [42] is committed to direct interaction with external constituencies and communities through the mutually beneficial exchange, exploration, and application of knowledge, expertise, and information. These interactions enrich and expand the learning and discovery functions of the academic institution while also enhancing community capacity. The work of the engaged campus is responsive to community-identified needs, opportunities, and goals in ways that are appropriate to the campus's mission and academic strengths. The interaction also builds greater public understanding of the role of the campus as a knowledge asset and resource. Mutual benefit is a critical attribute of engagement, as distinguished from service outreach. At least one major philanthropic foundation, the W.K. Kellogg Foundation, has developed major programs based on the concept of engaged institutions [43], although not necessarily mediated through information technology. Cogburn's work [44] on globally distributed collaborative learning, particularly the global seminar between Michigan, American University, and several universities in South Africa, illustrates one form of engagement supported by cyberinfrastructure. This seminar is not one-way distance education but rather an example of collaboration between teachers and learners on the topic of information technology and globalization. It enables participation by people from multiple disciplines, countries, and political perspectives and offers authentic opportunity to all participants that would not be possible outside a virtual community.

The impact of cyberinfrastructure-based knowledge environments on humanities research and teaching is also starting to be explored by the American Council of Learned Societies [45].

Developing nations are also beginning to explore a framework of cyberinfrastructure in the context of strategic directions for research and development. An example is South Africa's National Research and Development Strategy [46]. The report does not specifically mention cyberinfrastructure-enabled knowledge environments, but it calls for extensive national empowerment and international collaboration in research and higher education, which resonates strongly with the vision of the NSF report [32].

What Needs to Be Done

This chapter has surveyed opportunities for a next wave of coordinated investment in research, deployment, and application of cyberinfrastructure to revolutionize the how, what, and who of collaborative knowledge-based activities. Realizing this potential requires many stakeholders to come together around a complex agenda of research on both technical and social dimensions, defining, deploying, and operating cyberinfrastructure, and empowering communities of practice to use it to do new things in new ways. A discussion of what this requires is beyond the scope of this chapter, but the immediate, single most important next step is for the various national research sponsoring organizations to both strengthen and coordinate their various cyberinfrastructure initiatives along the lines suggested in Ref. [32]. Such a coordinated investment will have direct impact in the science–engineering community and likely will stimulate a larger set of complementary investments and activity in broader communities and fields.

Notes

1. Prepared for the First International Conference on the Economic and Social Implications of Information Technology, January 27–28, 2003. See http://cip.umd.edu/transform.htm.

2. The term *collaboratory*, playing on *collaborate*, was coined by William Wulf while NSF Assistant Director for CISE at the time of the 1989 collaboratory workshop. The word stuck and is now more often used than *co-laboratory*. It is sometimes confused with the more limited notion of *collaboration technology*—commercial applications such as Net Meeting, Centra, WebEx, or the Access Grid offering chat, video conferencing, shared viewing/pointing, and/or group editing. A collaboratory includes collaboration technology but is more than group communication tools—it is a virtual organization, (including a culture and norms of participation) of people, information, computational tools, and other facilities to support team-based knowledge discovery and dissemination—including but not limited to collaboration technology.

3. The terms *Grid* and *Access Grid* are sometimes confused. The Access Grid (AG) is a suite of applications to support human interaction over a grid network. The Access Grid supports large-scale distributed meetings, collaborative work sessions, seminars, lectures, tutorials, and training.

References

[1] Grudin, J., "CSCW: History and Focus." *IEEE Computer*, 1994. 27(5): 19–27.

[2] Wulf, W. A., "The National Collaboratory—A White Paper," Appendix A in *Towards a National Collaboratory*, the unpublished report of an invitational workshop held at Rockefeller University, March 17–18, 1989 (Joshua Lederberg and Keith Uncapher, co-chairs). 1989.

[3] NRC, *National Collaboratories: Applying Information Technology for Scientific Research*. 1993, Washington, D.C.: National Academy Press. pp. xi, 105.

[4] Space Physics and Aeronomy Research Collaboratory website, http://intel.si.umich.edu/sparc/.

[5] Olson, G. M., Atkins, D. E., Clauer, R., Finholt, T. A., Jahanian, F., Killeen, T. L., Prakash, A., and Weymouth, T., "The Upper Atmospheric Research Collaboratory (UARC)." *Interactions*, 1998. 5(3): 48–55.

[6] DOE, "U.S. Department of Energy National Collaboratories," http://doecollaboratory.pnl.gov/history.html. 2002.

[7] NIH Collaboratory Project website, http://collaboratory.psc.edu/.

[8] Arzberger, P., and Finholt, T. A., "Data and Collaboratories in the Biomedical Community: Report of a Panel of Experts," meeting held September 16–18, 2002, in Ballston, Va. Available at http://nbcr.sdsc.edu/ CollaboratoryPresentations.htm. 2002.

[9] Finholt, T. A., and Olson, G. M., "From Laboratories to Collaboratories: A New Organizational Form for Scientific Collaboration." *Psychological Science*, 1997. 8(1): 28–35.

[10] Finholt, T. A., "Collaboratories," in *Annual Review of Information Science and Technology*, ed. B. Cronin (available in pdf at http://intel.si.umich.edu/ crew/technical_reports.htm). 2003.

[11] Olson, G. M., Finholt, T. A., and Teasley, S. D., "Behavioral aspects of collaboratories," in *Electronic Collaboration in Science*, ed. S. H. Koslow and M. F. Huerta. 2000, Mahwah, N.J.: Lawrence Erlbaum Associates. pp. 1–14.

[12] Science of Collaboratories Project website, http://www.scienceofcollaboratories.org/.

[13] The Collaboratory for Research on Electronic Work (CREW) website, http://www.crew.umich.edu.

[14] Foster, I., and Kesselman, C. *The Grid: Blueprint for a New Computing Infrastructure.* 1999, San Francisco: Morgan Kaufmann Publishers. pp. xxiv, 677.

[15] Teragrid Project website, http://www.teragrid.org/.

[16] The Globus Project website, http://www.globus.org/about/.

[17] NSF Middleware Initiative website, http://www.nsf-middleware.org/.

[18] Shibboleth Project website, http://shibboleth.internet2.edu/.

[19] W3C, Web Services Activity website, http://www.w3.org/2002/ws/.

[20] W3C, Semantic Web Activity website, http://www.w3.org/2001/sw/.

[21] The DataGrid Project website, http://eu-datagrid.web.cern.ch/eu-datagrid/.

[22] Global Grid Forum website, http://www.gridforum.org/.

[23] Globus World website, http://www.globusworld.org/.

[24] U.K. Research Council e-Science Core Program website, http://www.research-councils.ac.uk/escience/.

[25] EU Sixth Framework Program (2002–2006) website, http://europa.eu.int/ comm/research/fp6/index_en.html.

[26] National Virtual Observatory website, http://www.us-vo.org/.

[27] The Network for Earthquake Engineering Simulation website, http:// www.nees.org/.

[28] Grid Physics Network website, http://www.griphyn.org/index.php.

[29] The National Ecological Observatory Network (NEON) website, http:// www.nsf.gov/bio/neon/start.htm.

[30] The ATLAS Experiment for the Large Hadron Collider website, http:// atlasexperiment.org/.

[31] NSF, "Complex Environmental Systems: Synthesis for Earth, Life, and Society in the 21st Century," available at http://www.nsf.gov/ geo/ere/ereweb/advisory.cfm. 2003.

[32] Atkins, D. E., Droegemeier, K. K., Feldman, S. I., Garcia-Molina, H., Klein, M. L., Messerschmitt, D. G., Messina, P., Ostriker, J. P., and Wright, M. H., "Final Report of the NSF Blue Ribbon Advisory Panel on Cyberinfrastructure: Revolutionizing Science and Engineering Through Cyberinfrastructure," available at http://www.cise.nsf.gov/evnt/reports/toc.htm. February 2003.

[33] NIH, The Biomedical Informatics Research Network (BIRN) website, http://www.nbirn.net/.

[34] Earth Simulation Center website, http://www.es.jamstec.go.jp/esc/eng/. 2002.

[35] UCAR, Windows to the Universe website, http://www.windows.ucar.edu/.

[36] Bugscope Project website, http://bugscope.beckman.uiuc.edu/.

[37] NRC, *Preparing for the Revolution: Information Technology and the Future of the Research University.* 2002, Washington, D.C.: The National Academies Press. p. 97.

[38] Duderstadt, J. J., Atkins, D. E., and VanHouweling, D., *Higher Education in the Digital Age: Technology Issues and Strategies for American Colleges.* American Council on Education/Praeger Series on Higher Education. 2002. Praeger Greenwood Publishing. pp. xi, 288.

[39] NRC, *Issues for Science and Engineering Researchers in the Digital Age* (http://www.nap.edu/catalog/10100.html). 2001, Washington, D.C: National Research Council of the National Academies.

[40] Raschke, C. A., *The Digital Revolution and the Coming of the Postmodern University.* 2003, New York: Routledge Falmer. p. 129.

[41] Cal-(IT)2_Project, Workshop on Cyberinfrastructure Research for Homeland Security, 25–27 Feb. 2003 (http://web.calit2.net/RiskReduction/). 2003.

[42] Holland, B. A., "Characteristics of Engaged Institutions and Sustainable Partnerships and Effective Strategies for Change" (http://www.oup.org/ researchandpubs/engaged.pdf). 2001.

[43] Kellogg Foundation, New Options for Youth Through Engaged Institutions Initiative website, http://www.wkkf.org/Programming/Overview.aspx?CID=166.

[44] Cogburn, D. L., "Globally-Distributed Collaborative Learning and Human Capacity Development in the Knowledge Economy," in *Globalization and Lifelong Education: Critical Perspectives*, ed. D. Mulenga. 2003, Mahwah, N.J.: Lawrence Erbaum Associates.

[45] American Council of Learned Societies website, http://www.acls.org/.

[46] The National Research and Development Strategy of South Africa at http://www.dst.gov.za/legislation_policies/strategic_reps/sa_nat_rd_strat.htm. 2002.

III

Networks and Organizations

7

Toward a Network Topology of Enterprise Transformation and Innovation

Panagiotis Damaskopoulos

Information and communication technology (ICT) is today recognized as the epicenter of a profound economic dislocation associated with what has come to be known as the transition to a new knowledge-driven economy. The capacity of organizations to engage in learning processes has increasingly come to be viewed as a crucial determinant of innovation, enterprise performance and economic development (Lunvall and Johnson, 1994; Nonaka and Hirotaka, 1995; OECD, 2001). In the emerging new economy innovation constitutes the foundation of the competitiveness and value-creation capabilities of economic organizations. Innovation has emerged as a strategic issue because of the disarticulation of established economic and social structures and processes that the knowledge-driven economy and society bring in their path. This disarticulation is the product of the interplay of technological, industrial, economic, and social transformations. The alignment and re-articulation of technological capabilities, especially ICT, through novel knowledge-creating organizational forms geared to constant innovation and value creation is the intangible quality that today determines the competitiveness of economic organizations and the national and regional environments within which they operate.

However, innovation is not something happening "inside" organizations but rather at the networked interfaces of organizations with the business, regulatory, and institutional environments within which they operate. The process of innovation is increasingly driven by open-source networks of cooperation and involves dynamic interrelationships between technological transformations, organizational capabilities of firms, and public institutional and regulatory structures supportive of innovation and entrepreneurship. In other words, for new ICT that powers the knowledge-driven economy to be able to spread throughout the whole economy, thus enhancing productivity growth, the organizational structure of business firms and institutions and the culture of the society need to undergo substantial change. This is why the agenda of research on the dynamics of adoption of new economy practices, innovation, and economic growth needs to be expanded beyond the level of the firm. It needs to be built around the dynamic interrelationships between technological transformations, firms' organizational knowledge-creating capabilities, emerging market and industry structures, and public institutions (Boyer and Saillard, 1995; Berger and Dore, 1996; Castells, 2000).

Locating the Strategic Importance of Networked Organizational Knowledge

One of the remarkable trends of the era of "irrational exuberance" was the almost exclusive emphasis of much business, academic, and professional commentary on the dot.com phenomenon and, in retrospect, the unrealistic valuations of high technology and Internet-based firms. Indeed, the proliferation of the "e" portions attached to economic activity, coupled to the rapid introduction of the Internet in established business processes, gave the impression that what was "new" in the emerging economic environment was the "transfer" of business processes online. In the wake of the collapse of the high-tech stock bubble, academic and business opinion is marked not only by uncertainty but also by skepticism as to whether the technological transformations associated with ICT and the Internet were the harbinger of a new phase in the development of the global economy or simply a temporary phenomenon that was brought about by rampant speculation. What is in question today is whether the technological, economic, and organizational changes associated with ICT amount to the formation of a new economic system, a new economy. In this context, it is imperative that research in the domain of e-business in particular and the knowledge-driven economy in general be underpinned by explicitly articulated operating assumptions and conceptual categories.

E-business is not an economic activity conducted through ICT-enabled electronic networks. E-business is a central component of an emerging economic system that is powered by ICT, is dependent on highly knowledgeable labor, and is organized around electronic and organizational networks. The historical specificity of this new economic system is that it is *knowledge driven*, it is *global* in its reach, and it is *networked* in its operation in terms of technology and organization. It is *knowledge driven* because the productivity and competitiveness of economic units depend upon their ability to create, process and convert information into knowledge geared to innovation and value creation. It is *global* because the core processes of production, consumption, and circulation are organized on a global scale through functional linkages among economic agents. It is *networked* because productivity and competition are organized through a global network of interaction between and across business networks.

These three central features do not mean that the emerging economic environment leads towards the convergence of economic systems. ICT broadens the scope of economic activity, which means that business systems interact on a global scale. In this context organizational forms diffuse across regulatory and institutional environments, borrow from each other, and create organizational amalgams that correspond to common patterns of business organization and competition, while adapting to the specific social environments within which they operate. In other words, forms of economic organization are mediated by antecedent organizational forms, institutional structures, and cultures. This mediation is of fundamental importance in the acceleration, or deceleration, of learning processes and processes of innovation (Castells, 2000; OECD, 2001).

One of the key drivers of change in the emerging economic environment is closely linked to two industries that not only introduced process and product/service innovations, but also applied such innovations to their own structures and processes, which resulted in higher growth and productivity, and, through competition, to the diffusion of new business models throughout the economy. These industries are *ICT* and *finance*. The global interconnection of financial markets, facilitated by ICT and regulatory reform, is one of the main features that make the new economy global. At the core of the new ICT industries are the Internet-centered firms and Internet-related components of "old economy" types of organizations. However, the centrality of Internet-related economic activity is not related to the until-recently exponential revenue growth and market capitalization value of Internet-related firms. Instead, their economic and business significance lies with the potentially dramatic impact of ICT on the way "old economy" business is conducted.

The financial component of the new economy is related to the successive rounds of innovation during the last quarter of the twentieth century that have resulted in a profound transformation of financial markets both organizationally and technologically. Financial markets are increasingly globalized and interdependent while they are one of the leading domains of application of new ICT. The global financial market is a central component of the emerging economic system. The ability of capital to flow in and out of securities and currencies across markets, and the hybrid nature of financial derivatives, are intertwining through regulatory changes. At the same time, ICT-enabled innovation is transforming the nature of financial transactions. The widespread use of ICT and the Internet have fundamentally changed financial trade between companies, between companies and the investment community, between sellers and buyers, and not least, the stock exchange markets. This change has important implications not only for financial markets but also for the entire economy. ICT-enabled transaction mechanisms reduce transaction costs, thus significantly increasing market volume because the globally interconnected financial markets are able to mobilize savings for investment on a planetary basis, while accelerating the turnover of investment (Strange, 1986; Canals, 1997; Orléan, 1999; Castells, 2000).

The dialectical interplay between ICT and finance has been in many ways the central axis, the *flywheel*, that accounts for the dynamism, global reach, and innovation potential of the emerging knowledge-driven economy. The technological infrastructure of financial markets allows for processes of financial innovation and the development of new financial products that create and allocate and destroy value on a global basis out of trade in securities. On the other hand, ICT-enabled financial innovation encompasses an increasingly larger sphere of economic life where almost any potential source of value can be converted into a security and traded in financial markets globally through ICT-enabled transaction systems.

This process of conversion of potential sources of value into financial securities, that is, securitization, is the driving force of the financial industry. Looking at the comparative performance of stocks of financial services firms and ICT firms between 1995 and 2002 the financial component of the new economy shows a 17 percent annual return. Financial services firms, the main providers of investment capital, have been the best performers, which reflects the growing centrality of the financial services sector in the economy (see table 7.1). Financial markets, in this respect, constitute a strategic network of the new economic environment. For it is there that value is assigned to economic activity as represented by its stocks, bonds, derivatives or any kind of security. The valuation of companies, and thus their capacity to attract capital, depends in a fundamental sense on the judgment of the financial market. The question of how this judgment is and should be formed is one of the most complex questions in contemporary economic analysis and is the subject of considerable debate. Nevertheless, recent experience and research suggest that *expectations* (on the part of financial markets) about the future growth projections of enterprises in terms of actual profitability and future financial value and *trust in the institutional environment* within which financial markets and enterprises operate are central determinants of investment (Castells, 2001). However, to reach the financial market, and to compete for higher value in it, firms have to go through innovation in technology, processes, product/service lines, management quality, and branding. Indeed, the ability to innovate in these domains becomes the cornerstone of competitiveness in the emerging economic environment (Tuomi, 1994; Shapiro, 2002).

Table 7.1
Percent increase in stock prices from January 3, 1995, through July 2, 2002

Financial services, banks, brokerages, insurance companies	224%
Health care, drugs, biotech, managed care	202%
Consumer staples, food, housewares, personal care	128%
Information technology, hardware, software, services	125%
Industrials, machinery, transportation, business services	115%
Consumer discretionary, autos, media, retailing, apparel	107%
Basic materials, aluminum, chemicals, steel, paper products	39%
Utilities, gas and electric companies	17%
Telecom services, telephone and wireless companies	1%

Note: Based on S&P 500 Sector Indexes. Data: Bloomberg Financial Markets.
Source: *Business Week*, July 15, 2002.

But the key to innovation lies in creative thinking and knowledge applied toward the identification of value-creating opportunities. It is leveraging these opportunities that leads to value creation. Indeed, today the connection between organizational knowledge and innovation has become so critical that many companies consider organizational knowledge, coupled to organizational processes geared to continuously improving information and communication channels, as risk management.

The reason is that sharing and transferring knowledge within and across organizations enables companies to increase organizational and operational transparency, which, in turn, helps to reduce risk. In other words, organizational knowledge is about access to timely and relevant information and the conversion of information into knowledge through open organizational channels of communication, which combine to improve judgment on the performance of a firm (Dore, 2001).

However, paths toward innovation are conditioned by three structural transformations associated with the new economy that have significant implications for the organizational structure of the firms operating in it. First, ICT centered on the Internet, in combination with globally integrated financial markets, tends to overcome one of the historic impediments to market transparency: geographical distance (Harvey, 1991). Transparency is a highly transforming condition that affects two dimensions of the business process. ICT can theoretically increase transparency in the operation of financial markets. Openness of corporations to financial markets is primarily a function of the financial disclosure regulations that govern public trading, i.e., access to capital markets. ICT increases transparency in that it can enhance the ability of shareholders and other stake-holding constituencies of organizations to track more intensely the performance of managers and align it more closely toward maximizing the value-creating capabilities of organizations (Goldman Sachs, 1999).

On the other hand, ICT increases price and process transparency. Pricing becomes more transparent as more transactions can be put to the test of auction. Customers can track the progress of their orders while suppliers can get information electronically out of their customers' databases. This kind of transparency affects every aspect of business operations. Small changes in things such as price, product quality, service, responsiveness, and even partnerships could, in theory, be rapidly registered in market share shifts. Putting a business process online has effects throughout a company, since it introduces more information and volatility into strategy. As a result, partnerships and customer relations that underpin existing business models are being reconfigured. In reality, excepting financial markets where they are negligible, switching costs for most industries still represent a significant element of friction. Nonetheless, the Internet contains the potential to move most industries closer to textbook transparency. As a recent authoritative report notes, the Internet is "the mother of all looking glasses" (Morgan Stanley Dean Witter, 2000).

The second implication of the new economy acts on the level of the spatial organization of firms. As information technology and the Internet become entrenched into corporate life, the economic foundation of the firm changes. Business theory on the spatial configuration of the firm has argued that the boundaries of firms are determined by the cost of transactions, especially the cost of communication (Coase, 1937). One of the central canons that guided business practice for much of the twentieth century was that an enterprise should aim for maximum integration as a key to competitiveness and efficiency. In the new economy, by contrast, disintegration and decentralization are becoming the new canon for business competitiveness. There are primarily two reasons for this. The first is that the knowledge needed for any economic activity has become highly specialized, which means that it is becoming increasingly costly and complex to maintain the necessary competencies for every major task within any given organization. And since knowledge is a quality that tends to be depleted unless it is used constantly, maintaining within an organization an activity that is used only intermittently leads to incompetence. The second reason that disintegration and decentralization are becoming important is that the physical cost of communication is becoming virtually nil, which means that in order to organize efficiently firms must search for the most economically optimum form of organization (Drucker, 2001).

Thus, the reduction of the information costs attached to transactions unleashes a process of reconfiguration of the internal and external boundaries of firms. The reduction of information costs enhances organizational capacity to link different operations within and between firms and outsource critical business process components. An important implication of this is the acceleration of the cycle from conception to rollout.

At the same time, the Internet is a fertile ground for the development of new ideas and hence competition, which reinforces the need for companies to develop mechanisms for "reading" and adjusting to the shifting conditions of competition. Within companies the implication is a greater need for collaboration in order to maximize synergies and increase efficiencies across all lines of the business process.

The third implication of the new economy is that it introduces a dialectic of centralization/decentralization in companies. This is largely a function of software standards required in order to enable the transfer of information within and between organizations with different software systems, naming conventions, procedural methodologies, etc. At the same time, standardization increases the capacity of all parties involved (management, employees, external partners, etc.) to "see through" the entire process. Transparency, in other words, though it significantly enhances the influence of shareholders, also increases the potential of other corporate stakeholders or partners to "see through" a company's activities. More specifically, it enables management to contribute to the activities at the frontlines of the company's operations. On the other hand, in the context of the pattern of economic change and heightened competition companies need information at the frontlines of their operations, hence the need for decentralized organizational forms that enhance the autonomy of employees not only in the generation of knowledge but also in terms of decision making and action, in order to acquire knowledge of developments at the front lines of their operations (i.e., the market touch-points with customers, suppliers, capital markets, etc.) and to constantly adjust to shifts in the competitive environment within which a firm operates (Cairncross, 2002).

The structural impact of this set of transformations is that the process of innovation is increasingly becoming a function of open-source networks of cooperation. Open-source networks are composed of teams of company employees and entrepreneurs within as well as across the formal boundaries of organizations. Innovation itself is driven by three main factors. The first is the generation of new knowledge in the form of scientific and technological know-why, know-how, know-what, and know-when and the practice of management. This presupposes the existence of well-developed public and private R&D systems able to provide the key ingredients of innovation. The second is the availability of highly educated, motivated, and autonomous labor, capable of applying new knowledge in innovative ways to increase productivity and improve business performance. The third factor is the existence of entrepreneurs. Entrepreneurial drive is a key element of innovation since it functions as a catalyst in the transformation of new business ideas and projects into innovation and improved business performance.

In the emerging economic environment timely access to information related to each market a company is operating in is critical for competitive success. However, such access in a constantly changing economic environment marked by highly diverse market dynamics is not feasible on the basis of inflexible and top-down organizational structures. ICT allows for the simultaneous decentralization of the information retrieval process from different spaces and for its integration into a flexible system. This technological structure spans different regulatory and institutional spaces that present the potential for large multinational firms to link with small and medium-size enterprises according to contingent project demands, forming networks that are able to innovate and adapt continuously. Business projects are implemented in diverse domains and can be directed to process, product and service line development and organizational tasks across different territorial areas. Successful business project implementation is a function of information that is generated and processed on the basis of ICT systems between and across companies, on the basis of knowledge acquired from each area. In other words, the key passages of information and knowledge that underpin the process of innovation run through networks: ICT and organizational networks within, between, and across companies (Castells, 2000).

Knowledge and Organizational Design: The "Network Enterprise"

It is this set of structural transformations associated with the emerging knowledge-driven economy that largely accounts for the ascending importance of intangible corporate assets in the process of value creation (Lev, 2001). The growing importance of intangibles can be appreciated in historical perspective. For much of the early twentieth century multinational firms were domestic firms organized internationally on the basis of a structure of subsidiaries that were operating quasi-autonomously within territorially defined institutional jurisdictions. During the closing decades of the twentieth century multinationals tended to become increasingly organized on a global basis that was defined by product and service lines. More recently corporate strategies underpinning foreign investment are geared toward the development of structured relationships between companies operating in different sectors and institutional environments. In the emerging context, it is alliances, joint ventures, know-how agreements, and minority stakes that are becoming the critical components of innovation strategies. At the same time, the organizational topology of the operations of multinational firms spans a global regulatory and institutional matrix. This means that the critical tasks of management are becoming balancing acts of conflicting demands between short-term profitability and long-term strategic growth made by the modern corporation's stakeholding constituencies: shareholders, i.e., financial markets, especially institutional investors and pension funds, customers, knowledge employees, and communities (Drucker, 2001).

In other words, the transition to the knowledge-driven economy involves a shift in the parameters of the valorization process that increases the value of the intangible assets of organizations and more specifically their "organizational capital." Organizational capital is not a "thing"; it is a relationship of different intra-organizational components or departments on the level of the firm itself and the relationship of these to the competitive, market, regulatory, and institutional environment within which the firm operates. Successful management of "organizational capital" depends on the knowledge-generating and learning capabilities of organizations and their deployment for innovation and value creation. The correlation between knowledge and organizational change and adaptation is a function of the fact that in the new economy though investment in technology is important, it is innovation in processes, and product and service lines that is the key determinant of the innovation capabilities and market capitalization of firms (Brynjolfsson, Hitt, and Yang, 2000; Bounfour and Damaskopoulos, 2001).

The term "organizational capital" refers to a nodal concept that is composed of several subcategories of intangible capital. It encompasses, but is not restricted to, the following: *Market capital* refers not to the physical qualities of the products a firm produces, but to the intelligence and know-how that go into creating and developing new products and services. It also includes intangible attributes that are closely related to products such as trademarks, patents, brand reputation, corporate reputation, and other insignia of corporate recognition. *Intellectual capital* includes the knowledge, skills, and competencies that the managers and employees of an organization possess. *Structural capital* includes any type of knowledge or innovation that affects ICT platforms or internal processes, which are critical to the formation of the processes that underpin the production and distribution of a firm's products and services. *Relationship capital* refers to a company's relationship with its customers and other stakeholders, including financial markets and the investment community, and government and community institutional structures. *Communications capital* includes the benefits of leveraging and communicating intangibles that may result in positive financial analysis recommendations, increased investor demand, premium pricing, more committed employees, and so on.

The growth of the strategic importance of the management of "organizational capital" can be understood as a shift that places increasingly higher value on the information assets, or more correctly, knowledge assets of corporations. The differentiation of *information* from *knowledge*, in this context, acquires strategic significance. The value of information generated by computer systems depends on human interpretation. Knowledge, by contrast, resides in a social intersubjective context and the human capacity for action based on that information. Thus, knowledge in a corporate organizational context can be distinguished from information since it is more directly linked to action and organizational performance. Organizations, of course, cannot manage knowledge *per se*. They can, however, create an environment that fosters the creation, continuity and sustained use of knowledge and its application within the organization (Davenport and Prusak, 1998; Von Krogh, Ichijo, and Nonaka, 2000).

One influential approach to the management of intangible corporate assets has proposed a model of the knowledge-creating company that is based on the organizational interaction between "explicit knowledge" and "tacit knowledge" at the source of innovation. This perspective argues that much of corporate knowledge is "tacit" and cannot be communicated under formalized management procedures. Yet a corporation's potential for innovation is significantly enhanced when it is able to build bridges that allow the conversion of "tacit" into "explicit" knowledge, "explicit" into "tacit" knowledge, "tacit" into "tacit," and "explicit" into "explicit" (Nonaka and Hirotaka, 1995; Nonaka and Nishiguchi, 2001). This conversion can be facilitated through the use of ICT tools. However, the creation of a knowledge-creating organization is not an issue of technology, it is an issue of organization. More precisely, the creation of a knowledge-creating organization is a question of creating process, which involves aligning technology, people, and organizational qualities toward the achievement of specific organizational goals. And this is primarily a process that involves skills, competencies, and commitment.

The quality of knowledge in the context of the new economy is not a function of the duration of formal education. Instead quality refers to the "type" and "relevance" of education to specific tasks involved in particular business projects. Labor in the knowledge-driven economy requires specific types of education that are characterized by continuous modification and expansion of the workers' knowledge throughout their working lives. The most important feature of this learning process is learning "how to learn," since in the context of accelerated economic and technological change most context-specific information is likely to be obsolete in short periods of time. Learning "how to learn" involves addressing the kind of learning that goes beyond mere acquisition of facts for the purpose of performing a specific task better. It involves developing the ability to forge meaningful connections that result in an awareness of different perspectives, and teaches one to ask the relevant questions in specific spatiotemporal contexts. It also involves the development of the ability to transform the information obtained from the learning process into knowledge and action geared to improving organizational performance. These abilities demand continuous education and decision-making autonomy, both of which have to do with a particular organizational structure and an organizational culture that instills commitment and encourages learning and autonomy.

Efficiently managing "organizational capital" depends in a fundamental sense on the development of organizational forms that generate mutually reinforcing dynamic interrelationships between ICT, organizational flexibility, and highly skilled and motivated labor (Bresnahan, Brynjolfsson, and Hitt, 2000; Bounfour and Damaskopoulos, 2001). There is a particular organizational form that has emerged as a critical component of competitiveness in the new economy: the "network enterprise" (Powell, 1990; Powell and Smith-Doerr, 1994; Applegate et al., 1999; Dutta and Evgeniou, 2002; Hagel and Seely Brown, 2001). In contrast to earlier vertically integrated hierarchical organizational structures, this is a flexible organizational form of economic activity, built around specific business projects and strategic objectives. The business projects themselves are set in motion through the cooperation of networks of various and flexible duration periods and diverse origins and compositions of skills and competencies. Indeed, such is the structural change associated with the transition to the new economy that the basic unit of economic activity and theoretical analysis is increasingly the network, not the firm. The firm continues, of course, to be the basic repository of property rights, strategic management, and the accumulation of capital. However, business practice is increasingly a function of *ad hoc* networks whose expertise is solicited for the achievement of specific business project goals. In terms of its internal organizational structure the "network enterprise" is characterized by several main trends: its organization is structured around process, not task; it has a flat organizational hierarchy; the work process is organized on the basis of teams; customer satisfaction is the primary measure of business performance; the structure of reward is based on team performance; the maximization of contacts with suppliers and customers is an integral part of the business process; and information and continuous training of employees at all levels are considered critical to business success (Castells, 2000).

Organizational Knowledge beyond the Boundaries of the Firm: "Clusters of Innovation"

It is synergy among these networked organizational components and their interaction with the business, regulatory, and institutional environment in which firms operate that decide the innovative capabilities and competitiveness of organizations in the knowledge-driven economy. ICT and the Internet have long been considered as bringing about "the end of geography" since the transparency they introduce into the economic process makes location less important—organizations have access anywhere and anytime. Yet, recent research demonstrates not only a remarkable geographical concentration of the production process of technologies that presumably annihilate geography but also the continuing concentration of significant ancillary services key to the operation of the knowledge-driven economy, services ranging from finance to legal services and advertising. Why is this happening? Research shows that spatial concentration and geographical proximity continue to hold a fundamental importance in fostering innovation. Innovation, in other words, is not something happening "inside" organizations but rather at the interface of organizations with the business, regulatory and institutional environment within which they operate (Saxenian, 1994; Porter, 1998; Gambardella and Malerba, 1999; Sassen, 2000; Cooke et al., 2000; Crouch, 2001; European Commission, 2000).

A key element in this spatial concentration has to do with "clusters of innovation" that denote organizational, social, and institutional constellations that underpin accelerated paces of technological uptake, organizational knowledge creation, and their deployment for innovation. These constellations incorporate specific sets of relationships of production and management, embedded in social and institutional structures that support a culture of entrepreneurship and encourage the development of new business processes geared to innovation. The central feature of the institutional infrastructure of these spatial concentrations is the synergistic network relationships they foster among and across firms and institutions of the public sector. Typical components of a "cluster" include companies that are networked within and through the cluster, venture capital firms, public institutions such as boards of trade and dedicated investment-attracting and promotion agencies (necessary for the creation of a business-friendly environment), universities and research institutes (necessary for the support of networked R&D activities and the generation of know-why, know-how, know-what, and know-when). The key in the competitive position of "clusters of innovation" is their ability to generate synergy, that is, the added value that results not from the cumulative economic impact of the critical elements present in the cluster but from their interaction in a way that fosters innovation and value creation (Castells and Hall, 1994; Morgan et al., 1999; OECD, 2001).

The processes of organizational learning that are central to continuous innovation themselves display a spatial logic of concentration, which increases the importance of new forms of comparative differentiation across regions. Spatial proximity between organizations is crucial to the exchange of information and knowledge through which organizational learning emerges (Storper, 1995). However, it is important to differentiate between "organizational proximity" and "spatial proximity." The former does not necessarily depend on the latter since the growing sophistication of ICT opens up new possibilities for the growth of effective learning networks among organizations based upon spatially dispersed interaction (Castells, 2000). However, the critical elements of organizational learning continue to take place within networks of organizations that are spatially proximate. Spatial proximity may create the conditions for organizational learning through channels of social interaction, for instance, by increasing the frequency of personal contacts among the knowledge agents within the innovation system. More fundamentally, however, at least some of the key elements of knowledge that are generated and disseminated through such social interaction are tacit, that is, they are embedded in particular local social systems. Access to this social field of knowledge, as a result, depends on participation in the local social system within which such knowledge is produced (OECD, 2001).

Thus, while there is accumulating evidence of structural changes that sustain trends toward the globalization of economic processes, this does not render the comparative difference among localities any less significant. On the contrary, a critical issue in the growing importance of locality has to do with the modalities and patterns of organizational learning that are implicated in the complex interactions between global and local processes. For instance, recent research indicates that the phenomenon of business incubation, that is, the creation of new businesses especially in regions that undergo industrial transformation, depends to a large extent on forging dynamic interrelationships between local business communities, structures of regional political authority and regional scientific know-how as this is produced in universities and research centers, and its commercialization through collaborative processes of knowledge and product or service development (CASIS, 2003).

The specific elements that structure the social and economic fabric of regions, that is, their economic structures, patterns of social and political relations, and cultural and institutional settings, are themselves critical factors that condition and shape emerging patterns of economic development and organizational forms. Hence, a key question regarding a locality's economic trajectory is the extent to which its social institutions can operate as frameworks enabling innovative responses to the challenges of the emerging knowledge-driven competitive environment (OECD, 2001). In other words, in order to reap the benefits associated with participation in a "cluster," firms, public bodies, and all the central elements that compose the cluster need to be knowledge-creating or "learning organizations" (Morgan, 1997). It is this amalgam of private enterprises and public institutions, conditioned by different historical contexts across national and regional economies that tend to attribute different roles to the public and the private, that is at the center of systems of innovation in the knowledge-driven economy.

Conclusion

ICT is one of the central parameters in the transition to the knowledge economy. The capacity of organizations to harness the potential of ICT and engage in learning processes has increasingly come to be seen as a crucial determinant of innovation, enterprise performance, and economic development. In the emerging knowledge-intensive economic environment the ability of organizations to generate new knowledge and apply it toward innovation in process, product, and service lines of the business process constitutes a critical determinant of competitiveness. However, as a consequence of the structural transformations associated with the transition to the new economy, innovation is migrating toward a more complex topology of networks that "pass through" the organizational structures of individual firms and the market, regulatory, and institutional environment within which they operate. The process of innovation is increasingly driven by open-source networks of cooperation and involves dynamic interrelationships between technological transformations, organizational capabilities of firms, and public institutional and regulatory structures supportive of innovation and entrepreneurship.

The diffusion of new ICT throughout the whole economy, which is a key condition for productivity growth, depends on broad-based mutually reinforcing transformations involving business organizations and the institutions and culture of society. The central challenge of modern management as well as public policy-making in this respect has to do with the development of strategic synergies between technological transformations, firms' organizational and knowledge-creating capabilities, emerging market and industrial structures, and public institutions. To put it another way, the transition to a sustainable knowledge-intensive economy depends on dynamics that extend beyond the domain proper of the firm. The sustainable profitability and social value of economic activity depends on the construction of linkages between technological capabilities, especially ICT, through novel knowledge-creating organizational forms geared to constant innovation and value creation in ways that support the competitiveness of economic organizations and the national and regional environments within which they operate.

References

Applegate, Lynda M., F. Warren McFarlan, and James L. McKenney (1999) *Corporate Information Systems Management: Text and Cases*. New York: McGraw-Hill.

Berger, Suzanne, and Ronald Dore (1996) *National Diversity and Global Capitalism*, Ithaca, N.Y.: Cornell University Press.

Bounfour, Ahmed, and Panagiotis Damaskopoulos (2001) "Managing organisational capital in the new economy: Knowledge management and organisational design." In *E-Work and E-Commerce: Novel Solutions and Practices for a Global Networked Economy*, volume 1, ed. Brian Stanford-Smith and Enrica Chiozza. Amsterdam: IOS Press.

Boyer, Robert, and Yves Saillard (1995) *Théorie de la regulation: l'état des saviors*. Paris: La Découverte.

Bresnahan, Timothy, Erik Brynjolfsson and Lorin Hitt (2000) "Information technology, workplace organization, and the demand for skilled labor: Firm-level evidence." Cambridge, Mass.: MIT-Sloan School Center for E-business, working paper.

Brynjolfsson, Erik, Lorin M. Hitt, and Shinkyu Yang (2000) "Intangible assets: How the interaction of companies and organizational structure affects stock market valuations." MIT working paper, July, at http://ebusiness.mit.edu/erik/.

Cairncross, Frances (2002) *The Company of the Future: Meeting the Management Challenges of the Communications Revolution*. London: Profile Books.

Canals, Jordi (1997) *Universal Banking: International Comparisons and Theoretical Perspectives*. Oxford: Oxford University Press.

CASIS (Centre for Advanced Studies in Innovation Systems) (2003) *Guiding Economic Innovation through Incubators: Critical Elements of Successful Business Incubation*, Paris. Available at www.casis.org.

Castells, Manuel (2000) *The Information Age: Economy, Society and Culture*. Volume I: *The Rise of the Network Society*, Volume II: *The Power of Identity*, Volume III: *End of Millennium*. Oxford: Oxford University Press.

Castells, Manuel (2001) *The Internet Galaxy: Reflections on the Internet, Business and Society*. Oxford: Oxford University Press.

Castells, Manuel, and Peter Hall (1994) *Technopoles of the World: The Making of Twenty-first Century Industrial Complexes*, London: Routledge.

Coase, Ronald, H. (1937) "The nature of the firm." *Economica*.

Cooke, Philip, et al. (2000) *The Governance of Innovation in Europe: Regional Perspectives on Global Competitiveness*. London: Pinter Publishers.

Crouch, Colin, ed. (2001) *Local Production Systems in Europe: Rise or Demise?* Oxford: Oxford University Press.

Davenport, H. Thomas, and Laurence Prusak (1998) *Working Knowledge: How Organizations Manage What They Know*, Boston: Harvard Business School Press.

Dore, Lucia (2001) *Winning Through Knowledge: How to Succeed in the Knowledge Economy*, Special Report by the *Financial World*, The Chartered Institute of Bankers in Association with Xerox. London: March.

Drucker, Peter (2001) "The next society: a survey of the near future." *Economist*, November 3.

Dutta, Soumitra, and Theodoros Evgeniou (2002) "CRM in a networked economy." INSEAD Working Paper.

European Commission (2000) *Innovation Policy in a Knowledge-Based Economy*; Maastricht Economic Research Institute on Innovation and Technology, Maastricht, Germany.

Gambardella, Alfonso, and Franco Malerba, eds. (1999) *The Organisation of Economic Innovation in Europe*. Cambridge: Cambridge University Press.

Goldman Sachs Investment Research (1999) *E-Commerce/Internet: B2B: 2B or Not 2B*, November.

Hagel, J., and John Seely Brown (2001) "Your next IT strategy." *Harvard Business Review*, March–April.

Harvey, David (1991) *The Condition of Postmodernity*. Oxford: Oxford University Press.

Lev, Baruch (2001) *Intangibles: Management, Measurement, and Reporting*, Washington, D.C.: Brookings Institute.

Lunvall, B. A., and G. Johnson (1994) "The Learning Economy." *Journal of Industry Studies*, no. 1.

Morgan, K. (1997) "The learning region: Institutions, innovation and regional renewal." *Regional Studies*, no. 31.

Morgan, K., G. Rees, and S. Garmise (1999) "Networking for local economic development." In *The Management of British Local Governance*, ed. G. Stoker. Basingstoke: Macmillan.

Morgan Stanley Dean Witter (2000) *The B2B Internet Report, Collaborative Research*, April.

Nonaka, Ikujiro, and Takeuchi Hirotaka (1995) *The Knowledge-Creating Company*. Oxford: Oxford University Press.

Nonaka, Ikujiro, and Toshihiro Nishiguchi, eds. (2001) *Knowledge Emergence: Social, Technical, and Evolutionary Dimensions of Knowledge Creation*. Oxford: Oxford University Press.

OECD (2001) *Cities and Regions in the New Learning Economy*, Paris.

Orléan, André (1999) *Le Pouvoir de la Finance*. Paris: Éditions Odile Jacob.

Porter, Michael (1998) "Clusters and the new economics of competition." *Harvard Business Review*, November–December.

Powell, W. W. (1990) "Neither market nor hierarchy: Network forms of organization." *Research in Organizational Behavior* 12.

Powell, W. W., and L. Smith-Doerr (1994) "Networks and economic life." In *Handbook of Economic Sociology*, ed. N. Smelser and R. Swedberg. Princeton, N.J.: Princeton University Press.

Sassen, Saskia (2000) *Cities in a World Economy*, 2d edition. London: Pine Forge Press.

Saxenian, Anna Lee (1994) *Regional Advantage*. Cambridge, Mass.: Harvard University Press.

Shapiro, Stephen M. (2002) *24/7 Innovation: A Blueprint for Surviving and Thriving in an Age of Change*. New York: McGraw-Hill.

Storper, M. (1995) "The resurgence of regional economies, ten years after: The region as a nexus of untraded interdependencies." *European Urban and Regional Studies*, no. 2.

Strange, Susan (1986) *Casino Capitalism*. London: Blackwell.

Tuomi, Ilkka (1994) *Corporate Knowledge: Theory and Practice of Intelligent Organizations*, Helsinki.

Von Krogh, Georg, Kazuo Ichijo, and Ikujiro Nonaka (2000) *Enabling Knowledge Creation: How to Unlock the Mystery of Tacit Knowledge and Release the Power of Innovation*. Oxford: Oxford University Press.

8

IT-Enabled Growth Nodes in Europe: Concepts, Issues, and Research Agenda

Ramon O'Callaghan

This chapter develops an agenda for research about knowledge management processes and the use of related information and communication technologies (ICTs) to foster "growth nodes" and emergent strategic growth opportunities within European regions. The project is based on research sponsored by the European Commission under the Information Society Technologies (IST) Programme.[1]

The working hypothesis is that the development of future competitiveness in the European Union will happen through the emergence of interconnected clusters ("growth nodes") with higher than average economic growth rates, including success in equity issues and social welfare. The term "growth node" was chosen instead of "cluster" or "growth pole" because the focus of the research is not only on the interrelatedness *within* different clusters but also on the interrelatedness *between* them.

From Clusters to Growth Nodes

Most experts define a cluster as a geographically bounded concentration of similar, related, or complementary businesses and other related organizations with active channels for business transactions, communication, and dialogue, that share a specialized infrastructure, labor markets, and services and that are faced with common opportunities and threats (Porter, 1998, 2001). The geographical boundary is determined largely by the distances and times that people are willing to travel for employment, meeting, and networking.

Clusters are based on systemic relationships among firms and related organizations. The relationships can be built on common or complementary products, production processes, core technologies, natural resource requirements, skill requirements, and/or distribution channels (Rosenfeld, 2002).

Porter (1998) posits that in a globalizing world the forces leading to cross-industry clustering and involving a knowledge base and social aspects have intensified. It is against this observation that the growth node concept is positioned. A key advantage of clustering is access to innovation and knowledge, but this generally is assumed to require geographic proximity. Such proximity is closely associated with the accumulation of social capital, strong social learning processes, and the advantages of networking.

A growth node (GN) is defined as *a high-performing geographical cluster of organizations and institutions, networked to other clusters, that is, other nodes, and amplified by ICTs.* A GN entails a local network of actors (i.e., organizations in a territory) collaborating and creating a growth path that generates "external economies" for a territory. This growth path is the result of an emergent process fostered by entrepreneurial action and appropriate enabling conditions.

In our paradigm, clusters are nodes in a wider network, and the development (i.e., growth) of a node arises from the interplay of internal and external interactions. In other words, growth nodes involve connections to organizations outside the territory in addition to connections to organ-

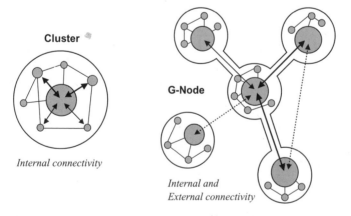

Figure 8.1
Illustration of cluster and growth node.

izations within the territory. Therefore, *nodality*[2] is a key feature of any growth node.

The use of diverse combinations of ICTs within and between growth nodes is assumed to have implications for the meaning of proximity. In traditional clusters, the need for physical proximity has led to regional agglomerations. The question is how will new ICTs affect traditionally perceived needs for physical proximity and introduce "virtual" proximity as a complement to physical proximity? Can growth nodes be expected to emerge and/or develop, in part, as a result of the widespread application of ICTs? What combinations of physically proximate and "virtual" arrangements best augment the social and economic performance of growth nodes? The next sections develop these issues and suggest associated research. The appendix summarizes the proposed research agenda.

Do Growth Nodes Exist?

Europe's existing clusters of industrial activity that are seeking to establish a stronger presence in European and global markets are likely to undergo a transition from relatively high degrees of internal connectedness to a situation in which they develop greater external connectedness. The synergistic effects of internal and external connectivity, augmented by ICTs, could give rise to growth nodes.

From a research perspective, the issue is how to identify existing growth nodes or clusters that have nodal potential. Constructs are needed for identifying, characterizing, and measuring growth nodes, their incidence, and their effects so that a typology can be used to classify the emergent properties of European regions.

Cluster attributes that exhibit the features of "nodality" and that may give rise to regional growth are likely to comprise the following:

• *Externality*: density of interactions with partners outside the growth node;
• *Reach*: geographic scope of the growth node—regional, national or international;
• *Knowledge intensity*: interactions that are strongly knowledge-based;
• *Employment structure*: high proportion of knowledge workers within total employees; and

• ICT *infrastructure*: an extensive network infrastructure for linking players internal and external to the node.

Systematic analysis of the above and other attributes is necessary to indicate whether a candidate agglomeration of economic activity or cluster exhibits nodality and, hence, can be considered to be in the transition toward growth node evolution. Thus, the proposed GN research agenda includes a survey of existing clusters in several regions in Europe to assess their nodality.

Growth Nodes and Regional Development

The importance of clusters is well established. Porter's identification of contemporary local agglomerations, based on a large-scale empirical analysis of the internationally competitive industries for several countries, has been especially influential, and his term "industrial cluster" has become the standard concept in this field (Porter, 1990, 1998, 2001). The work of Krugman (1991, 1996) has been more concerned with the economic theory of the spatial localization of industry. Both authors have argued that the economic geography of a nation is key to understanding its growth and international competitiveness.

Clusters lead to higher growth in several ways. Concentration, or clustering, gives businesses an advantage over more isolated competitors. It provides access to more suppliers and customized support services, to experienced and skilled labor pools. Clustering enables companies to focus on what they know and do best. In addition, clusters stimulate higher rates of new business formation, as employees become entrepreneurs in spin-off ventures, since barriers to entry are lower than elsewhere.

Among all of the advantages of clustering, however, none is as important as access to innovation, knowledge, and know-how. In the "knowledge-based economy," companies are expected to look for their main competitive advantages in terms of access to ideas and talent. This generally is assumed to require geographic proximity to professional colleagues, leading-edge suppliers, discriminating customers, highly skilled labor pools, research and development facilities, and industry leaders.

The following paragraphs briefly discuss three elements associated with clusters—social capital, social learning, and networking—that are also relevant to growth node performance.

The Social Factor

An important influence on cluster strategies has been the accumulation of "social capital" and its emergence as a factor in economic growth and social development. A region's stock of social capital resides in its civic and professional associations, and its economic value is deeply embedded in the functions of groups that bring people together to share ideas and knowledge. The contribution of social capital to economic development has roots in Europe in Northern Italy. An analytical framework for the social foundation of clusters was provided by Robert Putnam's research, which compared Northern and Southern Italy's economies in 1993, and by Annalee Saxenian's research, which compared Silicon Valley and Route 128 high-tech economies in 1994. Their widely cited analyses confirmed the importance of social infrastructure for competitiveness. An implication of their research is that regions should pay more attention to the roles of intermediaries and gatekeepers such as business associations, chambers of commerce, and community-based organizations.

Learning Systems

In the knowledge-based economy, growth depends on technology diffusion and knowledge spillover. Research shows that clusters can facilitate the transmission of knowledge—particularly tacit knowledge, which is embedded in the minds of individuals and the routines of organizations and which therefore cannot move as freely or easily from place to place as codified knowledge (Cortright, 2000).

Ideas about the importance of creating structures that support and accelerate learning have been translated in the context of the "new economy" in the form of strategies to create "learning cities" and "learning regions" (Boekema et al., 2000; OECD, 2001). Within clustered economies, there invariably is more interfirm mobility and thus more active transfer of information and knowledge among firms and workers.

Business Networks

Northern Italy is generally accepted as the prototypical economy of clusters. The region of Emilia-Romagna was first noticed because of its unusually small, flexible, and specialized firm structure. This was described by Piore and Sabel in the *Second Industrial Divide* (1984). But the success of Northern Italy was first attributed not to the clustering of companies but to the intensity of interfirm collaboration and to the specialized services created by the government and trade associations that gave the small companies access to external economies of scale.

It should be noted that whereas networking has been implicitly associated with the clustering concept, it can just as easily accommodate the concept of nodality and, hence, the relationship between clusters. What, then, does the growth node perspective add to the benefits of clusters?

Of particular importance for the growth node concept and for the associated research is the question of "experienced proximity" and "presence" in ICT-mediated environments. This topic is discussed later in more detail. By capitalizing on the potential offered by ICTs to achieve experienced proximity between points that are distant from each other in space, a growth node has the effect of making outside resources—cognitive, institutional, cultural, and material—available inside the growth node, thus stimulating its growth beyond that which could be achieved by employing local resources alone.

Concerns about Equity

Social networks are core assets of many clusters. They expose members to new processes and markets, nonpublic bid requests, and innovations. Companies outside the networks, however, may miss out on many economic opportunities. Clusters create a capacity to network and learn, but the more they are defined by formal membership, the higher the hurdles for outsiders to obtain the benefits of that knowledge. In conventional clusters, access to the network may be controlled by some large companies. This has traditionally been the problem for small and mid-sized enterprises (SMEs), which, as a result, have been slow to learn about and adopt new technologies or enter new markets.

Growth nodes, on the other hand, may be better equipped to tackle this issue. Such a proposition is based on the premise that knowledge in growth nodes is more freely available and not limited to the local resources. There seems to be some empirical evidence supporting it.

A number of regions in the European Union classified as "less favored" have sectors specializing in traditional industries with little innovation and a predominance of small family firms with weak links to external markets (Landabaso et al., 1999). The most successful clusters, on the other hand, include lead firms that are part of global networks and thus are exposed to global market opportunities and that employ people active in international professional associations and networks (Rosenfeld, 2002). These firms regularly benchmark themselves against the best practices anywhere.

Poorer regions and smaller companies have limited access to the benchmark practices, innovations, and markets. Without wider access, companies are limited to learning only within their regional borders and have a difficult time achieving any sort of competitive position. The question is whether, and how, growth nodes increase social inclusion. The research agenda should thus include the analysis and comparison of the structure and behaviors of a set of growth nodes against a set of traditional clusters.

Conditions to Sustain and Develop Growth Nodes

Research on economic growth suggests a number of factors that may also be relevant to growth node development: innovation, imitation and competition, entrepreneurship, networks, social capital ("connections"), specialized workforce, industry leaders, talent, and tacit knowledge (Rosabeth Moss Kanter, 1995).

Although the success of an individual firm depends on its ability to protect its proprietary knowledge, the success of a growth node depends on diffusion of innovations and information and spin-offs of new enterprises. Imitation is important because it is what circulates new concepts and practices among companies and fosters further innovation. Many imitators become innovators by improving on the practices they adopt, and this cycle of innovation and imitation develops the cluster/growth node.

An important operating principle of GNs is the ability to network extensively and form networks selectively. Networking is the process that moves and spreads ideas, information, and best practices throughout a cluster and imports them from other places. The limits or constraints to active participation are largely a function of "connections." Thus, the mechanisms and entities for collecting and disseminating knowledge, as well as the intermediaries that facilitate all forms of associative behavior, help build "social capital." The research agenda should include a longitudinal study of several growth nodes to analyze factors deemed relevant to the development and sustainability of growth nodes (such as those described above). The analysis would test whether changes in any of the factors over time translate into changes in the dependent variable (a growth measure). The research would also test additional explanatory variables unique to growth nodes, for example: extranode connectivity, availability of shared ICT infrastructure, and use of some key knowledge-management applications. A control group would be formed by a set of clusters with low external connectivity and low growth so as to differentiate GNs from traditional clusters.

Given the importance of networking, knowledge diffusion, and social learning in growth nodes, the next section discusses a knowledge-management framework to analyze growth nodes.

The I-Space: A Conceptual Framework to Assess Knowledge Management in Growth Nodes

This section introduces a conceptual framework to explore characteristics of information and knowledge flows relevant to growth nodes and how these might be affected by the information and communication technologies (ICTs) as well as their spatial implications. The framework, known as the *Information Space* or I-Space, is described in detail in Boisot (1995 and 1998).

The framework highlights the relation between the extent to which data can be structured and the extent to which they can be shared. The I-Space takes *codification* and *abstraction* as the two processes through which data are apprehended and structured. Well codified and abstract data economize on both senders' and receivers' data processing transmission resources. Well codified and abstract data will *diffuse* to more agents per unit of time than uncodified and concrete data. Thus, codification, abstraction, and diffusion constitute the three dimensions of the I-Space.

This three-dimensional space is used to position *learning* processes. The framework distinguishes six activities that contribute to learning: codification, abstraction, diffusion, absorption, impacting, and scanning all contribute to learning. Where they take place in sequence they make up the six phases of a social learning cycle (SLC), as illustrated in figure 8.2.

SLCs can have many shapes in the I-Space, reflecting the different blockages that can impede the learning process, and not all phases of an SLC will be immediately value-adding. In addition, learners will adopt

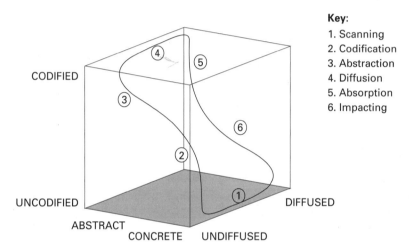

Key:
1. Scanning
2. Codification
3. Abstraction
4. Diffusion
5. Absorption
6. Impacting

CODIFIED

UNCODIFIED   DIFFUSED

ABSTRACT
CONCRETE   UNDIFFUSED

Scanning: Identifying threats and opportunities in generally available but often fuzzy data—i.e., weak signals.

Codification: The process of giving structure and coherence to such insights—i.e., codifying them.

Abstraction: Generalizing the application of newly codified insights to a wider range of situations.

Diffusion: Sharing the newly created insights with a target population.

Absorption: Applying the new codified insights to different situations in a "learning by doing" or a "learning by using" fashion.

Impacting: The embedding of abstract knowledge in concrete practices.

Figure 8.2
The social learning cycle in the I-Space.

different learning strategies. Learners or agents may engage in information hoarding or information sharing, or in some cases, in both strategies:

- *Information hoarding*: Recognizing that diffused information has no economic value, agents attempt to slow down the SLC by refraining from codifying or abstracting too much and by building barriers to the diffusion of newly codified abstract information—through patents, copyright, secrecy clauses, etc. Slowing down the SLC allows them to extract value from information in a controlled way.

- *Information sharing*: Recognizing that, through subsequent processes of absorption, impacting, and scanning, diffused information prepares the ground for further learning and knowledge creation, agents willingly share their information and watch how it is used by others. They gain first-mover advantages in being the first to initiate a new SLC and extract value from the process by participating in a succession of SLCs instead of dwelling as long as possible in a single one.

Network Institutional Orders

The framework introduces the role of institutions in the processes of social learning and knowledge management. Different institutional mixes embedded within a network may represent a variety of "strong" and "weak" cultures that interact in ways that influence the sharing of local and distant information resources and the extent to which learning progresses to enable efficient and effective knowledge management. The framework distinguishes four different types of institutional order: *bureaucracies, markets, clans,* and *fiefs.*

If codification and abstraction are taken as proxy measures for information *structuring*, and diffusion is taken as a proxy measure of information *sharing*, one can see how the different cultural features of actors connected within networks might help to institutionalize the social and economic transactions involved in the SLC.

The four different types of institutional order can be located in the I-Space as shown in figure 8.4. These institutions are potential attractors that exert a gravitational pull on transactions that are located within their field. Where these attractors are not too strong, they facilitate learning by speeding up the flow of the SLC through that region of the I-Space. However, where the pull that they exert becomes so powerful that

2. BUREAUCRACIES
- Information diffusion limited and under central control
- Relationships impersonal and hierarchical
- Submission to superordinate goals
- Hierarchical coordination
- No necessity to share values and beliefs

3. MARKETS
- Information widely diffused, no control
- Relationships impersonal and competitive
- No superordinate goals - each one for himself
- Horizontal coordination through self-regulation
- No necessity to share values and beliefs

1. FIEFS
- Information diffusion limited by lack of codification to face-to-face relationship
- Relationships personal and hierarchical (feudal/charismatic)
- Submission to superordinate goals
- Hierarchical coordination
- Necessity to share values and beliefs

4. CLANS
- Information is diffused but still limited by lack of codification to face-to-face relationships
- Relationships personal but nonhierarchical
- Goals are shared through a process of negotiation
- Horizontal coordination through negotiation
- Necessity to share values and beliefs

STRUCTURED INFORMATION

UNSTRUCTURED INFORMATION

UNDIFFUSED INFORMATION

DIFFUSED INFORMATION

Figure 8.3
Institutions in the I-Space.

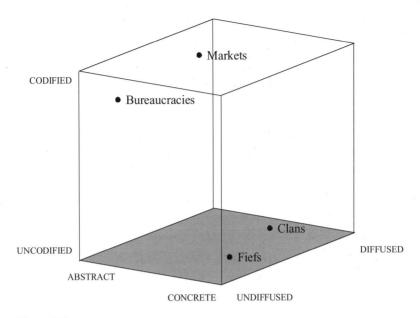

Figure 8.4
Location of institutions in the I-Space.

they can attract transactions from any part of the I-Space, they bring the SLC to a halt in their region of the space, blocking any further progress of the learning process.

This highlights the importance of fiefs, clans, and bureaucracies as governance structures. They can be as important as efficient markets in fostering effective collaboration, and social learning as well as competition.

The I-Space and the Effects of ICT

The new ICTs, by increasing data processing and transmission capacities at all levels of codification and abstraction, have the effect of shifting the whole diffusion curve to the right. Note that this shift has two quite distinct effects: (1) For any level of codification and abstraction, *more people can be reached* per unit of time with a given message. (2) A given size of population can now receive a message at a lower level of codification and abstraction. Another way of saying this is that the new ICTs allow *more of the context to be transmitted* with a message. But even with the new ICTs, the transmission of context (e.g., through video conferencing) remains costly. ICTs are likely to facilitate the transmission of data goods more than of information goods. In turn, this will facilitate the transmission of information goods more than that of knowledge goods. The implication is that ICT applications serve to differentiate knowledge-based services according to their *flow* characteristics. Data-intensive processing and transmission can be performed virtually anywhere, whereas the knowledge-intensive interactions remain rooted in time and place and continue to depend heavily on face-to-face interactions.

Competitive advantage will accrue to those who learn how to climb the ladder, firstly from data to information and then from information to knowledge. ICTs will get them on the bottom rung faster than they might have done on their own. But from then on, the critical assets will be an ability to learn fast rather than to access technology. Cities and regions, in effect, are engaged in learning races with each other, races in which competition and collaboration both have their place.

The predominance of a given institutional order is likely to be supported by a distinct set of ICTs. For instance, inward oriented, physically proximate networks of firms and organizations may be characterized by institutions such as "fiefs" that give very high priority to investment in high bandwidth intranets and extranets and proprietary software applications for decision support, putting less emphasis on interregional networks. Alternatively, externally oriented regions and firms and other organizations may develop many global network linkages to the neglect of the ICT applications that may enable and foster local learning cycles. The institution of the "market" may make it difficult or more costly to foster transactions within the regions in such cases.

The concepts introduced by this framework have been used to propose tools for identifying and analyzing growth nodes as discussed in the next section.

Identifying and Measuring Growth Nodes

A key premise of our research is that knowledge is a crucial asset that is shared and created though a process of "collective learning." Thus, in addressing growth node success factors, one needs to consider the role of knowledge diffusion processes, e.g., the formation of communities of practice, the conditions that enhance the exchange of ideas, and the processes that foster learning, innovation, and entrepreneurial culture.

To use the growth node concept as a policy instrument, regional policymakers will need practical tools to identify and assess some of the key intangible resources available to a given city, region, or country that could contribute to the development of growth nodes. The framework discussed earlier suggests three areas for measurement of growth nodes, i.e., knowledge assets, social learning cycles, and the different institutional orders in knowledge networks. The tools proposed are thus the following: (1) knowledge mapping, (2) learning mapping, and (3) institutional mapping. Taken together, they should allow different stakeholders in a given region to assess its dynamic capabilities and hence its potential for achieving growth node status. Taking each in turn:

• The knowledge mapping tool allows one to identify an organization's critical knowledge assets and to assess their strategic potential. Applied to a region, this tool will allow one to gauge a region's competitive position in the knowledge economy and will suggest possible avenues of strategic development.

• The learning mapping tool allows one to analyze the learning profile of a given organization in terms of its scanning, problem solving, diffusion, and absorption capacities. Applied to a region, this tool will allow one to assess how effectively the region is able to create and exploit knowledge assets.

• The institutional mapping tool will help to establish the extent to which local institutions and cultures facilitate or impede the learning processes that lead to the successful exploitation of knowledge resources.

The integrated use of these different knowledge management tools would allow one to explore the dynamic behavior of a region with respect to knowledge creation and sharing. Each of the knowledge management tools offers a distinct perspective on the knowledge resource of a given region, allowing one to progress, for example, from (1) an assessment of a given region's knowledge resources (knowledge mapping) to (2) an identification of the learning strengths and weaknesses that gave rise to such knowledge resources (learning mapping) as well as to (3) an understanding of the institutional context that facilitates or impedes such learning (institutional mapping).

Enabling ICT Infrastructure for Growth Node Development and Interaction

Much has been made of the potential of information and communication technologies to enable a despatialization of economic activity, but so far systematic analysis on how policy can stimulate European clusters in the twenty-first century by taking advantage of the characteristics of ICTs is lacking. To address this problem, future research should look at growth nodes as clusters of geographically proximate complex organizational systems of learning and economic and social activity that are globally networked with other systems and enabled by the effective use of ICTs. Such research could be structured around the three following issues:

The first issue has to do with the "tacit-codified" knowledge debate in knowledge management research. The understanding of growth nodes will be enhanced by a perspective on the potential contribution of ICTs that gives priority to socially mediated tacit skill sets and learning processes as prerequisites for the effective and efficient use of these technologies within a complex adaptive system. In this context, the conditions can be considered under which ICTs may enable new opportunities for codification that, in turn, may be expected to give rise to new clusters within the European economy. Research should also be particularly concerned with changes in the way that tacit skills sets and learning are mediated by ICTs. This aspect may be examined in terms of the extent to which these processes give rise to new hierarchies and forms of economic and social dominance that can lead to inequality and exclusion even within the context of relatively flat and decentralized network structures.

The second issue is related to debates about "mediated copresence." There is considerable empirical evidence suggesting that it is important to disaggregate ICTs and their applications. Different applications seem to contribute to varying degrees to the "social richness" of spatially proximate and distant encounters. Some may favor and support the growth of clusters of economic activity, while others are less favorable. Research should acknowledge that different modes of ICT-supported communication are likely to have different implications for the spatial distribution of economic activity and for agglomeration economies. Much work on the application of ICTs in the context of e-commerce, e-government, and intrafirm knowledge management systems neglects the social and culturally specific ways in which mediated environments give rise to new forms of communicative interaction and their consequences of networked relationships. The application of ICTs within and between organizational settings gives rise to the question of how to achieve an appropriate balance between on- and offline relationships and overlapping networked organizations. This question can be approached from the standpoint of the economics of technological change and the social phenomena that strengthen or weaken "experienced proximity."

The third issue is related to supporting architectures and infrastructures. Growth nodes are complex systems based on the networking of organizations, the cooperation of the players, and flexible access to resources. A growth node can be seen as a community that shares business, knowledge, and infrastructure in a highly dynamic way. Indeed, some researchers posit that the enterprise and other organizations of the future will be more fluid, amorphous, and, often, with transitory structures based on alliances, partnerships, and collaboration within clusters (or growth nodes). To support this scenario, a next generation of ICT interorganizational network infrastructure (i.e., business digital ecosystem) is envisaged that will provide a dynamic aggregation of network and software services to facilitate the dynamic interaction in both vertical and horizontal interorganizational relations. A question of particular interest is how new ICT infrastructures may serve as a catalyst that alters or transforms existing relationships between physical place and people's perceptions of the value of proximity.

The next sections explain in more detail each of these three themes.

Knowledge Sharing, Proximity, and ICT

The research community appears to have divided views with respect to knowledge diffusion. Some suggest that knowledge can be reduced ("codified") to messages that can then be sent and processed as information. This view suggests that the potential exists for the universal codification of knowledge. ICTs offer a means by which codified knowledge may be disseminated as information. These authors downplay the importance of geography. From this perspective, information flows are spatially unbounded in a world that is interlinked through the implementation of ICTs.

Others argue, instead, that knowledge cannot be considered independently from the processes through which it is generated. Comprehending and utilizing information encompass tacit skills that are intrinsically bound to social processes. These skills entail the cognitive capabilities of the agents and the organizational contexts within which they are interacting. The defining feature of this tacit knowledge is that it cannot be articulated (i.e., "codified") for the purpose of exchange. Tacit knowledge can refer to specific knowledge that is mainly held and shaped by individuals. It emerges from routines, conversations, memories, stories, and repeated interactions, instead of being encrypted in rules or in organizational design.

Focusing on the properties of the knowledge that is used in innovation-related activities and on the associated knowledge exchange, there are those who argue that the transmission of new knowledge occurs more efficiently among proximate actors within the cluster or growth nodes. This is the *knowledge spillover hypothesis.* Highlighting the complexity and tacit nature of the knowledge base, proponents of this view argue that proximity helps to reduce the costs of knowledge transmission by facilitating interpersonal contacts and the interfirm mobility of labor. The degree to which geographical proximity facilitates the sharing of knowledge, in turn, overlaps and combines with institutional, organizational and technical proximity in fostering effective processes of collective learning.

The debate about how the use of ICTs is likely to influence "experienced proximity" is related to whether these technologies support knowledge codification that provides new memory aids for individuals or facilitates collective recall within group exchanges. The use of ICTs may also provide a social memory device in environments where offline social processes for guiding knowledge codification are not available.

The key issue is whether a particular form of ICT use provides sufficient cognitive context for generally tacit knowledge to be transmitted explicitly when required so as to repair any problems that occur in applying that knowledge. This is a core issue in analyzing the spatial implications of "experienced proximity" and the potential of ICTs when they are used to support new network architectures and infrastructures of many different kinds.

"Presence" and ICT-Mediated Environments

It is important to examine the particular types of ICT applications that are in use or may be deployed in order to understand their implications for emergent growth nodes. Different ICT-supported networks are likely to contribute to varying degrees to the "social richness" of spatially proximate and distant encounters. Internet- and non-Internet-based ICTs will have different implications for the different phases of the social learning cycle. Thus, our proposed research is concerned with the several ways in which new ICTs differ from traditional modes of communication and information exchange (including face-to-face) in the configuration of constraints and processes available to those seeking to communicate in a growth node (not only in inward-focused transactions but also in outbound transactions, i.e., interactions with partners outside the growth node).

In investigating the role of ICTs in facilitating growth nodes, it may be helpful to conceptualize presence as "social richness." Thus, for both inter- and intraorganizational communication, "presence" is associated with whether a medium enables the reproduction of the capabilities for comprehending and utilizing information. ICTs may differ in the extent to which they (a) can overcome various communication constraints of time, location permanence, distribution, and distance, (b) transmit the social, symbolic, and nonverbal cues of human communication, and (c) convey useable information.

ICTs that are high in presence or social richness enable users to adjust more precisely to physical cues, i.e., facial expressions, gestures, vocal tones, etc., and to maximize the efficacy of interpersonal communication. Visual communication generally is believed to have more social presence than verbal (audio) communication, which, in turn, is expected to embody more presence than written text. These observations are important for understanding how ICTs may influence relationships between spatial proximity and "experience proximity" or "mediated presence."

The richness of information in face-to-face environments is important because usually this enables actors to quickly repair problems or gaps that arise in information exchanges. Depending on the social conventions operating within virtual networks, repair is also easy, or even much easier (as in open source development networks where very high levels of shared skills can be assumed). But this may not be the case, or at least not in the same way, with many other types of knowledge that might in principle be shared virtually, e.g., non-ICT product designs, marketing strategies, organizational management and negotiating skills, etc.

Research has shown that higher resolution images in a video conferencing system elicited reports of greater "communicative" presence. Many presence-evoking ICTs are promoted as enabling people to more efficiently accomplish specific tasks (i.e., transmit knowledge). However, despite the fact that these technologies often enable tasks to be completed in a new way, there is relatively little research available to indicate the extent to which these new ways are more effective or efficient in transmitting knowledge than older, more traditional methods. Although one of the most important groups of tasks for which presence evoking media have been designed and used involves specialized skills training (i.e., flight simulation), more research is required to identify the characteristics of tasks for which presence actually enhances knowledge transfer and acquisition.

Key questions for the research agenda are, therefore, for what types of knowledge are virtual networks more likely to be sustained as effective means of sharing, developing, and repairing knowledge and for which types of knowledge are virtual networks more likely to need supplementing or replacing by face-to-face forms of information exchange? Answering these questions calls for comparative studies of practices within and between organizations in order to elicit principles that might influence the emergence of growth nodes. It also requires an analysis of different types of ICT applications and their appropriation by users including firms, universities, civil society organizations, etc. within their various social networks.

Shared ICT Infrastructure

The logic for concentrating and sharing resources in a cluster can be extended to the ICT infrastructure and related ICT services in growth nodes. In this case, the potential benefits go beyond local access to technology and know-how. Indeed, achieving some interorganizational ICT architecture, with standardized interfaces, flexible access, and shared elements, can lead to significant benefits in terms of interoperability and flexibility. This should be of special interest to SMEs since it would lower their barriers to adopt ICT and allow them to fully engage in growth node network(s).

The working hypothesis is that enterprises in growth nodes will be able to form ad hoc, temporary alliances, partnerships, and collaboration with partners within and beyond clusters (or growth nodes). Consequently, the need for ad hoc and flexible interorganizational information exchange is paramount. To support this scenario, a next-generation ICT interorganizational network infrastructure is likely to be required. Such a network infrastructure would provide a dynamic aggregation of network and software services to facilitate such dynamic interaction between business partners and other institutions in the growth node.

The adoption of Internet-based technologies for business, where business services and software components are supported by a pervasive software environment, can be designated as a *business digital ecosystem (BDE)*. The key elements of the BDE structure are software components and agents, which show evolutionary and self-organizing behavior, i.e., they are subject to evolution and to self-selection based on their ability to adapt to the local business requirements (Nachira, 2002).

Recognizing the need for such an infrastructure, a growth node research agenda would include the design and development of a DBE platform as well as its subsequent testing in several regions that exhibit growth node potential. Such a project would take into consideration the three aspects: technology, e-business models and services, and interorganizational knowledge sharing. The holistic approach requires fundamental research on self-organization in complex adaptive systems (CASs) and on network architectures. The following topics could be considered:

• Pervasive, adaptive, self-configuring, and self-healing network software architectures
• Semantic discovery and registering applications
• Distributed security and federated network identity

• Dynamic component composition, software component and knowledge sharing on the network
• Interoperability
• Multiagents, behavior of complex systems and agent communities
• Grid technology
• Semantic web, knowledge sharing and cooperation mechanisms, ontologies, business process modeling, and integration.

The approach would be evolutionary, based on continuous development and adoption of DBE architectures. The project would likely entail design, simulation, development, and field tests. The testing should be articulated as a series of field experiments. These would be used to test hypotheses relating to the role of ICT and the development of growth nodes.

The Policy Perspective

Over the past few years, the cluster concept has found a ready audience among policymakers at all levels. The argument is not that governments can create clusters or growth nodes but that they can help to provide the business, innovative, and institutional environments necessary for their success.

The first step is usually to identify the clusters in the region or the country. Organizations and government agencies that view their regions as clustered production systems are predisposed to tailoring existing policies and programs to that model and, in some instances, to creating new strategies. The most common policy levers are those that alter the way agencies organize and deliver their services, work with employers, recruit businesses, and allocate resources. But the most popular goals are to market a political region and attract businesses and highly educated and skilled people. The following are examples of cluster-based policy levers:

• *Promotion.* Giving "official" recognition to a cluster represents a form of collective marketing of the cluster and its products and creates avenues for more effective lobbying efforts.
• *Investment.* Often the public sector is interested in identifying clusters because it represent a way to increase the odds of attracting investment.

• *Education.* Governments adapt the appropriate degree of specialization in higher education to meet the needs of clusters and regional economies.

• *Social cohesion.* Clusters offer ways to restructure equity policies to more effectively serve less-advantaged regions and lower income and less-educated populations.

• *Collective awareness.* A common intervention to strengthen clusters is to form and empower "cluster councils." These get companies to articulate a collective vision, foster cooperation, and create awareness that they represent a larger regional economic entity.

• *Organization of services.* A solution to help SMEs is to integrate the services within a cluster either by creating a hub (*one-stop shop*) or creating a set of intermediaries (*knowledge brokers*) to serve as linking agents.

Toward Growth Node Policies

To apply the concept of growth nodes to policy, one must believe not just that growth nodes exist but also that they can be brought into existence. The question is how to identify existing growth nodes or clusters that have nodal potential.

As discussed earlier, growth node research should develop constructs for identifying, characterizing, and measuring growth nodes and their effects. These instruments would then be used to identify growth nodes in some European countries or regions and to establish whether the regional/national economies can be effectively examined through the growth node lens, and if so, whether policymakers can more accurately identify market imperfections, find pressure points, envisage or pinpoint systemic failures, and determine what interventions can have the greatest impacts.

Assuming that growth nodes can exist in reasonable numbers (i.e., that they are significant for policy analysis), the second question is whether the traditional cluster-based policies apply to growth nodes. Are different policies required when regions are examined from the perspective of growth nodes?

If public policymakers proactively integrate advanced ICTs to link local geographically clustered firms and other organizations beyond their immediate regional surroundings, there may be substantial opportunities for a departure from the conventional pattern. Global, national, regional, and local ICT links and information flows may, in fact, fuel an "innovative milieu" and help to provide the catalyst for the social learning cycle that gives rise to successful and enduring growth nodes.

Growth nodes differ from clusters in their nodality and the enabling role of ICTs. ICTs provide a new means of linking up local places and regions within networks of organizations. Inclusion in the network requires an adequate local technological infrastructure, a system of ancillary firms and other organizations providing support services, a specialized labor market, and a system of services required by the professional labor force. Thus, another set of relevant policies refers to actions to facilitate the adoption and usage of ICTs by SMEs. As discussed above, an interorganizational infrastructure based on the BDE principles might lower the barriers to ICT usage by SMEs and facilitate access to the wider business network in a cluster or growth node. If this potential exists, a policy that sponsors the development, testing, and deployment of a BDE-like infrastructure may prove to be a more effective way to support SMEs than the take-up actions sponsored by earlier initiatives.

Another policy topic that should be addressed has to do with growth node dynamics. Porter (1990) has argued that fast-growing, innovative, geographically clustered firms—"hot spots"—often turn into "'blind spots." More recently, other researchers have shown how rapidly the fortunes of "hot spots" can be reversed (Pouder and St. John, 1996), leading to the deterioration of formerly vibrant and innovative regions including both urban and rural agglomerations. Firms first begin to cluster and to forge a "hot spot" identity, but convergence of clustered firms ultimately leads to a "hot spot" failure.

Growth node research should provide a conceptual framework that helps researchers and policymakers to clarify the conditions under which emergent growth node outcomes might be expected that are departures from the hot spot/blind spot cycle. The perspective on emergent complex adaptive systems opens the possibility for the discovery of key factors and policies that encourage divergence from historical pathways that are believed to characterize regional and local clusters.

Conclusion

This chapter has introduced the idea of a "growth node" and has positioned it as a further evolution of the cluster concept that emphasizes external networking dimensions as well as the cross-industry knowledge transfer and social learning conventionally associated with clusters. A series of research questions have been raised: How does one identify and measure growth nodes? How do growth nodes contribute to regional development? What are the conditions that sustain and develop them? What are the desired properties of the enabling IT infrastructure? What policies are likely to foster emergent growth nodes in Europe?

In addressing these questions, existing literature on the theory and practice of clusters has been reviewed, and new research areas have been identified in order to advance the understanding and implications of growth nodes. The process has led to the identification of key research issues that constitute the basis for a preliminary research agenda on ICT-enabled growth nodes. This research agenda is summarized in the appendix.

The proposed research agenda addresses the question of how the deployment and use of new ICTs might modify our understanding of what constitutes a viable and sustainable growth node. There is relatively clear evidence that the spread of global and local networks is creating the potential for a new dynamic designated as "nodality." What is not clear is what mix of ICT production and application in any given cluster will give rise to the emergent properties of growth nodes. The research agenda seeks to clarify how new insights in this area can be generated. It calls for investment in producing appropriate data sets and in further conceptualizing the foundation principles of growth nodes as augmented clusters.

Advanced ICTs provide a new means of linking up local places and regions within a "network of networks." Inclusion in these networks requires an adequate local technological infrastructure, a system of ancillary firms and other organizations providing support services, a specialized labor market, and a structure of the services required by the professional labor force. Positive synergies are generated partly by the dynamics of social networks within a given territorially bounded place and partly by the global interconnectedness of that place with many other places. Both social and technical networks seem to play essential roles in whether agglomeration economies emerge out of networking synergies and their interactions with the features of a given cluster of economic and social activity.

In addition, the research agenda is designed to consider the new rules of the game that will be required to equitably foster the development of ICT-enabled growth nodes under a regime of intensifying interregional competition. Regional and local policymakers as well as those working at national and European levels are searching for new means of governance that will not only foster competitiveness in global markets but also achieve sustainable growth within the framework of the European social model.

Many regions are confronting the decline of longstanding industrial sectors without having identified the means to dynamize "new economy" developments on a scale sufficient to ensure sustained growth and a favorable process of social cohesion that is equitable for all. There is a need for finding mechanisms to counter the threats of rising unemployment and increasing divides within regions that could see their efforts to encourage successful clustering eroded by the cycle of hot spot/blind spot development. The research agenda calls for a rethinking of past strategies that is consistent with what many now believe to be an era of increasing uncertainty, heightened risk, and a new economic dynamic that makes winners out of those who can find the resources to compete and collaborate effectively not only within clusters but also far beyond the boundaries of localized economic activity.

Appendix: Research Agenda for ICT-Enabled Growth Nodes in Europe

1. Analysis of existing clusters in several member states or geographical areas in Europe to assess their "nodality," and, consequently, their potential for designation as growth nodes.

A composite measure based on a combination of growth node attributes (e.g., external links, reach, knowledge intensity, employment structure, ICT infrastructure) would indicate the degree and/or type of "nodality" of the target clusters (i.e., the potential growth node candidates). Such a project would help refine the definition of growth nodes and develop an appropriate typology.

2. *Development of practical mapping tools to identify and measure growth nodes for a given region or national economy.* These tools would include the following:

I. A knowledge mapping tool would allow identification of a growth node's critical knowledge assets and assessment of their strategic potential. This tool would allow a region's competitive position in the knowledge economy to be gauged and suggest possible avenues for strategic development.

II. A learning mapping tool would support analysis of the learning profile of a growth node in terms of its scanning, structuring, diffusion, and absorption capacities. This tool would allow assessment of how effectively the region is able to create and exploit knowledge assets.

III. An institutional mapping tool would help establish the extent to which local institutions and cultures are facilitating or impeding the learning processes that lead to the successful exploitation of knowledge resources.

3. *Assessment of usability and utility of the tools to identify growth nodes, in order to establish whether selected regional/national economies can be examined usefully through the growth node lens.*

The tools to measure growth nodes (agenda item No. 2) should be tested by regional authorities in some European countries or regions. The idea is to ascertain whether policymakers find them useful tools for policy analysis and design. An important output of such research would be a refined policy toolbox. In this regard, the project could explore, test, and propose complementary measures/instruments that might be relevant indicators from a policy perspective.

4. *Analysis of designated growth nodes to validate the proposition that outside resources can be made available inside the growth node, thereby stimulating growth.*

Growth nodes capitalize on the potential offered by ICTs to achieve "experienced proximity" and "presence" between points that are distant from each other in space. The hypothesis is that a growth node has the effect of making outside resources—cognitive, institutional, cultural, and material—available inside the growth node, thus stimulating its growth beyond that which could be achieved by employing local resources alone.

5. *Identification of factors contributing to the development and sustainability of growth nodes.*

A longitudinal study is proposed to track the evolution of a set of growth nodes. A multivariate analysis could be used to determine which factors are significant. The analysis would test whether changes in any of the factors over time translate into changes in the dependent variables (growth node performance measures).

6. *Assessment of virtual network properties and typologies to establish what network types are more likely to be sustained as effective means of sharing, developing, and repairing knowledge.*

This implies an analysis of different types of existing and emerging ICT applications (including new broadband services) and their appropriation by users including firms, universities, civil society organizations, etc. within their various social networks. The goal is to establish what types of knowledge-based virtual networks are more likely to need supplementing or replacing by face-to-face forms of information exchange.

7. *Design, development, and test of an infrastructure providing a dynamic aggregation of network and software services (business digital ecosystem) to facilitate dynamic interaction between business partners and other institutions within and beyond the growth node.*

Research on the network infrastructure is likely to cover topics such as self-configuring architectures, dynamic component composition, multi-agents, behavior of complex systems, knowledge sharing mechanisms, business process modeling, and integration. The implementation approach would be based on the iterative process of prototyping, development, and adoption. The evolving ecosystem will allow for trials and early adoption in test-bed growth nodes.

8. *Policies for growth node: Establish whether traditional cluster-based policies also apply to growth node development, and also whether new policies are required.*

The project would entail (1) the adoption of such policies on an experimental basis and (2) analysis of their effects over time, (3) testing the hypothesis that internodal ICT links enable a virtual "innovative milieu" and its necessary social learning cycle. If public policymakers proactively integrate advanced ICTs to link local geographically clustered firms and other organizations beyond their immediate regional surroundings, there may be substantial opportunities for a departure from the conventional pattern.

9. *Test whether the availability and use of a BDE interorganizational infrastructure may lower the barriers to ICT adoption by SMEs and facilitate SME access to the wider business network in a cluster or growth node.*

If this potential exists, a policy that sponsors the development, testing, and deployment of a DBE infrastructure may prove to be a more effective way to support SMEs than some of the take-up actions sponsored by earlier initiatives.

10. *Ascertain the conditions under which one might expect emergent growth node outcomes that depart from the "hot spot"/"blind spot" cycle (i.e., rapid success followed by rapid failure).*

This research is concerned with growth node dynamics, their emergence, and their potentially chaotic behavior as complex systems. The perspective on emergent complex adaptive systems opens the possibility for the discovery of key factors and policies that encourage divergence from historical pathways that are believed to characterize regional and local clusters.

Notes

1. Code named G-NIKE (for Growth Nodes in a Knowledge Based Europe), the project studies regional and interregional IT-enabled growth nodes in order to understand their role in regional development. It is a collaborative effort of Universitat Oberta de Catalunya (UOC), Spain; the London School of Economics, UK; European Institute for the Media, Germany; Tilburg University, Netherlands; and the University of Tampere, Finland.

2. The concept of a node in a regional context was initially developed by the French geographer Vidal de la Blanche (1845–1918), who borrowed the concept

of *nodality* from the British geographer Mackinder to indicate the major cross-roads that generate change of all kinds and that, as a result, have the greatest power of organization.

References

Boekema, F. W. M., K. Morgan, S. H. P. Bakkers, and R. P. J. H. Rutten (2000) *Knowledge, Innovation and Economic Growth. The Theory and Practice of Learning Regions.* Cheltenham: Edward Elgar Publishing.

Boisot, M. (1995) *Information Space: A Framework for Learning in Organizations, Institutions and Cultures.* London: Routledge.

Boisot, M. (1998) *Knowledge Assets: Securing Competitive Advantage in the Information Economy.* Oxford: Oxford University Press.

Cortright, J. (2000) "New Growth Theory, Technology, and Learning: A Practitioner's Guide to Theories for the Knowledge Based Economy." Draft report to the Economic Development Administration, Washington, D.C.: U.S. Department of Commerce.

Douglas, M. (1973) *Natural Symbols.* Middlesex: Penguin Books.

Krugman, P. (1991) *Geography and Trade.* Cambridge, Mass.: MIT Press.

Krugman, P. (1996) "The localisation of the global economy." In *Internationalism.* Cambridge, Mass.: MIT Press.

Landabaso, M., C. Oughton, and K. Morgan (1999) "Innovation networks—Concepts and challenges in the European perspective." Paper presented at the Fraunhofer Institute in Karlsruhe, Germany, November 18, 1999.

Moss Kanter, R. (1995) *World Class: Thriving Locally in the Global Economy.* New York: Simon & Schuster.

Nachira, F. (2002) "Towards a network of digital business ecosystems fostering local development." European Commission, IST Programme, Brussels, September. www.opencontent.org/openpub.

OECD (2001) *Cities and Regions in the New Learning Economy.* Paris: Organization for Economic Cooperation and Development (OECD).

Piore, M., and C. Sabel (1984) *The Second Industrial Divide.* New York: Basic Books.

Porter, M. (1990) *The Competitive Advantage of Nations.* London: Macmillan.

Porter, M. (1998) "Clusters and the new economics of competition." *Harvard Business Review,* November/December.

Porter, M. (2001) "Clusters of innovation: regional foundations of U.S. competitiveness." Washington, D.C.: Council on Competitiveness.

Pouder, R., and C. St. John (1996) Hot spots and blind spots: Geographical clusters of firms and innovation. *Academy of Management Review* 21(4): 1192.

Putnam, R. (1993) "Social capital and institutional success." In *Making Democracy Work: Civic traditions in modern Italy*, ed. R. Putnam, R. Leonardi, and R. Nanetti. Princeton: Princeton University Press, pp. 163–185.

Rosenfeld, S. A. (1997) "Bringing business clusters into the mainstream of economic development." *European Planning Studies* 5(1): 3–23.

Rosenfeld, S. A. (2002) *Creating smart systems: A guide to cluster strategies in less favored regions*. Carrboro, N.C.: Regional Technology Strategies.

Sassen, S. (2002) *Global Networks, Linked Cities*. New York: Routledge.

Saxenian, A.-L. (1994) *Regional Advantage: Culture and Competition in Silicon Valley and Route 128*. Cambridge, Mass.: Harvard University Press.

9

Supernetworks: Paradoxes, Challenges, and New Opportunities

Anna Nagurney

Background

Throughout history, networks have served as the foundation for connecting humans to one another and their activities. Roads were laid, bridges built, and waterways crossed, so that humans, be they on foot, animal, or vehicle, could traverse physical distance. The airways were ultimately conquered through flight. Humans, separated by physical distance, communicated with one another, in turn, using the available means of the period, from smoke signals, drum beats, and pigeons to the telegraph, telephone, and computer networks of today.

Today network systems provide the infrastructure and foundation for the functioning of our societies and economies. They come in many forms and include physical networks such as transportation and logistical, communication, and energy and power networks, as well as more *abstract* networks comprising economic and financial, environmental, social, and knowledge networks.

For example, transportation networks give us the means to cross physical distance in order to conduct our daily activities. They provide us with access to both food and consumer products and come in a myriad of forms: road, air, rail, or waterway. According to the U.S. Department of Transportation (1999), the significance of transportation in dollar value alone as spent by U.S. consumers, businesses, and governments was $950 billion in 1998.

Communication networks, in turn, allow us to communicate with friends and colleagues and to conduct the necessary transactions of life. They, through such innovations as the Internet, have transformed the manner in which we live, work, and conduct business today.

Communication networks allow the transmission of voice, data/ information, and/or video and can involve telephones and computers as well as satellites and microwaves. The trade publication *Purchasing* (2000) reports that corporate buyers alone spent $517.6 billion on telecommunications goods and services in 1999.

Energy networks, in addition, are essential to the very existence of the *network economy* and help to fuel not only transportation networks but in many settings also communication networks. They provide electricity to run the computers and to light our businesses, oil and gas to heat our homes and to power vehicles, and water for our very survival. In 1995, according to the U.S. Department of Commerce (2000), the energy expenditures in the United States were $515.8 billion.

Financial networks supply businesses with the resources to expand, to innovate, and to satisfy the needs of consumers. They allow individuals to invest and to save for the future for themselves and for their children and for governments to provide for their citizens and to develop and enhance communities.

Information technology has transformed the ways in which individuals work, travel, and conduct their daily activities, with profound implications for existing and future networks. Moreover, the *decision-making process* itself has been altered due to the addition of alternatives and options that were not, heretofore, possible or even feasible. The boundaries for decision making have been redrawn since individuals can now work from home or purchase products at work. Indeed, we now live in an era in which the freedom to choose is weighted by the immensity of the number of choices and possibilities: Where should one live? Where should one work? And when? How should one travel? Or communicate? And with whom? Where should one shop? And how?

Managers can now locate raw materials and other inputs from suppliers through information networks in order to maximize profits while simultaneously ensuring timely delivery of finished goods. Financing for their businesses can be obtained online. Individuals, in turn, can obtain information about products from their homes and make their purchasing decisions accordingly. How should businesses avail themselves of new opportunities made possible through information technology? What kind of supply chain network structures will allow for greater productivity, efficiencies? How can firms more effectively compete and when and with whom should they cooperate? Finally, what are the ramifications of the decisions made in the new networked economy for the environment and its sustainability?

The reality of today's networks includes a large-scale nature and complexity, increasing congestion, and alternative behaviors of users of the networks, as well as interactions between the networks themselves, notably between transportation and telecommunication networks. Indeed, recent historical events have dramatically and graphically illustrated the interconnectedness, interdependence, and vulnerability of organizations, business, and other enterprises on one another and on such critical network infrastructure systems as transportation and telecommunications. The decisions made by the users of the networks, in turn, affect not only the users themselves but others, as well, in terms of profits and costs, timeliness of deliveries, the quality of the environment, etc.

In this chapter, we argue that new paradigms are needed to capture the complexities of decision making in the Information Age. In particular, we believe that the concept of *supernetworks is* sufficiently general and yet elegantly compact to formalize such decision making. "Super" networks are networks that are "above and beyond" existing networks, which consist of nodes, links, and flows, with nodes corresponding to locations in space, links to connections in the form of roads, cables, etc., and flows to vehicles, data, etc. Supernetworks are conceptual in scope, graphical in perspective, and, with the accompanying theory, predictive in nature.

In particular, the supernetwork framework captures, in a unified fashion, the decision making facing a variety of economic agents including consumers and producers as well as distinct intermediaries in the context of today's networked economy. The decision-making process may entail weighting trade-offs associated with the use of transportation versus telecommunication networks. The behavior of the individual decision makers is modeled as well as their interactions on the complex network systems with the goal of identifying the resulting flows and prices.

The origins of supernetworks can be traced to the study of transportation and telecommunication networks, as well as economic and financial networks, and, interestingly, to biology. Here we take the synthetic approach promulgated by Nagurney and Dong (2002). In figure 9.1, we provide a conceptualization of supernetworks that emphasizes the interdependence of distinct network systems.

Classical Networks

For definiteness, we first present in table 9.1 some basic *classical* networks and the associated nodes, links, and flows. By *classical* network is meant a network in which the nodes correspond to physical locations in space and the links to physical connections between the nodes.

We note that the topic of networks and the management thereof dates to ancient times with examples including the publicly provided Roman road network and the "time of day" chariot policy, whereby chariots were banned from the ancient city of Rome at particular times of day (Banister and Button, 1993). The formal study of networks, consisting of *nodes*, *links*, and *flows*, in turn, involves how to model such applications (as well as numerous other ones) as mathematical entities, how to study the models qualitatively, how to design algorithms to solve the resulting models effectively to enable the ultimate prediction of the

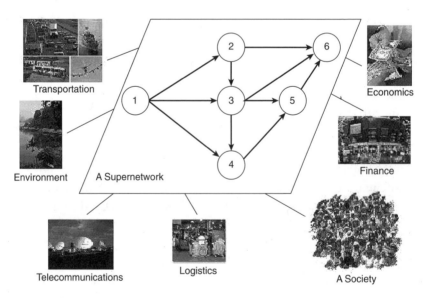

Figure 9.1
Conceptualization of a supernetwork.

Table 9.1
Examples of classical networks

Network system	Nodes	Links	Flows
Transportation			
Urban	Intersections, homes, places of work	Roads	Autos
Air	Airports	Airline routes	Planes
Rail	Rail yards	Railroad track	Trains
Manufacturing and logistics	Distribution points, processing points	Routes Assembly line	Parts, products
Communication	Computers	Cables	Messages
	Satellites	Radio	Messages
	Phone exchanges	Cables, microwaves	Voice, video
Energy	Pumping stations	Pipelines	Water
	Plants	Pipelines	Gas, oil

underlying variables, and, finally, how to design appropriate policy instruments. The study of networks is necessarily *interdisciplinary* in nature due to their breadth of appearance and is based on techniques from applied mathematics, computer science, and engineering with applications as varied as finance and even biology. Network models and tools are widely used by businesses and industries as well as governments today (Ahuja, Magnanti, and Orlin, 1993; Nagurney and Siokos, 1997; Nagurney, 1999, 2000a; and references therein).

Basic examples of network problems are the *shortest path* problem, in which one seeks to determine the most efficient path from an origin node to a destination node; the *maximum flow* problem, in which one wishes to determine the maximum flow that one can send from an origin node to a destination node, given that there are capacities on the links that cannot be exceeded; and the *minimum cost flow* problem, where there are both costs and capacities associated with the links and one must satisfy the demands at the destination nodes, given supplies at the origin nodes, at minimal total cost associated with shipping the flows and subject to not exceeding the arc capacities. Applications of the shortest path problem are found in transportation and telecommunications, whereas the maximum flow problem arises in machine scheduling and network reliability settings, with applications of the minimum cost flow problem ranging from warehousing and distribution to vehicle fleet planning and scheduling.

Networks also appear in surprising and fascinating ways for problems, which initially may not appear to involve networks at all, such as a variety of financial problems and in knowledge production and dissemination. Hence, the study of networks is not limited only to physical networks, where nodes coincide with locations in space, but applies also to abstract networks. The ability to harness the power of a network formalism provides a competitive advantage since

• Many present-day problems are concerned with flows, be they material, human, capital, or informational, over space and time and, hence, ideally suited as an application domain for network theory,
• One may avail oneself of a graphical or visual depiction of different problems,
• One may identify similarities and differences in distinct problems through their underlying network structure, and
• One may apply efficient network algorithms for problem solutions to predict the vehicular, commodity, financial, and informational flows.

The Realities of Today's Networks

The characteristics of today's networks include a large-scale nature and complexity of network topology; congestion; alternative behavior of users of the network, which may lead to paradoxical phenomena; and the interactions among networks themselves such as in transportation versus telecommunications networks. Moreover, policies surrounding networks today may have a major impact not only economically but also socially.

Large-Scale Nature and Complexity

Many of today's networks are characterized by both a large-scale nature and complexity of the underlying network topology. For example, in Chicago's Regional Transportation Network, there are 12,982 nodes, 39,018 links, and 2,297,945 origin/destination (O/D) pairs (see Bar-Gera, 1999), whereas in the Southern California Association of Governments model there are 3,217 origins and/or destinations, 25,428 nodes, and 99,240 links, plus 6 distinct classes of users (Wu, Florian, and He, 2000).

In terms of the size of existing telecommunications networks, AT&T's domestic network has 100,000 origin/destination pairs (Resende, 2000), whereas in their detail graph applications in which nodes are phone numbers and edges are calls, there are 300 million nodes and 4 billion edges (Abello, Pardalos, and Resende, 1999).

Congestion

Congestion is playing an increasing role not only in transportation networks but also in telecommunication networks. For example, in the case of transportation networks in the United States alone, congestion results in $100 billion in lost productivity, whereas the figure in Europe is estimated to be $150 billion. The number of cars is expected to increase by 50% by 2010 and to double by 2030 (see Nagurney, 2000a).

In terms of the Internet, with over 275 million present users, the Federal Communications Commission reports that the volume of traffic is doubling every 100 days, which is remarkable given that telephone traffic has typically increased only by about 5% a year (Labaton, 2000). As individuals increasingly access the Internet through wireless communication such as handheld computers and cellular phones, experts fear that the heavy use of airwaves will create additional bottlenecks and congestion that could impede the further development of the technology.

System Optimization versus User Optimization and the Braess Paradox

In many of today's networks, not only is congestion a characteristic feature leading to nonlinearities, but the behavior of the users of the networks themselves may be that of noncooperation. For example, in the case of urban transportation networks, travelers select their routes of travel from an origin to a destination so as to minimize their own travel cost or time, which although "optimal" from an individual's perspective (user optimization) may not be optimal from a societal one (system optimization), where one has control over the flows on the network and, in contrast, seeks to minimize the total cost in the network and, hence, the total loss of productivity (see, e.g., Wardrop, 1952; Beckmann, McGuire, and Winsten, 1956; Dafermos and Sparrow, 1969; and Nagurney, 1999). Consequently, in making any kind of policy decisions in such networks one must take into consideration the users of the particular network. Indeed, this point is vividly illustrated through a famous example known as the Braess paradox, in which it is assumed that the underlying behavioral principle is that of user optimization. In the Braess (1968) network, the addition of a new road with no change in the travel demand results in all travelers in the network incurring a higher travel cost and, hence, being worse off! The increase in travel cost on the paths is due, in part, to the fact that in this network two links are shared by distinct paths and these links incur an increase in flow and associated cost. Hence, Braess's paradox is related to the underlying topology of the networks. One may show, however, that the addition of a path connecting an O/D pair that shares no links with the original O/D pair will never result in Braess's paradox for that O/D pair.

Interestingly, as reported in *The New York Times* by Kolata (1990), this phenomenon has been observed in practice in the case of New York City when in 1990, 42nd Street was closed for Earth Day and the traffic flow actually improved. Just to demonstrate that it is not purely a New York or U.S. phenomenon concerning drivers and their behavior, an analogous situation was observed in Stuttgart where a new road was added to the downtown but the traffic flow worsened and, following complaints, the new road was torn down (see Bass, 1992).

This phenomenon is also relevant to telecommunications networks (see Korilis, Lazar, and Orda, 1999) and, in particular, to the Internet, which is another example of a "noncooperative network," and therefore network tools have wide application in this setting as well, especially in terms of congestion management and network design (see also Cohen and Kelly, 1990).

Network Interactions

Clearly, one of the principal facets of the network economy is the interaction among the networks themselves. For example, the increasing use of electronic commerce especially in business-to-business transactions not only is changing the utilization and structure of the underlying logistical networks but is also revolutionizing how business itself is transacted and the structure of firms and industries. Cellular phones are being used as vehicles move dynamically over transportation networks, resulting in dynamic evolutions of the topologies themselves. The unifying concept of supernetworks with associated methodologies allows one to explore the interactions among such networks as transportation and telecommunication networks, as well as financial networks.

More Paradoxes: Transportation and Telecommunications versus the Environment

The demand for transportation on the one hand with a growing realization of the associated negative externalities due, for example, to congestion and pollution is raising questions of sustainability of the transportation infrastructure. For example, 15% of the world's emissions of carbon dioxide are due to motor vehicles, as are 50% of the emissions of nitrogen oxide and 90% of the carbon monoxide. The necessity of identifying the behavior of the users of such networks coupled with the interactions between transportation and environmental networks is vividly illustrated through several transportation/environmental paradoxes identified by Nagurney (2000a, b). For example, the addition of a new link (road) to a transportation network may result in an increase in vehicular emissions with no change in travel demand; a decrease in travel demand associated with a particular origin/destination pair of nodes of travel may result in an increase in emissions; and a reallocation of travelers from a mode of higher emissions to that of one with lower emissions may actually result in an increase in total emissions!

Recently, Nagurney and Dong (2001) identified paradoxes in networks with zero emission links such as telecommunication networks. In particular, they showed through simple examples how the addition of a zero emission link may result in an increase in total emissions with no change in demand and how a decrease in demand on a network with a zero emission link may result in an increase in total emissions. Hence, one must incorporate the network topology, the relevant cost and demand structure, as well as the behavior of the users of the particular transportation/telecommunication network into any policy aimed at pollution abatement! These paradoxes further illustrate the interconnectivity among distinct network systems and that they cannot be studied simply in isolation.

Supernetworks and Applications

Supernetworks may be comprised of such networks as transportation, telecommunication, logistical, and financial networks, among others. They may be *multilevel*, as when they formalize the study of supply chain networks, or *multitiered*, as in the case of financial networks with intermediation. Furthermore, decision makers on supernetworks may be faced with multiple criteria and, hence, the study of supernetworks also includes the study of multicriteria decision making. In table 9.2, some specific applications of supernetworks are given, upon which we elaborate below.

In particular, the supernetwork framework allows one to formalize the alternatives available to decision makers, to model their individual behavior, typically characterized by particular criteria that they wish to optimize, and, ultimately, to compute the flows on the supernetwork, which may consist of product shipments, travelers between origins and destinations, and financial flows, as well as the associated "prices." Hence, the concern is with *human decision making* and how the supernetwork concept can be utilized to crystallize and inform in this dimension.

Telecommuting versus Commuting Decision Making

According to Hu and Young (1996), person-trips and person-miles of commuting increased between 1990 and 1995, both in absolute terms and as a share of all personal travel. Constituting 18% of all person-

Table 9.2
Examples of supernetwork applications

Telecommuting/commuting decision making
Teleshopping/shopping decision making
Supply chain networks with electronic commerce
Financial networks with electronic transactions

trips and 22% of all person-miles in 1995, commuting is the single most common trip purpose. Furthermore, as argued by Mokhtarian (1998; see also Mokhtarian, 1991), it is very likely that a greater proportion of commute trips rather than other types of trips will be amenable to substitution through telecommunications. Consequently, telecommuting most likely has the highest potential for travel reduction of any of the telecommunication applications. Therefore, the study of telecommuting and its impacts is a subject worthy of continued interest and research. Furthermore, recent legislation that allows federal employees to select telecommuting as an option (see United States, 2000), underscores the practical importance of this topic.

The decision makers in the context of this application are travelers, who seek to determine their *optimal* routes of travel from their origins, which are residences, to their destinations, which are their places of work. Note that, in the supernetwork framework, a link may correspond to an actual physical link of transportation or an abstract or virtual link corresponding to a telecommuting link. Furthermore, the supernetwork representing the problem under study can be as general as necessary and a path may consist of a set of links corresponding to physical and virtual transportation choices such as would occur if a worker were to commute to a work center from which she could then telecommute. In figure 9.2, a conceptualization of this idea is provided.

Observe that, in figure 9.2, nodes 1 and 2 represent locations of residences, whereas node 6 denotes the place of work. Work centers from

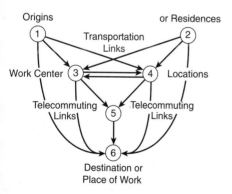

Figure 9.2
A supernetwork conceptualization of commuting versus telecommuting.

which workers can telecommute are located at nodes 3 and 4, which also serve as intermediate nodes for transportation routes to work. The links (1,6), (3,6), (4,6), and (2,6) are telecommunication links depicting virtual transportation to work via telecommuting, whereas all other links are physical links associated with commuting. Hence, the paths (1,6) and (2,6), consisting of individual single links, represent "going to work" virtually, whereas the paths consisting of the links (1,3), (3,6) and (2,4), (4,6) represent first commuting to the work centers located at nodes 3 and 4, from which the workers then telecommute. Finally, the remaining paths represent the commuting options for the residents at nodes 1 and 2. The conventional travel paths from node 1 to node 6 are as follows: (1,3), (3,5), (5,6); (1,3), (3,4), (4,5), (5,6); (1,4), (4,5), (5,6); and (1,4), (4,3), (3,5), (5,6). Note that there may be as many classes of users of this network as there are groups who perceive the trade-offs among the criteria in a similar fashion.

Of course, the network depicted in figure 9.2 is illustrative, and the actual network can be much more complex, with numerous paths depicting the physical transportation choices from one's residence to one's work location. Similarly, one can further complexify the telecommunication link/path options. Also, we emphasize that a *path* within this framework is sufficiently general to also capture a choice of mode, which, in the case of transportation, could correspond to buses, trains, or subways (that is, public transit) and, of course, to the use of cars (i.e., private vehicles). Similarly, the concept of path can be used to represent a distinct telecommunications option.

In this framework, since the decision makers are travelers, the path flows and link flows by class would correspond, respectively, to the number of travelers of the class selecting a particular path and link.

We now turn to a discussion of the criteria that one can expect to be reasonable in the context of decision making in this particular application. The first multicriteria traffic network models, due to Schneider (1968) and Quandt (1967), considered two criteria, travel time and cost. Of course, telecommuting was not truly an option in those days. Dafermos (1981), Leurent (1993), Marcotte (1998), and Nagurney (2003) also considered those two criteria but handled congestion on the networks as well. Nagurney, Dong, and Mokhtarian (2000), in turn, focused on the development of an integrated multicriteria network equilibrium model, which was the first to consider telecommuting versus commuting trade-offs. They considered three criteria: travel time, travel cost, and an opportunity cost to trade off the opportunity cost associated with not being able to physically interact with colleagues. Further developments, including the incorporation of additional decision-making criteria, including safety, and discussion of the associated analytical methodologies, can be found in Nagurney and Dong (2002) and in Nagurney, Dong, and Mokhtarian (2002).

The behavioral assumption is that travelers of a particular class are assumed to choose the paths associated with their origin/destination pair so that the generalized cost on that path, which consists of a weighting of the different criteria (which can be different for each class of decision maker and can also be link dependent), is minimal. An equilibrium is assumed to be reached when the multicriteria network equilibrium conditions are satisfied whereby only those paths connecting an O/D pair are employed such that the generalized costs on the paths, as perceived by a class, are equal and minimal.

Modeling Teleshopping versus Shopping Decision Making

Here a multicriteria network equilibrium model for teleshopping versus shopping is described. The model generalizes the model proposed in Nagurney, Dong, and Mokhtarian (2001). For further details, including numerical examples, see Nagurney and Dong (2002).

Although there is now a growing body of transportation literature on telecommuting (Mokhtarian, 1998), the topic of teleshopping, which is a newer concept, has received less attention to date. In particular, shopping refers to a set of activities in which consumers seek and obtain information about products and/or services, conduct a transaction transferring ownership or right to use, and spatially relocate the product or service to the new owner (Mokhtarian and Salomon, 2002). Teleshopping, in turn, refers to a case in which one or more of those activities is conducted through the use of telecommunication technologies. Today, much attention is focused on the Internet as the technology of interest, and Internet-based shopping is, indeed, increasing. In this setting, teleshopping represents the consumer's role in B2C electronic commerce.

Although the model is in the context of Internet-based shopping, the model can apply more broadly.

Note that outside the work of Nagurney, Dong, and Mokhtarian (2001, 2002), there has been essentially no study of the transportation impacts of teleshopping beyond speculation (e.g., Gould, 1998; Mokhtarian and Salomon, 2002). Assume that consumers are engaged in the purchase of a product that they do so in a repetitive fashion, say, on a weekly basis. The product may consist of a single good, such as a book, or a bundle of goods, such as food. Assume also that there are locations, both virtual and physical, where the consumers can obtain information about the product. The virtual locations are accessed through telecommunications via the Internet, whereas the physical locations represent more classical shopping venues such as stores and require physical travel to reach.

The consumers may order/purchase the product, once they have selected the appropriate location, be it virtual or physical, with the former requiring shipment to the consumers' locations and the latter requiring, after the physical purchase, transportation of the consumer with the product to its final destination (which we expect, typically, to be his residence or, perhaps, place of work).

Refer to the network conceptualization of the problem given in figure 9.3. We now identify the above concepts with the corresponding network component. The idea of such a shopping network was first proposed by Nagurney, Dong, and Mokhtarian (2001).

Observe that the network depicted in figure 9.3 consists of four levels of nodes, with the first (top) level and the last (bottom) level corresponding to the locations (destinations) of the consumers involved in the purchase of the product. We emphasize that each location may have many consumers. The second level of nodes, in turn, corresponds to the information locations (and where the transactions also take place), with the first set of such nodes representing the virtual or Internet-based locations and the second such set denoting the physical locations of information corresponding to stores, for example. The third level of nodes corresponds to the completion of the transaction, with the first set of such nodes corresponding to Internet sites where the product could have been purchased (and where it has been assumed that information has also been made available in the previous level of nodes) and the second set of such

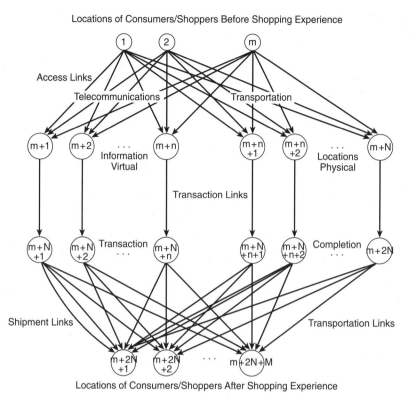

Figure 9.3
A supernetwork framework for teleshopping versus shopping.

nodes corresponding to the completion of the transaction at the physical stores.

We now discuss the links connecting the nodes in the network in figure 9.3. A link connecting a top-level node (consumer's location) to an information node at the second level corresponds to an *access* link for information. The links terminating in the first set of nodes of the second level correspond to telecommunication access links, whereas those terminating in the second set of nodes correspond to (aggregated) transportation links.

As can be seen from figure 9.3, from each second-tier node there emanates a link, which corresponds to a completion of a transaction node. The first set of such links corresponds to virtual orders, whereas the subsequent links denote physical orders/purchases. Finally, there are links emanating from the transaction nodes to the consumers' (final) destination nodes, with the links emanating from the first set of transaction nodes denoting shipment links (since the product, once ordered, must be shipped to the consumer) and the links from the second set representing physical transportation links to the consumers' destinations. Note that, in the case of the latter links, the consumers (after purchasing the product) transport it with themselves, whereas in the former case, the product is shipped to the consumers. Observe that in the supernetwork framework, we explicitly allow for alternative modes of shipping the product, which is represented by an additional link (or links) connecting a virtual transaction node with the consumer's location.

The above network construction captures the *electronic dissemination* of goods (such as books or music, for example) in that an alternative shipment link in the bottom tier of links may correspond to the virtual or electronic shipment of the product.

An origin/destination pair in this network corresponds to a pair of nodes from the top tier in figure 9.3 to the bottom tier. In the shopping network framework, a path consists of a sequence of choices made by a consumer. For example, the path consisting of the links $(1, m + 1)$, $(m + 1, m + N + 1)$, $(m + N + 1, m + 2N + 1)$ would correspond to consumers located at location 1 accessing virtual location $m + 1$ through telecommunications, placing an order at the site for the product, and having it shipped to them. The path consisting of the links $(m, m + N)$, $(m + N, m + 2N)$, and $(m + 2N, m + 2N + M)$, on the other hand, could reflect that consumers at location m (which could be a work location or home) drove to the store at location $m + N$, obtained the information there concerning the product, completed the transaction, and then drove to node M. Note that a path represents a sequence of possible options for the consumers. The flows, in turn, reflect *how many* consumers of a particular class actually select the particular paths and links, with a zero flow on a path corresponding to the situation that no consumer elects to choose that particular sequence of links.

The criteria that are relevant to decision making in this application are time, cost, opportunity cost, and safety or security risk, where, in contrast to the telecommuting application, time need not be restricted simply to *travel* time and, depending on the associated link, may include transaction time. In addition, the cost is not exclusively a travel cost but depends on the associated link and can include the transaction cost as well as the product price, or shipment cost. Moreover, the opportunity cost now arises when shoppers on the Internet cannot have the physical experience of trying the good or the actual sociableness of the shopping experience itself. Finally, the safety or security risk cost now can reflect not only the danger of certain physical transportation links but also the potential of credit card fraud, etc.

For example, an article in *The Economist (2001)* notes that "websites are not much good for replicating the social functions of shopping" and that "consumers are often advised against giving their credit-card numbers freely over the Internet, and this remains one of the most-cited reasons for not buying things online."

Assuming weights for each class, link, and criterion, a generalized link cost for each class and link can then be constructed, as well as a generalized path cost for a class of consumer (Nagurney and Dong, 2002). The behavioral assumption is that consumers of a particular class are assumed to choose the paths associated with an O/D pair so that their generalized path costs are minimal.

Using the methodologies discussed in Nagurney and Dong (2002), one can then solve the model and obtain the number of decision makers who select the different options and the incurred generalized costs. One can then ascertain the relative popularity of the various options.

Supply Chain Networks

The study of supply chain network problems through modeling, analysis, and computation is a challenging topic due to the complexity of the relationships among the various decision makers, such as suppliers, manufacturers, distributors, and retailers, as well as the practical importance of the topic for the efficient movement of products. The topic is multidisciplinary by nature since it involves particulars of manufacturing, transportation and logistics, and retailing/marketing, as well as economics.

The introduction of electronic commerce has unveiled new opportunities in terms of research and practice in supply chain analysis and management since electronic commerce (e-commerce) has had a huge effect on the manner in which businesses order goods and have them transported, with the major portion of e-commerce transactions being in the form of business to business (B2B). Estimates of B2B electronic commerce range from approximately $0.1 trillion to $1 trillion in 1998 and with forecasts reaching as high as $4.8 trillion in 2003 in the United States (see Federal Highway Administration, 2000; Southworth, 2000). It has been emphasized that the principal effect of business-to-business commerce, estimated to be 90% of all e-commerce by value and volume, is in the creation of new and more profitable supply chain networks.

In figure 9.4, we depict a four-tiered supply chain network in which the top tier consists of suppliers of inputs into the production processes used by the manufacturing firms (the second tier), who, in turn, transform the inputs into products that are then shipped to the third tier of decision makers, the retailers, from whom the consumers can then obtain the products. Here we allow not only for physical transactions to take place but also for virtual transactions, in the form of electronic transactions via the Internet to represent electronic commerce. In the supernetwork framework, both B2B and B2C can be considered, modeled, and analyzed. The decision makers may compete independently

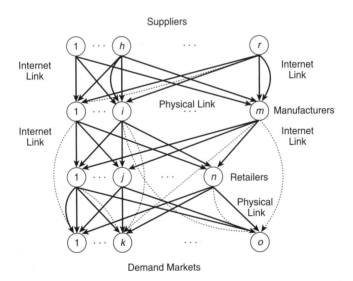

Figure 9.4
The supernetwork structure of the supply chain network with suppliers, manufacturers, retailers, and demand markets and electronic commerce.

across a given tier of nodes of the network and cooperate between tiers of nodes.

In particular, Nagurney, Loo, Dong, and Zhang (2002) have applied the supernetwork framework to supply chain networks with electronic commerce in order to predict product flows between tiers of decision makers as well as the prices associated with the different tiers. They assumed that the manufacturers as well as the retailers are engaged in profit maximizing behavior, whereas the consumers seek to minimize the costs associated with their purchases. The model therein determines the volumes of the products transacted electronically or physically.

As mentioned earlier, supernetworks may also be multilevel in structure. In particular, Nagurney, Ke, Cruz, Hancock, and Southworth (2002) demonstrated how supply chain networks can be depicted and studied as multilevel networks in order to identify not only the product shipments but also the financial flows as well as the informational ones. In figure 9.5, we demonstrate how a supply chain can be depicted as a

Supply Chain-Transportation Supernetwork Representation

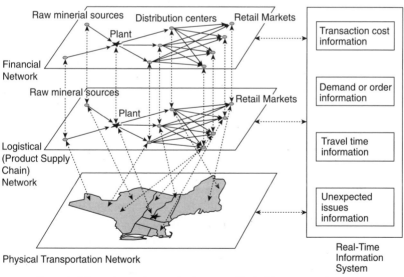

◀┄┄▶ Two-way information exchanges between specific decision-makers

Figure 9.5
A multilevel supply chain supernetwork.

multilevel supernetwork in which the financial network as well as the actual physical transportation network are also represented. For example, in the supernetwork depicted in figure 9.5, the logistical network affects the flows on the actual transportation network, whereas the financial flows are due to payments as they proceed up the chain and as the transactions are completed. The information flows, in turn, are in the form of demand, cost, and flow data at the instance in time.

Obviously, in the setting of supply chain networks and, in particular, in global supply chains, there may be much risk and uncertainty associated with the underlying functions. Some research along those lines is being undertaken (Dong, Zhang, and Nagurney, 2002; Nagurney, Cruz, and Matsypura, 2002). Continuing efforts to include uncertainty and risk in modeling and computational efforts in a variety of supernetworks and their applications is of paramount importance given the present economic and political climate.

In addition, we emphasize that the inclusion of environmental variables and criteria is also an important topic for research and practice in the context of supply chain networks (Nagurney and Fuminori, 2002). In particular, we note that additional effort needs to be extended on the topic of reverse logistics and the recycling of electronic wastes.

Financial Networks

Financial networks have been utilized in the study of financial systems since the work of Quesnay in 1758, who depicted the circular flow of funds in an economy as a network. His conceptualization of the funds as a network, which was abstract, is the first identifiable instance of a supernetwork. Quesnay's basic idea was subsequently applied in the construction of flow of funds accounts, which are a statistical description of the flows of money and credit in an economy (Board of Governors, 1980). However, since the flow of funds accounts are in matrix form, and hence two-dimensional, they fail to capture the behavior on a micro level of the various financial agents/sectors in an economy, such as banks, households, insurance companies, etc. Moreover, the generality of the matrix tends to obscure certain structural aspects of the financial system that are of continuing interest in analysis, with the structural concepts of concern including those of financial intermediation.

Advances in telecommunications and, in particular, the adoption of the Internet by businesses, consumers, and financial institutions have had an enormous effect on financial services and the options available for financial transactions. Distribution channels have been transformed, new types of services and products introduced, and the role of financial intermediaries altered in the new economic networked landscape. Furthermore, the impact of such advances has not been limited to individual nations but, rather, through new linkages, has crossed national boundaries.

The topic of *electronic* finance has been a growing area of study (Claessens, Glaessner, and Klingebiel, 2000, 2001; Allen, Hawkins, and Sato, 2001; and references therein), due to its increasing impact on financial markets and financial intermediation, as well as related regulatory issues and governance. Of particular emphasis has been the conceptualization of the major issues involved and the role of networks in the transformations (see McAndrews and Stefanidis, 2000; Banks, 2001; Allen, Hawkins, and Sato, 2001; Economides, 2001; Nagurney and Dong, 2002).

Nevertheless, the complexity of the interactions among the distinct decision makers involved, the supply chain aspects of the financial product accessibilities and deliveries, the availability of physical as well as electronic options, and the role of intermediaries have defied the construction of a unified, quantifiable framework in which one can assess the resulting financial flows and prices.

Here we briefly describe a supernetwork framework for the study of financial decision making in the presence of intermediation and electronic transactions. Further details can be found in Nagurney and Ke (2001, 2002). The framework is sufficiently general to allow for the modeling, analysis, and computation of solutions to such problems.

The financial network model consists of agents or decision makers with sources of funds, financial intermediaries, and consumers associated with the demand markets. In the model, the sources of funds can transact directly electronically with the consumers through the Internet and can also conduct their financial transactions with the intermediaries either physically or electronically. The intermediaries, in turn, can transact with the consumers either physically in the standard manner or electronically. The depiction of the network at equilibrium is given in figure 9.6.

It is assumed that the agents with sources of funds as well as the financial intermediaries seek to maximize their net revenue (in the presence of transaction costs) while, at the same time, minimizing the risk associated with the financial products. The solution of the model yields the financial flows between the tiers as well as the prices. Here we also allow for the option of having the source agents not invest a part (or all) of their financial holdings. More recently, Nagurney and Cruz (2002) have demonstrated that the financial supernetwork framework can also be extended to model international financial networks with intermediation in which there are distinct agents in different countries and the financial products are available in different currencies.

Summary and Conclusions

In this chapter we have described the realities surrounding networks today and the challenges and complexities posed for their analysis and study. In particular, we have argued for new paradigms to capture decision making in the Information Age. We have focused on the concept of supernetworks and have discussed a variety of applications that come under this umbrella, ranging from telecommuting versus commuting

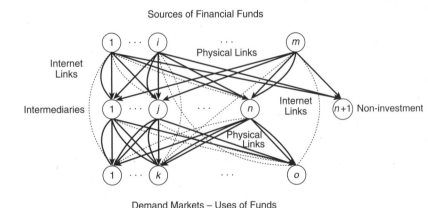

Figure 9.6
The structure of the financial network with electronic transactions.

decision making to financial networks with electronic transactions and intermediation. In addition, we have emphasized possible new directions for research. The advances in information technologies not only have enabled new connections and applications but have, at the same time, allowed for the implementation of powerful analytical methodologies for the solution of complex network problems that underlie our economies and societies today.

References

Abello, J., P. M. Pardalos, and M. G. C. Resende (1999) "On Maximum Clique Problems in Very Large Graphs." In *External Memory Algorithms*, ed. J. Abello and J. Vitter. AMS Series on Discrete Mathematics and Theoretical Computer Science, 50, pp. 119–130.

Ahuja, R. K., T. L. Magnanti, and J. B. Orlin (1993) *Network Flows: Theory, Algorithms, and Applications*. Upper Saddle River, N.J.: Prentice Hall.

Allen, H., J. Hawkins, and S. Sato (2001) "Electronic Trading and its Implications for Financial Systems." Bank of International Settlements Paper no. 7, November, Bern, Switzerland.

Banister, D., and K. J. Button (1993) "Environmental Policy and Transport: An Overview." In *Transport, the Environment, and Sustainable Development*, ed. D. Banister and K. J. Button. London: E. & F.N, pp. 130–136.

Banks, E. (2001) *e-Finance: The Electronic Revolution*, John Wiley & Sons, New York.

Bar-Gera, H. (1999) "Origin-Based Algorithms for Transportation Network Modeling." National Institute of Statistical Sciences, Technical Report #103, Research Triangle Park, North Carolina.

Bass, T. (1992) "Road to Ruin." *Discover*, May, 56–61.

Beckmann, M. J., C. B. McGuire, and C. B. Winsten (1956) *Studies in the Economics of Transportation*. New Haven: Yale University Press.

Braess, D. (1968) "Uber ein Paradoxon der Verkehrsplanung." *Unternehmenforschung 12*, 258–268.

Claessens, S., T. Glaessner, and D. Klingebiel (2000) "Electronic Finance: Reshaping the Financial Landscape Around the World." Financial Sector Discussion Paper no. 4, September, The World Bank.

Claessens, S., T. Glaessner, and D. Klingebiel (2001) "E-Finance in Emerging Markets: Is Leapfrogging Possible?" Financial Sector Discussion Paper no. 7, June, The World Bank.

Cohen, J. E., and F. P. Kelly (1990) "A Paradox of Congestion on a Queuing Network." *Journal of Applied Probability* 27: 730–734.

Dafermos, S. (1981) "A Multicriteria Route-Mode Choice Traffic Equilibrium Model." Lefschetz Center for Dynamical Systems, Brown University, Providence, Rhode Island.

Dafermos, S. C., and F. T. Sparrow (1969) "The Traffic Assignment Problem for a General Network." *Journal of Research of the National Bureau of Standards* 73B, 91–118.

Dong, J., D. Zhang, and A. Nagurney (2002) "A Supply Chain Network Equilibrium Model with Random Demands." Forthcoming in *European Journal of Operational Research.*

Economides, N. (1996) "The Impact of the Internet on Financial Markets." *Journal of Financial Transformation*, 8–13.

The Economist (2001) "We Have Lift-Off." February 3.

Federal Highway Administration (2000) "E-Commerce Trends in the Market for Freight. Task 3 Freight Trends Scans." Draft, Multimodal Freight Analysis Framework, Office of Freight Management and Operations, Washington, D.C.

Gould, J. (1998) "Driven to Shop? Role of Transportation in Future Home Shopping." *Transportation Research Record* 1617, 149–156.

Hu, P. S., and J. Young (1996) *Summary of Travel Trends: 1995 Nationwide Personal Transportation Survey*, December, U.S. DOT, FHWA, Washington, D.C.

Kolata, G. (1990) "What If They Closed 42d Street and Nobody Noticed?" *New York Times*, December 25.

Korilis, Y. A., A. A. Lazar, and A. Orda (1999) "Avoiding the Braess Paradox in Noncooperative Networks." *Journal of Applied Probability* 36: 211–222.

Labaton, S. (2000) "F.C.C. to Promote a Trading System to Sell AirWaves." *New York Times*, March 13.

Leurent, F. (1993) "Cost versus Time Equilibrium over a Network." *European Journal of Operations Research* 71: 205–221.

McAndrews, J., and C. Stefanidis (2000) "The Emergence of Electronic Communications Networks in U.S. Equity Markets." *Current Issues in Economics and Finance* 6: 1–6, Federal Reserve Bank of New York.

Mokhtarian, P. L. (1991) "Telecommuting and Travel: State of the Practice, State of the Art." *Transportation* 18: 319–342.

Mokhtarian, P. L. (1998) "A Synthetic Approach to Estimating the Impacts of Telecommuting on Travel." *Urban Studies* 35: 215–241.

Mokhtarian, P. L., and I. Salomon (2002) "Emerging Travel Patterns: Do Telecommunications Make a Difference?" In *In Perpetual Motion: Travel Behavior Research Opportunities and Application Challenges*, ed. H. S. Mahmassani. The Netherlands: Elsevier Science, pp. 143–181.

Nagurney, A. (1999) *Network Economics: A Variational Inequality Approach.* Second and revised edition. Dordrecht: Kluwer Academic Publishers.

Nagurney, A. (2000a) *Sustainable Transportation Networks.* Cheltenham, England: Edward Elgar Publishing Company.

Nagurney, A. (2000b) "Congested Urban Transportation Networks and Emission Paradoxes." *Transportation Research D 5*, 145–151.

Nagurney, A. (2000c) "A Multiclass, Multicriteria Traffic Network Equilibrium Model." *Mathematical and Computer Modelling 32*: 393–411.

Nagurney, A., and J. Cruz (2002) "International Financial Networks with Intermediation: Modeling, Analysis, and Computation." Forthcoming in *Computational Management Science;* see also: http://supernet.som.umass.edu.

Nagurney, A., and J. Dong (2001) "Paradoxes in Networks with Zero Emission Links: Implications for Telecommunications versus Transportation." *Transportation Research D 6*, 283–296.

Nagurney, A., and J. Dong (2002) *Supernetworks: Decision-Making for the Information Age*. Cheltenham, England: Edward Elgar Publishers.

Nagurney, A., and T. Fuminori (2003) "Supply Chain Supernetworks and Environmental Criteria." *Transportation Research D*, 185–213.

Nagurney, A., and K. Ke (2001) "Financial Networks with Intermediation." *Quantitative Finance 1*: 441–451.

Nagurney, A., and K. Ke (2003) "Financial Networks with Electronic Transactions: Modeling, Analysis, and Computations." Revised and resubmitted to *Quantitative Finance.*

Nagurney, A., and S. Siokos (1997) *Financial Networks: Statistics and Dynamics*, Springer-Verlag, Berlin, Germany.

Nagurney, A., J. Cruz, and D. Matsypura (2002) "Dynamics of Global Supply Chain Supernetworks." *Mathematical and Computer Modelling 37*: 963–983; see also: http://supernet.som.umass.edu.

Nagurney, A., J. Dong, and P. L. Mokhtarian (2000) "Integrated Multicriteria Network Equilibrium Models for Commuting Versus Telecommuting." Isenberg School of Management, University of Massachusetts, Amherst, Massachusetts; see http://supernet.som.umass.edu.

Nagurney, A., J. Dong, and P. L. Mokhtarian (2001) "Teleshopping Versus Shopping: A Multicriteria Network Equilibrium Framework." *Mathematical and Computer Modelling 34*: 783–798.

Nagurney, A., J. Dong, and P. L. Mokhtarian (2002) "Multicriteria Network Equilibrium Modeling with Variable Weights for Decision-Making in the Information Age with Applications to Telecommuting and Teleshopping." *Journal of Economic Dynamics and Control 26*: 1629–1650.

Nagurney, A., K. Ke, J. Cruz, K. Hancock, and F. Southworth (2002) "A Multilevel (Logistical/Informational/Financial) Network Perspective." *Environment & Planning B 29*, 795–818.

Nagurney, A., J. Loo, J. Dong, and D. Zhang (2002) "Supply Chain Networks and Electronic Commerce: A Theoretical Perspective." *Netnomics 4*, 187–220.

Purchasing (2000) "Corporate Buyers Spent $517.6 Billion on Telecommunication." 128, 110.

Quandt, R. E. (1967) "A Probabilistic Abstract Mode Model." In *Studies in Travel Demand VIII*. Princeton: Mathematica, pp. 127–149.

Quesnay, F. (1758) *Tableau Economique*, reproduced in facsimile with an introduction by H. Higgs by the British Economic Society, 1895.

Resende, M. G. C. (2000) personal communication.

Schneider, M. (1968) "Access and Land Development." In *Urban Development Models*, Highway Research Board Special Report 97, pp. 164–177.

Southworth, F. (2000) "E-Commerce: Implications for Freight." Oak Ridge National Laboratory, Oak Ridge, Tennessee.

United States (2000) Public Law #106-346, Washington, D.C.

United States Department of Commerce (2000) Statistical Abstract of the United States, Bureau of the Census, Washington, D.C.

United States Department of Transportation (1999) "Guide to Transportation." Bureau of Transportation Statistics, BTS99-06, Washington, D.C.

Wardrop, J. G. (1952) "Some Theoretical Aspects of Road Traffic Research." *Proceedings of the Institute of Civil Engineers*, part 2, pp. 325–378.

Wu, J. H., M. Florian, and S. G. He (2000) "EMME/2 Implementation of the SCAG-II Model: Data Structure, System Analysis and Computation." Submitted to the Southern California Association of Governments, INRO Solutions Internal Report, Montreal, Quebec, Canada.

IV

Sector Transformation

10
Transforming Production in a Digital Era

John Zysman

How revolutionary is the digital revolution? In the spring of 2000, we would have asked: Are the arrival of the Internet, the pervasive spread of digital networks, and the extraordinary expansion of computing intelligence the edge of a historical revolution, a transformation? Today, in the summer of 2003, the question is different: Was this the revolution that never happened, the dreams evaporating with stock values, first with the dot-com collapse and then in the telecoms debacle?

"Information technology builds the most all-purpose tools ever, tools for thought. . . . These tools for thought amplify brainpower in the way the technologies of the Industrial Revolution amplified muscle power."[1] Steam substituted energy and mechanical power for muscles of animals and men. Capabilities created to process and distribute digital data multiply the scale and speed with which thought and information can be applied. The tool may be a personal computer, a student's calculator, or the microprocessor in a car engine or a refrigerator. Thought and information can be applied to almost everything, almost everywhere.[2]

To assess the consequences of these tools and the character of the digital revolution, we turn to the classic process of the industrial economy: manufacturing. What happens to manufacturing in a digital era? Certainly these digital tools permit the reorganization of production as communication and data exchange become easier. But how do these tools alter the significance of manufacturing in a firm's strategic choices?

Does manufacturing—production—continue to matter in the digital era? Perhaps "manufacturing" is best used to refer to the physical production of an industrial era, and the broader category of "production" can apply to an array of digital information goods as well. In answering this question, we will discover that just as the emergence of a service economy did not mean the end of manufacturing, so the emergence of a digital era does not mean the end of production. Put in historical perspective, we have passed from an electromechanical industrial era, through a transition period I call "Wintelism," into a digital age. In that digital age, things continue to be made, whether they are textiles or consumer durables manufactured in factories or software developed in very different kinds of facilities.

I will conclude that manufacturing in a digital era can be either a strategic asset or a vulnerable commodity, for both companies and countries. For companies, the questions are: When can production serve to generate and maintain advantage? Under what circumstances is the lack of in-house world-class manufacturing skills a strategic vulnerability? When is it simpler and easier to just buy production as a commodity service? For the nation, or the region perhaps, the questions are: What can be done to make this country or region an attractive location for world-class manufacturing? How can it be made an attractive place for companies to use production to create strategic advantage?

In conclusion, I will consider the role of production in value creation and market position for three different sectors.

Exit or Transformation: The Future of Manufacturing

The conviction that we are living through a digital revolution suggests that as a national economy, we can safely exit manufacturing.[3] That is, there will be a secure economic life developing software and other digital applications and providing services—a whole array of activities that do not involve making physical things. This logic is an extension, almost a translation into the twenty-first century, of the argument of 20 years ago about services and manufacturing. Then, it seemed that because the United States was supposedly moving from an industrial economy to a postindustrial or service economy, it was acceptable for the economy to lose manufacturing production and jobs, just as it had when it moved from agriculture to industry.

But the agriculture-into-industry model was never literally correct. The agricultural sector didn't disappear. Farm production was reorganized, and the process of growing things in Nebraska, where I grew up, and California, where I now live, evolved. Throughout agriculture, labor shifted from the land into production inputs in the form of fertilizer and machinery. Similarly, the manufacturing story was not about an exit into services. It was rather about the reorganization of production—specifically, the emergence of the production supply chains, or global value chains, as a substitute for in-house production and the adaptation of production to distribution channels. The statistics showed that manufacturing as a portion of the economy had dropped precipitously, and the portion included in the category *services* had risen.[4] The conventional categories showed private goods-producing industries in the United States declining toward 20 percent. Durable goods manufacturing fell below 8 percent, while private service-producing industries rose to over 67 percent. How government is included—although some would argue that no government expenditures are services—determines the precise balance of services in the economy as a whole. But instead of illuminating matters, the statistics simply mislead us.[5] The statistical game often confuses more than it enlightens. The real story lies within the numbers. Let us see why.

As we analyze the numbers, the evidence that a service economy is replacing an industrial economy slowly dissipates. Let us initially separate business services from personal and social services. Then, let us divide business services into two categories: those services downstream from the point of production and those upstream. What is the difference between downstream and upstream services? An auto dealer is downstream from production and does not care where the car was made, whether the Ford was produced in Brazil or Michigan, or the Toyota in Japan or the United States. By contrast, the upstream services are the ones feeding into manufacturing, supporting the production activities. The question is whether the services are tightly linked to the manufacturing operation or whether they can be separated from the production and moved elsewhere. There are services that go directly into production line activity that are an integral part of manufacturing. Then there are ancillary services, such as window washing, and supportive services such as customer relation phone services or back office services.

There are two points worth noting as we disassemble these categories: statistics and linkages. First, consider the statistics. If the window washers, or phone service personnel, or billing service personnel work for General Motors, they are classified as manufacturing sector employees. If they work for Ace Window Washers, Phone Service Outsourcing, or Back Office Temp Services, then they are considered service sector employees. Although the employees in both examples are engaged in the same activities, they fall in different statistical categories based on their employer. So the statistic *services* is a confused measure that focuses less on the activity than on the business of the employer.

Next, consider the tightness of the linkages between the services and the underlying manufacturing activity. If General Motors moves to Brazil, the window washers will not go with it. On the other hand, many back office services can now be performed overseas. The back office and customer support services are much more mobile than window washing; window washing is location dependent. Even before the manufacturing moved to Brazil, the back office might have moved to South Dakota and the phone services to Bangalore.

Hence, we must ask: What links these activities together? What strengthens or weakens these linkages? For this discussion, we must distinguish between strong and weak locational and organizational linkages, between which activities must geographically or organizationally stay together. And what is the glue that binds them? Indeed, in a digital era with easy communications, including data document transfer, these various back office and customer support services become even more mobile. Is a mastery of English and a sophisticated telecom infrastructure with global links, even if it has limited local ties, all that is needed? Certainly, the ability to communicate fluidly and collaboratively over distances loosens the locational linkages, alters appropriate organizational structure, and changes control structures among other kinds of activities. This is evident in the evolution of supply chains into very complex, tightly controlled production systems as well as in the development of distributed innovation and production evidenced in open source software that involves the distributed efforts by seemingly undirected groups of volunteer workers.

The Digital Era in Historical Perspective

The evolving model of production and competition follows a historical sequence that goes from American dominance with mass manufacture through challenges to mass manufacture in the form of Japanese lean production and European flexible specialization or diversified quality production.[6] The transition from a mechanical or electromechanical age comes with Wintelism.[7]

American Dominance: Fordism and Mass Manufacture

Mass manufacture, epitomized by Henry Ford and the Model T, was the first twentieth-century production revolution. With far-reaching societal implications, this civilian production innovation, mass manufacture, made possible the large-scale deployment of the tanks and planes that provided American and Allied forces an advantage in the Second World War. Mass manufacture is broadly understood to mean the high-volume output of standard products made with interchangeable parts connected using machines dedicated to particular tasks and manned by semiskilled labor.[8] Features of this basic definition include:

• The separation of conception from execution—managers design systems, operated by workers in rigidly defined roles that match them to machine function;
• The "push" of product through these systems and into the market; and
• Large-scale integrated corporations, whose size and market dominance reflect mass manufacture's economies of scale.

In this system, large-scale manufacture implied rigidity. Fixed costs in the production line and design were high; consequently, changes in products or reductions in volume were difficult and expensive. That meant the national economy was rigid; drops in demand would be difficult for mass production companies to absorb. An initial downturn in demand could cumulate into sharper economic downturns. Hence, the national economic policy management counterpart of that corporate rigidity became the policy question of how to avoid business cycles. Demand management policies, associated with the label of Keynes, were born. And alongside the technical policy issues was a political one. Booms and busts implied worker dislocations. As unemployment surges it often initiates political debate about how to use a public policy to cushion not only the economic dislocations but also the political dislocations that would come from mass unemployment. In any case, Fordist mass manufacture, the label taking its name from Henry Ford's automobile production lines but being a method for organizing assembly production with large numbers of parts, was associated with American industrial development, military success, and postwar hegemony.

Challenges from Abroad: Lean Production and Flexible Specialization
Producers abroad, often with the support of their governments, tried to imitate the American mass manufacture model. While most producers abroad failed in competition against American companies, some of these efforts generated new rounds of production innovation, spawning a second phase in twentieth-century manufacturing.

These challenges to American manufacturing came from two different directions. The more important was the interconnected set of Japanese production innovations loosely called *flexible volume production* or *lean production*.[9] The mechanisms and sources of the Japanese flexible volume manufacturing system attracted intense attention because of the stunning world market success of the Japanese companies in consumer durable industries requiring complex assembly of a large number of component parts. Japan's automobile and electronics firms burst onto world markets in the 1970s and consolidated into powerful conglomerates in the 1980s. The innovators were the core auto and electronics firms who, in a hierarchical manner, dominated tiers of suppliers and subsystem assemblers; the production innovation was the orchestration and reorganization of the assembly and component development process. The core Japanese assembly companies of the lean variety have been less vertically integrated than their American counterparts, but they have been at the center of vertical Keiretsu. A Keiretsu is, loosely speaking, a Japanese conglomerate conventionally understood to be headed by a major bank or one consisting of companies with a common supply chain linking wholesalers and retailers, that have tightly linked the supplier companies to their clients.[10] The Japanese lean production system provides flexibility of output in existing lines as well as rapid introduction of new products, which permits rapid market response. High quality has come hand-in-hand with lower cost.

The developmental strategies of Japan were essential to its production innovation. The distinctive features of the Japanese lean production system were a logical outcome of the dynamics of Japanese domestic competition during the rapid growth years, and this system was firmly in place by the time of the first oil shock in the early 1970s.[11] Indeed, protected domestic markets and exports were decisive in Japanese success in export markets. Moreover, those closed markets were critical to the emergence of the innovative and distinctive system of lean flexible volume production.[12] While the Fordist story highlights national strategies for demand management, this Japanese story of lean production and developmentalism highlights the interaction among the markets and producers of the advanced countries in international competition. Lean production was the focus of policy and corporate attention because it represented a direct challenge to both mass manufacturing and assumptions of American global economic policy.

The second challenge to the classical American mass production model had little to do with the volume production strategies emerging in Japan. Different accounts of its development variously labeled this collection of innovations as *diversified quality production* and *flexible specialization*.[13] The "Third Italy" and the Germany of Baden-Wurttemberg were the first prominently displayed examples of an approach in which craft production, or at least the principles of craft production, survived and prospered in the late twentieth century. The particular political economy of the two countries gave rise to distinctive patterns of company and community strategies.[14] Firms in these countries often competed in global markets on the basis of quality not price; they used production methods involving short runs of products that had higher value in the marketplace because of distinctive performance or quality features. Competitive position rested on skills and flexibility, not low wages. These challenges—often in high value-added niche markets—came from small- and middle-sized firms rooted in particular industrial districts. "Craft production or flexible specialization," argue Paul Hirst and Jonathan Zeitlin, "can be defined as the manufacture of a wide and changing array of customized products using flexible, general purpose machinery and skilled, adaptable workers."[15] Communities of groups of small companies arose, organized in what are perceived as twentieth century versions of industrial districts. These communities are able, in at least some markets and circumstances, to adapt, invest, and prosper in the radical uncertainties and discontinuities of global market competition more effectively than larger, more rigidly organized companies. "These districts escape ruinous price competition with low-wage mass producers," Charles Sabel explains, "by using flexible machinery and skilled workers to make semicustom goods that command an affordable premium in the market."[16] The emphases in these discussions are the *horizontal* connections, the connections within the community or region of peers, as distinct from the *vertical* or *hierarchical* connections of the dominant Japanese companies.

The Transition to a Digital Age and the American Comeback: Wintelism

Wintelism[17] is the transition period out of an electromechanical era into a digital age. Twenty years ago, it seemed that American firms were being dominated in international markets, when a flood of innovative entertainment products such as the Sony Walkman and the VCR joined traditional electronic products such as televisions. The reason for this dominance was not simply low wages but the fact that firms outside the United States also had the capacity to turn ideas into competitive products. As the semiconductor industry joined consumer electronics and automobiles as sectors under intense competitive pressure in the late 1980s, it seemed that the fabric of advanced electronics was coming unraveled. That is, the competitive position of equipment suppliers to the semiconductor industry was eroding, making it more difficult for American semiconductor producers to hold market position. With weakening position of the semiconductor makers it seemed less likely that final product producers would have access to the most innovative chip designs needed in their final products. Then, suddenly, it seemed that American producers rebounded. They had not reversed the decline of production in electromechanical products but, rather, a new sort of electronics product had emerged, defining a new segment of the industry.

The "new" consumer electronics, as Michael Borrus has argued, are networked, digital, and chip-based.[18] They involve products from personal computers to mobile devices. The nature of manufacturing and the sources of functionality change dramatically. The core engineering skills moved to chip-based systems given functionality by software.

More or less at the same time, products that were thought to spin off from technology investment in military goods into civilian products began to seem less significant. Leading edge civilian technologies contained more advanced technologies and components than their military counterparts. Technologies began to spin on from the civilian to the military spin-on technologies.[19]

The process of creating value and the role of production were beginning to change as well. Consider the PC, the personal computer. What part of the value chain confers the most advantage? It is not the producer of the final product, the metal box we call the PC, even if, like Gateway's or Hewlett Packard's, the box carries the company logo. Is it the producer of the constituent elements, the components of the system such as the chip, the screen, and the operating system? The added value is in the components or subsystems. Those components and subsystems are built to generally agreed upon standards that emerge in the marketplace, and thus part of their value lies in the standards. Much of the value is in the intellectual property (IP), formally in the components, often in partially open but owned standards that create de facto IP-based monopolies or dominant positions. A large portion of intellectual property resides in the chip, and a large portion in the screen.

Modularization, as it came to be called, facilitated a vertical disintegration of production. Outsourcing, a tactical response usually aimed at cost savings with a decision to procure a particular component or service outside the organization, evolved into cross-national production networks (CNPNs) that could produce the entire system or final product. Then what began as an academic discussion of CNPNs transformed into a broader business debate of how to manage the supply chain.

Let us state it formally: *Wintelism* is the word Michael Borrus and I coined to reflect the shift in competition away from final assembly and vertical control of markets by final assemblers.[20] Competition in the Wintelist era is a struggle over setting and evolving de facto product market standards, with the possibility of exerting market power lodged anywhere in the value chain, including product architectures, components, and software. Each point in the value chain can involve significant competition among independent producers of the constituent elements of the system (e.g., components, subsystems)—not just among assemblers—for control over the evolution of technology and final markets. As these fundamentals of Wintelism have evolved, the constituent elements of the product became modules. Even if distinctive intellectual property remains in the modules, production becomes modularized as the knowledge about the elements and components they interconnect becomes codified, that is, formally stated and expressed in code, and then diffused.

CNPN is a label we apply to the consequent disintegration of the industry's value chain into constituent functions that can be contracted out to independent producers wherever those companies are located in the global economy. This strategic and organizational innovation, what we might now call supply chain management, means that even production of complex products can become a commodity service that can be purchased in the market. The nature of those chains, now often labeled *global value chains*, varies with the complexity of the transactions, the codifiability of the knowledge involved, and the competence of the suppliers.[21] The strategic weapon for companies such as Dell moves from the factory to the management of the supply chain. And the supply chain itself is extended both forward into the marketplace and backward into development.

Competing in a Digital Age

Wintelism was the beginning of the transition from an electromechanical era into a digital age, into a digital era in which tools for thought—broadly, communications and computing—are central. Our question here is: How do these tools alter the significance of manufacturing in a firm's strategic choices? Does manufacturing become a commodity easily purchased and of little strategic significance or a critical weapon in the fight for market position?

We must begin by considering how digital tools affect a firm's core process of creating and sustaining value. A three-pronged explanation is necessary to suggest the array of strategic choices that are opened by these tools: (1) market segmentation and product functionality, (2) the ever-shifting line between services and goods, and (3) the very character of marketplaces themselves.

First, digital tools permit markets to be segmented, and then permit the segments to be attacked with functionally varied product. A fundamental feature of the digital era is that analytic tools of database management permit the consumer community to be segmented into subcomponents, each with distinct needs and wishes. At an extreme, individuals and their particular needs can be targeted. Early on, the insurance industry moved from using computers exclusively for back office operations to using them to create customized products for particular consumers.[22] Thus, collecting consumer information in a variety of forms—credit cards or grocery store purchases are obvious examples—is a critical matter. The result, of course, is a policy struggle about what information can be gathered, shared, and combined. The desire of companies and governments to assemble information from diverse sources into consumer profiles or threat assessments is offset by individuals' expectations and rights of privacy.

At the same time, digital tools permit ever greater functional variety in products, which permits firms to address these now defined or created market segments. A standard product can now be given diverse functionality. The coffee maker that automatically turns on at a particular time in the morning depends on simple digital functionality. The difference between many higher speed, higher priced printers and their slower, lower priced counterparts is in the software that tells the printer how to operate.[23]

Let us overstate the conclusion. Electromechanical functionality of the Sony Walkman or a Bang & Olufsen high-end CD stereo system rests on proprietary manufacturing skills. The digital functionality of the coffee maker or an MP3 player rests largely on commodity microchips in products that can be assembled by commodity production services. This combination of market segmentation and digitally based functionality would seem to make production into a commodity and supply chain management into a critical corporate asset.

In any case, new problems are created in the struggle over value creation. Let us consider just one, intellectual property. Market advantage rests increasingly on proprietary product and market knowledge. Consequently, protecting that knowledge as intellectual property becomes a central issue. Digital information makes product and process knowledge explicit and permits it to be stored in easily replicated forms. This is the case whether the firm is a media company, a company building routers, or Microsoft. When surgical technique can be formally expressed, the surgeon can be replaced by a robot. The surgical program becomes essential as hip surgery becomes a form of high-end machining. It is easier to transfer, or lose control of, formalized knowledge than intuitively held know-how. Often, what might have previously been embedded in organizational know-how, as the accumulation of individual understandings shrouded from view in the final product, is now potentially transferable as a data file. Suddenly, intellectual property, a creation of law and social agreement if there ever was one, becomes central to company strategy. Not surprisingly, who owns, or can construct the right to own, which intellectual property becomes a central business problem and policy question. Consider that if you redefine copyright law, one thereby changes who controls the use and distribution of media products such as music. By redefining the control of use and distribution one alters the value of many existing media products. In so redefining copyright law, one redefines the valuation of an entire swath of media companies.

Second, the distinction between *service* and *product*, which concerned us at the beginning of this chapter, blurs even more in a digital era. A matter of aggregate accounting in the first part of our story, now it becomes a matter of strategic importance. Consider accounting: Accounting is a personal service provided by accountants utilizing tools from the original double-entry bookkeeping system but now using computers. But if you create a digital accounting program and put it on a CD, put it in a box, call it Quicken, and allow its unlimited use by the purchaser, then you have a product.[24] If you put the program on the Web for access with support for use on a fee basis, then you likely offer a service, as an ASP, or application service provider. Next, consider pharmaceuticals. If NextGenPharma sells a drug to be dispensed by a doctor or hospital or sold in a pharmacy, it is producing a product. With gene mapping and molecular analysis, we are moving toward the possibility of a service model of therapies adapted to particular physiologies. If NextGenPharma really is a database company with a store of detailed molecular-level drug information and genome functionality, it could sell an online service to customize drugs or therapy. Slowly the distinction between product and service empties of meaning; we are left instead with the question with which we began. If what is being sold is a service of defining a customized drug, then does it matter who produces the drug, the product? Does it matter to the enduring competitive position of the custom drug service company if it sources the product, the drug, as a commodity in the marketplace?

Third, marketplaces themselves are being redefined. The "tools for thought" that define the digital age are not only transforming the dynamics of traditional marketplaces, but they are also enabling the emergence of marketplaces operated by natural competitors, so-called *collaborative markets*. Dramatic increases in computer processing speed, encryption capabilities, and intellectual property protection are permitting the rapid and secure interchange of data—including price quotations, product information, and credit information—necessary to effect multiple and instantaneous transactions in a wide range of established marketplaces, including equity and bond markets and commodity markets. Advances in data networking are democratizing the spread of the established markets beyond the richest institutions, bringing price transparency and market liquidity to smaller entities and extending even to individuals, as in the case of equity trading. In addition, the new tools for thought have enabled companies to develop supply chain management software and complex risk and credit management procedures which, in combination with shifts in government policy toward the deregulation of finance, telecommunications, energy, and other industries, have emboldened corporate leaders to organize markets collectively.

It is not that such collaborative markets never existed before. Industry interests (e.g., OPEC) have a long history as market makers, though it is perhaps more common for them to seek to control prices or output rather than to operate the markets per se. Competing agricultural interests created the Chicago Produce Exchange in 1874, which evolved into the Chicago Mercantile Exchange. Similarly, rival dairy merchants created the Butter and Cheese Exchange of New York in 1872, which eventually became the New York Mercantile Exchange. In the case of cotton, an entire system of private law has evolved dating from the mid-1800s, whereby industry representatives both write and police the rules. Today, collaborative markets take the form of business-to-business and private commodity exchanges, transacting currencies, metals, energy products, automobile parts, chemicals, and dozens of other products.[25] Digital tools make such markets easier to create and construct. They represent a novel type of marketplace requiring a fundamentally different means of analysis from established markets operated by traders or neutral entities and overseen by state regulators.[26]

Production: Strategic Asset or Commodity?

Clearly, products continue to be made; production does not disappear. But production's place in creating value shifts and evolves. When is production a strategic asset, and when is it a commodity that can be purchased in the marketplace? There will not be a single answer, but rather answers that are specific to, or most evident in, particular industries. Let us consider three different sectoral groupings, based on the sector's relation to digital tools and to production.[27]

Digital Goods/Digital Markets[28]

At an extreme, some products can be at once digital and exchanged in entirely online marketplaces. Sectors such as media and finance are sectors where the product can be represented digitally *and* the marketplace, even delivery of the product, can be online. If production still matters in this extreme case, then we know the production questions will endure into the digital era.

What does it mean to make or produce an entertainment or financial product for delivery? There is the creation of the underlying entertainment content or financial instrument and then the digital construction, the programming or development of the digital product. Even pure software products, be it a Windows operating system or the web structure for delivering an accounting service, are built.

Moreover, that digital product is part of a system; it resides on a server and is delivered via a network of digital equipment. More generally, for computers or telecom equipment, the core functionality is the information or data processing. The hardware is simply an instrument for delivering the digital material. The digital functionality expressed through the hardware differentiates the products. The issue, which is distinct from our pure software products, is what hardware knowledge is required to effectively implement the software solutions.[29] Is the semiconductor a commodity, as it is for Dell in a PC, or a proprietary chip, as it may be for some telecommunications applications, or a specialty chip shared with other producers? That answer, commodity or proprietary house for digital IP, depends on the particular product and the particular hardware environment, and there is no consistency to the answers. Dell outsources its actual manufacturing, making its supply chain management into a strategic weapon. Dell's market link is the key; it has limited distinct product knowledge. Cisco likewise outsources production, but its distinct product knowledge is in the development of generations of equipment in which functionality is expressed through electronic hardware but determined by software instructions.

While manufacturing implies manipulating things and materials, its dictionary definition more generally refers to "the organized action of making goods and services for sale" and of putting something together from components and parts.[30] Certainly, our example of Quicken qualifies as "manufacturing" according to this definition, as does the creation of the Yahoo web site and the assembly of the software tools that allow that web site to function.

But the word "manufacturing" implies smoke and factories. We require a new word, stripped of the grime of nineteenth century industry. It may not be possible to apply the concepts we are developing to a word already loaded with centuries of accumulated meaning. So why not just talk of *production* as the general case and *manufacturing* as the specific case of physical production? This means that production—the know-how, skills, and mastery of the tools required—is absolutely central to the products in the digital sector. All the arguments about the linkages and mastery of groups of activities that we developed in the first section of the chapter then would simply be revisited.

In sum, we must broaden the meaning of a production worker from someone who works in a factory to someone who works in an array of other activities. But when we do, the traditional questions of what should be produced or built in-house and what can be outsourced do not disappear. What skills are required to produce the digital product? Is the quality influenced by outsourcing? The same questions remain, they are just posed in a new context.[31]

This new context raises entirely new issues. Cross-national production networks were precursors of global value chains, and supply chain management emerged alongside factory management. Data networks permit and facilitate these varied networked production systems.[32] But the most dramatic evolution comes with distributed product development of software. It is not simply collaboration across distances by traditional software developers, but rather the emergence of entirely new production systems in the open source community.[33]

Indeed, open source software may be the archetype of the digital era, a system of distributed innovation where tasks are self-assigned and where even the management of the innovation is voluntary.[34] It is quite a contrast to the archetype of the industrial era, division of labor, as exemplified in Adam Smith's pin factory, where the production of the classic good, the pin that had been made by a craftsman, is now made by an industrial process. Smith's approach to production would set the process and the divisions of labor, assigning tasks that subdivide the process.

These two systems of political economy—division of labor and open source—rest, moreover, on quite different notions of property, each defined by aspects of its era. In the industrial era, property gave the right to exclude others from using what you possessed, such as the creation of the private use of land from what had been a commons. By contrast, in the digital age, property in the form of open source software is the antithesis of exclusion. Steve Weber writes:

Property in open source is configured fundamentally around the right to distribute, not the right to exclude. If that sentence feels awkward on first reading, it is a testimony to just how deeply embedded in our intuitions and institutions the exclusion view of property really is.[35]

The two eras are each characterized by distinctive notions of property, and perhaps by evolutions in the notions of property as well.

Sectors Based on New Processes and Materials

At the other extreme from digital functionality, let us consider, as a separate case, emerging sectors based on making things using new processes and new materials. Nanotechnology and biotechnology are examples of these emerging sectors. And, indeed, we would include here the semiconductor industry as well, in which the underlying production process and materials evolve radically as transistor size shrinks. In these sectors the issues of production, product innovation, value creation, and market control remain entangled.[36] A generation ago, the industry was threatened when its ability to develop and source leading-edge production equipment was weakening. The capacity to retain an innovative edge in product seemed endangered. Now, the cycle comes full circle after a generation in which design has often become separated from production, with foundries producing for pure design houses. Once again, the question is whether product position can be held if the underlying technologies and their implementation in production systems cannot be maintained.

The strategic place of production is evident if we ask, Who will dominate the new sectors? Will those who generate or even own, in the form of intellectual property rights, the original science-based engineering on which the nanotechnology or biotechnology rests be able to create new and innovative firms that become the significant players in the market? Or will established players in pharmaceuticals and materials absorb the science and science-based engineering knowledge and techniques, by purchasing firms that have spun off from a university or alternately by parallel internal development by employees hired from those same universities?[37]

There is an ongoing, critical interaction among (1) the emerging science-based engineering principles, (2) the reconceived production tasks, and (3) the interplay with lead users that permits product definition and debugging of early production. Arguably, that learning is more critical in the early phases of the technology cycle. Can a firm capture the learning from that interplay if it outsources significant amounts of production?

Twenty years ago, as the American industry generally faced new competition in international markets, the risk was that the industry could not maintain a competitive cost and quality position. The question of course became whether a weakened production position would ultimately influence company abilities to sustain product innovation. For the firm, the question is whether that interaction between development and production, which generates learning, is more effective within the firm, or possible at all through arms-length marketplaces. As new processes or materials emerge, it is harder to find the requisite manufacturing skills as a commodity. Certainly, with new processes and materials, new kinds of production skills become essential. Will outsourcing risk transferring core product/process knowledge, developing in others strategically critical intellectual assets? For the nation or region, the question is whether ongoing production activity is needed to sustain the knowledge required to implement the new science and science-based engineering. In other words, a regional or national government may not care if the learning goes on within a specific firm, as long as the learning is captured in technology development within its domain. Those intimate interplays have traditionally required face-to-face, and hence local and regional, groupings. With the new tools of communication, what happens to the geography of the innovation node is an open question. If a firm, or a national sector, loses the ability to "know how to make things," to use production as a strategic capacity, then it will lose the ability to capture value. Whatever goes on in the labs at Berkeley, if you can't capture it in a product you can make and defend, then the science is not going to translate into a defensible position in terms of jobs and production.

Conventional Products with Digital Functionality and a Physical Function

Certainly, traditional markets will be altered by market segmentation addressed with digital functionality, as we noted above. Let us consider for a moment products that remain physical, that are usually best evaluated in person (textiles and cars) and must be delivered in our physical world. In the case of a car or refrigerator the IT instrumentality creates distinct controls and adds value to the product. Yet, the underlying purpose and the source of functionality, transportation or refrigeration, is something physical and not digital.

Digital tools permit new answers to the fundamental question of how much people are willing to pay for which products. Firms have new ways to identify who will pay how much for what, ways to create products people are willing to pay more money for. But the story goes beyond that.

Digitally rooted online sales/marketing and supply chain management alter the links between a firm and its customers as well as suppliers. The Dell story illustrates how innovative uses of the net that ties customers from sales to product-build can create dramatic advantage.[38] And, as development and production processes merge, resulting in decreased time to market and improved design choices, the divisions among production, design, and development blur even more. Then, because the firm is constructing and evolving a complex evolutionary system, not just procuring a set of defined components, more of the system—a larger portion of the value added—must be kept in-house and not outsourced. More generally, if production becomes characterized by rapid turnaround and custom activity, has the decision about where to locate production within the firm changed? Do the lessons of diversified quality production and flexible specialization, that custom production and rapid turnaround, imply tighter geographical and organizational links between development and production?

The range of conventional physical products is too great to be put into a single set. The critical question is whether production is a strategic asset to be hoarded or a commodity outsourced at the best available market price. The answer depends on a number of issues.

1. As markets are segmented, products must have variable functionality to address each specific segment. Is a proprietary product position, that is, a product and functionality that cannot be simply reverse engineered and then built easily with commodity components, required in the market? Can a proprietary position be developed with outsourced digital development of hardware and software? If the functionality is achieved by the application of commodity digital components to a relatively standard product, then the production process is likely to be a commodity. If variability rests on distinctive knowledge of the application, be it a car or a refrigerator, and requires distinctive skills embedded in the knowhow of making something, then production may be a strategic asset. Or, better, some parts of the production process embody critical knowledge. The production chain must be itself divided between commodity and strategic steps.

2. Can your supplier, source, become your rival? How much product knowledge is now derived from production? Where does market and product learning take place? Is it possible for rivals to enter the market based on their learning from producing?

3. How radical might changes in production be? How confident can we be that another lean production revolution is not around the corner? What skill sets and in-house knowledge of production would be required to respond if there were a radical break?

4. Without production, how is innovation in the core product affected? How much production knowledge is required for next-generation efforts?

But even these questions are conventional. We might ask an altogether different set of questions: When do the new tools alter fundamentally the underlying business models on which firms operate? When do market knowledge and new communication tools transform a product business into a service business?

Conclusion: What Have We Learned?

The digital era is defined by a set of tools for thought—data communication and data processing technologies—that manipulate, organize, transmit, and store information in digital form, with information defined as a data set from which conclusions can be drawn or control exercised. The emerging digital tool set and networks mean that information in digital form becomes critical to a firm's strategies to capture value and market position.

Business strategies and organization, the business models that define the links between objectives and implementation, have all evolved in response to these tools. With that evolution, the meaning, not just the role, of manufacturing has evolved as well. The term *production*, as the act or process of producing something, can encompass a range of products, digital as well as physical, and also the delivery platforms that provide services. Clearly, one implication is that matters of both software and supply-chain management must be understood as elements of production as much as of service.

For a company, the question is how to use production as a strategic weapon. For a country, the question is how to be the most attractive location for strategic production. When production changes very rapidly, jobs can be dislocated or altered. However, if production doesn't change, then those jobs become commodities and are vulnerable to innovation abroad or to moving abroad. For both company and country the question, differently framed for each, is how to adapt to the changing logics of production.

Does production matter? Absolutely, but production can either be a commodity that is vulnerable to relocation or closure, or it can become a strategic asset. Corporate strategists and national policymakers alike must ensure that production capability is a strategic asset that they control, not one that is used against them.

Notes

1. John Zysman, Steve Cohen, and Brad Delong, *Tools for Thought: What is New and Important about the "E-conomy"?* (Berkeley: BRIE, 2001).
2. Ibid, pp. 3–4.

3. Stephen Cohen and John Zysman, *Manufacturing Matters: The Myth of the Post-Industrial Economy* (New York: Basic Books, 1987).

4. Robert E. Yuskavage and Erich H. Strassner, "Gross Domestic Product by Industry for 2002" http://www.bea.gov/bea/ARTICLES/2003/05May/0503GDPbyIndy.pdf (May 2003).

5. Ibid.

6. Japanese lean production is a term associated with the work of Jim Womack, *The Machine that Changed the World* (New York: HarperPerennial, 1991). Flexible specialization is a term widely used by Charles Sabel and Michael Piore, generating a veritable industry of studies. See *The Second Industrial Divide* (New York: Basic Books, October 1990). Diversified quality production, Wolfgang Streeck's term for a similar phenomena, is developed in "On the Institutional Conditions of Diversified Quality Production," in Egon Matzner and Wolfgang Streeck, *Beyond Keynesianism* (Aldershot: Elgar Publishing, 1991).

7. John Zysman and Michael Borrus, "Globalization with Borders: The Rise of Wintelism as the Future of Industrial Competition," *Industry and Innovation*, Vol. 4, Number 2, Winter 1997. John Zysman, "Production in a Digital Era: Commodity or Strategic Weapon?" BRIE Working Paper 147 (Berkeley: BRIE, September 2002).

8. James P. Womack, Daniel T. Jones, and Daniel Roos, *The Machine that Changed the World* (New York: HarperPerennial, 1991). See also Paul Hirst and Jonathan Zeitlin "Flexible Specialization: Theory and Evidence in the Analysis of Industrial Change," in *Contemporary Capitalism: The Embeddedness of Institutions*, ed. J. Rogers Hollingsworth and Boyer (Cambridge: Cambridge University Press, 1997).

9. Stephen Cohen and John Zysman, *Manufacturing Matters: The Myth of the Post Industrial Economy* (New York: Basic Books, 1987). Benjamin Coriat, "The Revitalization of Mass Production in the Computer Age," paper presented at the UCLA Lake Arrowhead Conference Center, Los Angeles, CA, March 14–18 1990. Ramchandran Jaikumar, "From Filing and Fitting to Flexible Manufacturing: A Study in the Evolution of Process Control," Working Paper 88–045 (Boston: Division of Research, Graduate School of Business Administration, Harvard University, c1988). James P. Womack, Daniel T. Jones, and Daniel Roos, *The Machine that Changed the World* (New York: HarperPerennial, 1991). Chalmers Johnson, Laura Tyson, and John Zysman, eds., *Politics and Productivity: The Real Story of How Japan Works* (Cambridge, Mass.: Ballinger, 1989).

10. Taken from the online dictionary Word Web: http://wordweb.info/2/lookupfreefail.pl?keiretsu.

11. John Zysman and Laura Tyson, "The Politics of Productivity: Developmental Strategy and Production Innovation in Japan," in *Politics and Productivity: The Real Story of How Japan Works*, ed. Chalmers Johnson, Laura Tyson, and John Zysman (Cambridge, Mass.: Ballinger, 1989).

12. John Jay Tate, "Driving Production Innovation Home: Guardian State Capitalism and the Competitiveness of the Japanese Automotive Industry"

(Berkeley: BRIE, 1995). The argument is simple. The relationships of production and development in these production systems are, at best, delicate. Just-in-time delivery, subcontractor cost/quality responsibility, and joint component development push onto the subcontractor considerable risk in the case of demand fluctuations. True, there were techniques to continuously reappraise demand levels and indicate to "client" firms their allocations so that the client firms could in turn plan. This reduced unpredictability throughout the system. But if demand moved up and down abruptly, those techniques would not have mattered. True, government and corporate programs to reduce the capacity break-even point in small firms helped. Nonetheless, imagine that Japan's emerging auto sector had to absorb continuously the stops and starts of the business cycle that typified Britain in the 1950s and 1960s. Would the trust relationships that are said to characterize Japan have held up? Could the fabric of small firms have survived to support just-in-time delivery and contractor innovation? Simply a smooth and steady expansion of demand typified the Japanese market in sectors such as autos and facilitated these arrangements and developments. The high growth rates— combined with the need to re-equip Japan in the postwar years—created the basis of the continuous expansion. But domestic growth did fluctuate and the rivalries for market share led consistently to overinvestment, or excess capacity, in the Japanese market. The story about Japan told by Yammamura and Murakami, Tsuru, Zysman, and Tyson, and by Tate in the case of the auto industry, shows that the excess capacity was "dumped" off onto export markets. Seen differently, these exports permitted a steady and smooth expansion without which the production innovations outlined here would not have emerged. The developmental strategies of Japan were essential to its production innovation.

13. Wolfgang Streeck, "On the Institutional Conditions of Diversified Quality Production," in Egon Matzner and Wolfgang Streeck, *Beyond Keynesianism*, pp. 21–61 (Aldershot: Elgar, 1991). Michael Piore and Charles F. Sabel, *The Second Industrial Divide: Possibilities for Prosperity* (New York: Basic Books, 1990). Robert Boyer and J. Rogers Hollingsworth, *Contemporary Capitalism: The Embeddedness of Institutions* (New York: Cambridge University Press, 1997). Robert Boyer and Yves Saillard, *Regulation Theory: The State of the Art* (New York: Routledge Press, 2002).

14. Charles F. Sabel, Horst Kern, and Gary Herrigel, *Collaborative Manufacturing: New Supplier Relations in the Automobile Industry and the Redefinition of the Industrial Corporation* (Cambridge, Mass.: International Motor Vehicle Program, Massachusetts Institute of Technology, 1989). Charles Sabel, *Work and Politics* (Cambridge: Cambridge University Press, 1982). Suzanne Berger and Michael J. Piore, *Dualism and Discontinuity in Industrial Societies* (New York: Cambridge University Press, 1980).

15. Paul Hirst and Jonathan Zeitlin, "Flexible Specialization: Theory and Evidence in the Analysis of Industrial Change," in *Contemporary Capitalism: The Embeddedness of Institutions*, ed. J. Rogers Hollingsworth and Robert Boyer (Cambridge: Cambridge University Press, 1997).

16. Charles Sabel, "Flexible Specialization and the Re-Emergence of Regional Economies," in *Post-Fordism: A Reader*, ed. Ash Amin (Oxford: Blackwell Publishers, 1994).

17. Michael Borrus and John Zysman, " 'Wintelism' and the Changing Terms of Global Competition: Prototype of the Future?" BRIE Working Paper 96B (Berkeley: BRIE, 1997).

18. Ibid.

19. Steve Vogel et al., eds., *The Highest Stakes: The Economic Foundations of the Next Security System* (New York: Oxford University Press, 1992). Jay Stowsky, "Secrets to Shield or Share? New Dilemmas for Dual Use Technology Development and the Quest for Military and Commercial Advantage in the Digital Age," BRIE Working Paper 151 (Berkeley: BRIE, April 2003).

20. By vertical control we mean both *vertical integration* from inputs through assembly to distribution, as in the case of American auto producers, and the *"virtual" integration* of Asian enterprise groups, as when Japanese producers of consumer durables effectively dominate market relations with semi-independent suppliers through the Keiretsu group structure. See Masahiko Aoki, *The Japanese Firm as a System of Attributes: A Survey and Research Agenda* (Stanford, CA: Center for Economic Policy Research, Stanford University, 1993). Masahiko Aoki and Ronald Dore, eds., *The Japanese Firm: The Sources of Competitive Strength* (New York: Oxford University Press, 1994). Masahiko Aoki, *Information, Incentives, and Bargaining in the Japanese Economy* (New York: Cambridge University Press, 1988). Michael L. Gerlach, *Alliance Capitalism: The Social Organization of Japanese Business* (Berkeley: University of California Press, 1992).

21. Global Value Chain Initiative, http://www.globalvaluechains.org/.

22. Barbara Baran, "The Technological Transformation of White Collar Work: A Case Study of the Insurance Industry," dissertation (Berkeley: University of California, 1986).

23. Carl Shapiro and Hal R. Varian, *Information Rules: A Strategic Guide to the Network Economy* (Boston: Harvard Business School Press, 1999) (p. 59 refers to the versioning of IBM printers).

24. Certainly downloading the program would also be sale of a product, but it confuses the presentation.

25. Lisa Bernstein, "Private Commercial Law in the Cotton Industry: Creating Cooperation Through Rules, Norms, and Institutions," Olin Working Paper No. 133 (Chicago: University of Chicago Law & Economics, August 2001).

26. Andy Schwartz's work on Enron produces this insight. Schwartz et al., "Enron's Missed Opportunity: Enron's Refusal to Build a Collaborative Market Turned Bandwidth Trading into a Disaster," BRIE Working Paper 152 (Berkeley: BRIE, February 2003).

27. Francois Bar, "The Construction of Marketplace Architecture," in BRIE-IGCC E-conomy Task Force (Ed.), *Tracking a Transformation: E-commerce and*

the Terms of Competition in Industries (Washington, DC: Brookings Institution Press, 2001).

28. This categorization follows Bar's, previously cited (note 27).

29. Clearly, the meaning of manufacturing, or production, changes as software becomes more important. At one point a central office switch cost tens of millions of dollars to develop and several thousand workers to manufacture. Then, by the early 1990s, the development costs became a billion dollars, but with semiconductor, board stuffing, and automated assembly the manufacturing could be done with a few hundred people. Early versions of routers and Internet access equipment were really honed when the product was already in the field in the hands of very sophisticated early users, universities and early Internet service providers. And there were serious mistakes with stories of early product catching fire because heating problems were not resolved. In any case, if the product must work first time out for more conventional users such as telecom companies, the lines between development, production, distribution, and support vanish. Consequently, the manufacturing solution can be workable at the beginning. Is assuring the product will work at the beginning of the cycle a design and development problem that can then be handed over to contract manufacturing folks, or does that design and development expertise require hands-on internal production of the hardware?

30. It seems appropriate to use the definition from an online dictionary: WordWeb Online Dictionary, http://wordweb.info/.

31. The critical question, once we acknowledge that software production is a form of manufacturing, is: What are the most effective ways of organizing software production? For this discussion, the list begins with the conventional questions of whether to outsource, of where, geographically, to locate software development. The story becomes interesting when we ask whether to choose conventional hierarchical production structures typified by Microsoft or new alternatives such as the commercialization of Linux products developed in an open source model.

32. Niko Waesche, *Internet Entrepreneurship in Europe: Venture Failure and the Timing of Telecommunications Reform* (Northampton: Edward Elgar, 2003).

33. Steven Weber, *The Success of Open Source* (Boston: Harvard University Press, 2004, in press).

34. Ibid.

35. Ibid.

36. Michael Borrus, Jim Millstein, and John Zysman, "US–Japanese Competition in the Semi-Conductor Industry" (Berkeley: Institute of International Studies, 1982). Michael Borrus, Dieter Ernst, and Stephan Haggard, eds., *International Production Networks in Asia: Rivalry or Riches?* (London: Routledge, 2000).

37. What happened in semiconductor development was that at a moment of new technology development, the then two major dominant established players—IBM and ATT—were restricted by antitrust competition concerns from producing

semiconductor products for sale in the merchant markets. But the antitrust ruling was critical to that outcome, and to the emergence of the merchant semiconductor firms. That merchant sector changed the course of the information technology evolution worldwide.

38. Gary Fields, "From Communications and Innovation, to Business Organization and Territory, The Production Networks of Swift Meat Packing and Dell Computer," BRIE Working Paper 149 (Berkeley: BRIE, March 2003). Martin Kenney and David Mayer, "Economic Action Does Not Take Place in a Vacuum: Understanding Cisco's Acquisition and Development Strategy," BRIE Working Paper 148 (Berkeley: BRIE, September 2002).

11

Automotive Informatics: Information Technology and Enterprise Transformation in the Automobile Industry

John Leslie King and Kalle Lyytinen

This chapter examines the role of information technology (IT) in the transformation of the automobile industry during the twentieth century, including the complements, enablers, and constraints that make the industry one of the largest and most influential enterprises in history. It adopts a view of IT as *deep infrastructure* capable of producing the vital *information capability* necessary for transformation. The focus on infrastructure entails consideration of factors often considered obvious, mundane, and pervasive compared to the exciting accounts of new IT applications involving the Internet and the World Wide Web. Yet, as the analysis reveals, this powerful industry transformation occurred as a result of the slow accretion of new capability enabled by information technologies over a long period of time.

The objective of the chapter is two-fold. Most obvious is the effort to tell a story about the historical role of IT in the industry and to use the insights gained to predict important transformations yet to come. In this particular instance, the historical account argues that the most powerful influences arise from government and industry record-keeping systems and process control devices, while relatively little effect has been seen from the Internet and the Web. The prediction is that these influences have so dramatically shifted the liability and market characteristics of the industry that the industry itself is in the process of shifting from one that sells its products to a service industry that never sells what it manufactures. The prediction is deliberately provocative but is justified by the analysis. Nevertheless, the history and prediction are merely illustrative; whether they are "right" or "wrong" is less important than the analytical strategy they reveal.

The second objective is the explication of an analytical strategy for considering transformation arising from use of IT. A common failing of transformational accounts is the lack of a sufficiently broad and systematic view of the industry in context and the role of that context in shaping the industry's destiny. Admittedly, it is difficult to capture the complexity of even simple industries; it is much harder with something as large and complicated as the automobile industry. Still, we strive to be comprehensive on all the essential factors of transformation.

The Industry in Context

The automobile industry is technologically intensive, systemic, and institutionalized.[1] It has been technology intensive since 1885 when Gottlieb Daimler perfected a high-speed internal combustion engine with a sufficiently wide power band to enable controlled acceleration of a light vehicle. It has been highly systemic since 1907 when Henry Ford revolutionized production through large-scale parts standardization and process engineering.[2] The industry has from its inception been tied irrevocably to local and national governments, as well as to other industries that are themselves highly institutionalized.

The automobile industry today is arguably the world's largest coordinated production system. Some statistics are shown in table 11.1.

The automobile industry has extraordinary global reach. There is approximately one automobile for every 11 people on Earth, although the distribution is uneven: The ratio in the United States is one for every two or three people. The new vehicle industry is economically important; the value of new auto production and sales is between one-tenth and one-eleventh of the total GNP. New auto manufacturing as measured under NAICS code 33611 employs about 0.5% of the U.S. population, and adding the parts supply industry raises that figure to about 2.5%. In fact, the automobile industry, if considered broadly to include everything that makes the use of automobile transportation functional, is of enormous importance in the United States. It has been estimated that one out of every seven jobs here is tied to the automobile through its effects on manufacturing of new vehicles and parts, fuel, service, insurance, and so on.

Table 11.1
Car industry indicators

Global autos in use 2001	About 550 million
Global autos produced in 2001	40 million
U.S. family cars in use in 2001	135 million cars (24% of global); 215 million with SUVs, minivans, and pickups
U.S. new vehicle assembly companies in 2001	11 in the U.S.
U.S. new auto production in 2001	About 5.5 million
U.S. auto parts suppliers	About 5,000
U.S. auto parts supply employment	About 550,000
U.S. NAICS code 33611 "Automobile Manufacturing"	
Companies	About 200
Employees	About 120,000
Revenue	About $100 billion
Value-added	About $30 billion

The data further show that the number of automobiles in use dominates global production of new automobiles by a factor of nearly 14:1. If the "automobile industry" incorporates both new vehicles and those already on the road, it is overwhelmingly a *used vehicle* industry. The average service life of an automobile has been rising as the market's global growth has been decelerating. The average age of a car is now about eight years in the United States, the highest in history. The main reason for this is increased global competition and the resulting radically improved vehicle quality, part of which has been the result of intensified uses of IT. The inevitable consequence of this trend is that production will continue to face constrained growth in future even while the total population of vehicles expands and some promising new markets such as China emerge. This change is directly reflected in several contextual measures. Ownership of automobiles per household is rising, as seen in the trend toward construction of U.S. housing with garages for three or more vehicles. The manufacturing side of the industry is also experiencing significant growth, with some OEMs reporting spare parts sales accounting for between 6% and 10% of total revenue. There also has been almost explosive growth in the so-called "specialty equipment" aftermarket that provides equipment and services to customize and augment used vehicles. New automobile dealers have increased their involvement in used vehicle markets that they used to despise, while OEMs actively seek to gather a larger fraction of those markets through institution of manufacturer-certified used vehicle programs and other marketing strategies.

Curiously, the industry is not following the pattern of the typical "maturing" industry, in which demand is saturated and manufacturing becomes essentially a contest among efficient producers. Even among OEMs a number of surprises in recent years have upset the market saturation hypothesis: the sudden popularity of minivans and SUVs, the surge in light truck sales, and the success of upscale brands representing image and status. And automobile use continues to rise in both developed and developing countries. The industry, if one looks beyond the new vehicle sector, is not "mature" at all; it is in a continual state of flux and could change in dramatic new directions as a result of innovations in underlying technologies (e.g., fuel cells), production and manufacturing technologies such as build-to-order and componentization (see, e.g., Helper and MacDuffie 2001; Holweg and Pil 2001), and changes in distribution and service networks (e.g., intelligent cars). This turbulence has been a hallmark of the industry since its inception (see, e.g., Abernathy and Clark 1985) and is likely to continue in the future. The industry has also gone through more than one cycle of rapid expansion. In 1900 there were more than 50 automobile OEMs in the United States; by 1930 there were only a handful. That trend has repeated itself on the global scale: The number of OEMs rose during the industrial expansion after WWII and then entered a period of consolidation over the past two decades.

Finally, the key purpose of the industry—the automobile itself—has become interwoven into many infrastructures of modern life (see, e.g., Edwards 2001).

The automobile industry was the first systematic global industry and arguably launched the modern era of enterprise through its exploitation of mass production and the invention of both continuous production lines and Alfred P. Sloan's M-form organization. The industry was instrumental in the creation of the U.S. middle class by enabling production of vast quantities of goods at constantly decreasing prices and by stimulating the economy in ways that required steady wage increases for industrial workers. The industry was extraordinary in the speed with which it generated wealth for entrepreneurs, but it also spread the wealth throughout the society rather than leaving it concentrated in the hands of a few. Automobile design and use has become an essential feature of cultural identity. The shared mission of both OEMs and buyers is to make automobile choices a mechanism for expressing social status and individual personality. The realm of the auto extends to the creation and support of a number of large, complementary industries such as insurance, fuel supply, parts and service provision, road building, transportation planning, trucking, and so on. The auto is deeply embedded in regulatory and other institutional sectors of society at the local, national, and global levels. It is no wonder that Peter Drucker has called it "the industry of industries."

IT as an Engine of Industry Transformation

This analysis requires a broad view of both the automobile industry and IT. For our purposes, the industry includes all of the activities and infrastructure required to make automobile transportation possible. Beyond OEM and sales, this includes all of the tiers of supply in original production, and the servicing of automobile transport thereafter (fuel, insurance, streets/highways, regulation, resale, disposal). Information technology includes all modes of information collection, processing, storage, and dissemination. In addition to modern digital technologies, this includes manual methods that have been part of the industry from the earliest days.

Our focus is on *embedded* information capability. This means that the capability has evolved and been adopted and adapted to play vital and ongoing roles in the industry. Three kinds of capability are examined: (1) enhanced product platforms, (2) production and distribution systems, and (3) use and service monitoring. Each of these can be broken into specific regimes defined by either product or service boundaries or specific institutional or regulatory authority, as shown in table 11.2. There is no inherent ordinality to this list—none of the regimes is presumed to precede the others in time or importance. There are also overlaps among the regimes: For example, telematic services might include entertainment (downloading music or offering e-mail services), passenger safety

Table 11.2
Regimes of IT impact in car industry

IT impact realm	Product, process, or regulation regimes	Examples of IT applications and impacts
Use and service monitoring	Property regulation, risk mitigation, and complementary asset provision	Car and driver registration services, insurance
Enhanced product platforms	Atmospheric emissions control	Embedded digital fuel and emission control systems
	Passenger safety	Air bag and collision detection systems
	Entertainment, conviviality, and control	Sound systems, hands-free phones, remote control
Production and distribution systems	Expediting and coordinating production and distribution	MRP, ERP systems, supply-chain management (EDI, e-collaboration, e-markets)
	Manufacturer–customer relationship construction and maintenance	Car service history, customer profiling/service (On-Star), intelligent service management

(automatic call back to a service center using location-based data and analysis of vehicle status), customer relationship management (profiling user habits; producing service reminders automatically), and risk mitigation (setting car insurance premiums based on usage and driving habits). This scheme is not a definitive classification structure; it merely serves to provide order to an otherwise unwieldy topic.

Use and Service Monitoring with IT

Property Regulation, Risk Mitigation, and Complementary Asset Provision

In the earliest days of the automobile, ownership was restricted to the upper classes and to organizations that could afford expensive specialty-built chassis and coaches. When the auto came within reach of common people, it had a profound effect on the economics of households and communities. For most households an automobile remains the second most expensive piece of personal property, after a residence. Residential property law had evolved over centuries and was well established by the automobile era. No similar evolution had taken place to deal with autos. There was no established registry of ownership, and given their extraordinary theft potential (both valuable and made to be driven away), it became necessary very early to develop an official registration system so that thieves could be prosecuted and stolen property identified and restored to rightful owners (Bonnier 2001). It also soon became clear that automobiles in the hands of inexpert operators could cause mayhem, resulting in death, injury, and destruction of property. It thus became necessary to regulate operation, which meant creating a registry of legitimate operators and licensing those who met the criteria. These registries were the first systematic records systems involving individuals and households since the development of vital statistics registration (birth, marriage, death) and land records tied to individual owners (Bonnier 2001). Moreover, the fact that autos changed hands and individuals changed residences meant that these records systems required updating much more frequently than earlier systems (automobile and operator registration in many locales predated metered utility services—water, electric, gas, telephone—that later necessitated ubiquitous records systems). These automobile-related records were instrumental in ways that soon became evident.

The potential for loss quickly demonstrated the need for risk mitigation in the form of insurance, which required accurate historical records on both automobiles and operators so that premiums could be set appropriately. This meant that insurance records had to be interoperable with the official registries maintained by the government. This interaction made it possible to deliver rewards and sanctions for operator behavior: "Good drivers" could be given discounts while "bad drivers" paid higher premiums or were refused service altogether. As the systems grew more sophisticated, the disparate records of various locales were linked such that accidents or violations occurring anywhere would be known to the insurance provider as well as to law enforcement. Given the important role of automobiles in the commission of crimes, vehicle and operator registries became important tools of law enforcement generally. While not directly part of the automobile realm, such uses of automobile-related records have provided powerful incentives for governments and other institutions to improve records systems (Bonnier 2001).

The growth of automobile use quickly spawned demand for improved roads. Road construction is enormously expensive, and with the exception of a few limited-access facilities (turnpikes, bridges, tunnels), roads must be provided as public goods, open to all users. Making the entire population pay for an expensive infrastructure that only some would use is economically dysfunctional, so the early vehicle records systems provided a ready mechanism for collecting revenues. Fees to help pay for roads could be imposed not only on sale or resale but annually for operation. These revenue collection mechanisms were essential complements to the excise taxes levied on motor fuel, tires, batteries. and other supplies.

The extensive record keeping related to automobiles resulted in time-series data bases that facilitated systematic analysis. The data were valuable for market analysis and advertising, for assessment of road use patterns for highway planning, and for the study of population mobility in land use planning, which raised the question of privacy long before it became a Web issue (Bonnier 2001). In time, they became essential to the implementation of regulatory requirements related to atmospheric emissions, passenger safety, and a variety of other problems. Most of the early records systems were manual and paper based, but the industry quickly adopted new technologies as they became available, systems that today are continuously being upgraded to new levels of reliability, accuracy, and usability.

Enhanced Product Platforms

Atmospheric Emissions Control

Automobiles work by mixing fuel with air, compressing the mixture in a cylinder, and capturing the energy released by its combustion. The inevitable result is atmospheric emission of gases and other materials: carbon dioxide, carbon monoxide, oxides of nitrogen, unburned hydrocarbons, and other gases and particulates. This was not considered a problem for many decades; indeed, smoke in the air was often equated with industrial progress. Also, the pollution from mobile sources dissipated rapidly, and there simply were not that many vehicles in use. That all changed around the middle of the century when residents of communities such as Los Angeles began to notice a persistent cloud of brownish-gray gas that seemed like a mix of smoke and fog. This "smog" proved to be a revolutionary issue for the automobile industry.

Researchers at Caltech demonstrated in 1956 that smog was not merely collected emissions but was actually "produced" by sunlight acting on precursor chemicals contained in automobile exhaust and other sources. Photochemical smog became a political issue in the early 1960s in the Los Angeles area largely because the region's combination of mountains, prevailing winds, and atmospheric "inversion" trapped the gases and allowed the smog to concentrate. The Surgeon General's 1964 report on the health risks of smoking coincided with a growing sense that smog might have serious health consequences. In 1968 Congress passed the Clean Air Act mandating air quality standards, and California set its own even stronger standards. The U.S. auto industry initially fought the standards, arguing that meeting them would be either impossible or prohibitively expensive, but foreign competition produced new technologies to meet them, and this plus other factors shifted the focus of the entire industry toward finding solutions to the emissions problem.

A number of technologies had been developed to alter combustion temperatures and re-circulate exhaust gases through the combustion process, but these were not terribly effective. The big breakthrough was the development of the three-way catalytic converter, whose potential had been demonstrated in theory and in highly controlled test conditions, where it converted more than 98% of harmful gases into harmless byproducts. It had a serious drawback, however: The combustion process had to operate very near to stoichiometric conditions. Even minor deviations resulted in a catastrophic decline in cleaning efficiency. The only way to obtain the right operating conditions was to make micro adjustments in the fuel/air ratio on a continuous basis, a huge obstacle. The Lambda-sond solution developed in the mid-1970s by Volvo consisted of a special sensor between the exhaust manifold and the catalytic converter chamber that would sense oxygen partial pressure so sensitively it could detect even minor deviance from stoichiometric optimum. The sensor relayed its information to a computer that contained state tables directing the carburetor or fuel injection system to make adjustments in the fuel/air ratio in real time to keep the output gas in stoichiometric balance. This was the first application of modern IT in an automobile engine for emissions control and was incorporated in Volvo's 240 model sold in California in 1976.

This general strategy was exploited aggressively in the following years as the demands for reduced emissions were joined with expectations of improved fuel economy following the energy crises in the 1970s.[3] The general trend was toward incorporation of a larger variety of sensors and variables to more finely tune engine performance, which over the years has resulted in standardized interfaces for computer-controlled performance analysis and emissions control. Simultaneous improvements in combustion chamber design, valve performance, fuel injection, and other technologies eventually resulted in automobiles that were remarkably clean while still developing considerable power as a function of their fuel consumption level. The success of this technological development is seen clearly in the dramatic decreases in air pollution throughout metropolitan regions wherever it has been applied.

The story of atmospheric emissions control highlights one critical application of IT, but there is another, more subtle implication. California's aggressive pursuit of lower emissions proved the need not only for the manufacture of new autos with lower emissions but for the maintenance of those low emissions. The OEMs had repeatedly complained that the emissions control equipment would fail or wear out in time, presenting the customer with an expensive repair bill. The prospect of angry customers turning on the regulators might have caused the regulators to back off, but that is not what happened. They simply ordered the OEMs to implement warranties on all components related to the emissions control system for five years or 50,000 miles, thus forcing the OEMs to take responsibility for used vehicle performance as well. The rise of automobile-related records played a crucial role in the subsequent evolution of this mandate and produced surprising results when considered in light of other factors.

Passenger Safety

Early automobiles were extremely unsafe by today's standards. They were basically carts powered by an engine instead of draft animals. They handled poorly, their brakes were inadequate, they were prone to mechanical failures, and they provided passengers almost no protection.

Design and technical progress improved passenger safety. John and Horace Dodge introduced the first all-steel car bodies in mass production in 1914. Polyvinyl acetate laminated safety glass was introduced in 1938. Many minor improvements in brakes, tires, steering gear, windshield wipers, and other components appeared over the years. The OEMs in Europe had started to focus on passenger safety in the 1950s: Volvo began incorporating the now universal three-point shoulder/lap seat belt in its vehicles by 1960, and several European manufacturers were building their models with disc brakes at least on the front wheels.

The U.S. OEMs were less attentive to safety until the publication of Ralph Nader's path-breaking book *Unsafe at Any Speed* (Nader 1965). Nader's book went far beyond claims that many automobiles were unsafe for passengers; it accused the OEMs of knowingly making unsafe vehicles when they had the means to do much better. The OEMs reacted at first through denial and public relations efforts while making modest efforts to improve safety. Their slow approach came to an abrupt end in 1972 when a seemingly routine lawsuit erupted into one of the largest product liability judgments ever handed down. The case involved a 1972 Pinto hatchback that was rear-ended, rupturing its gas tank against its differential housing and spraying fuel throughout the passenger compartment. The fuel ignited and the two occupants were terribly burned, one dying within days and the other surviving with permanent disfigurement and handicaps. The trial revealed that the highest level of Ford's management knew that this specific design flaw could be remedied at a cost of $4–$8 per vehicle, but management proceeded with production without remediation because the likely losses from such accidents were low enough to make the fix uneconomical. The jury returned a verdict for the plaintiffs of $3 million in direct damages and $125 million in punitive damages.

During the 1970s and 1980s U.S. insurance companies aggressively lobbied for safety regulations for new automobiles, and the OEMs began to take passenger safety more seriously. Information technology played an important role through automatic braking systems that reduced skidding, deceleration-detecting air bag systems for supplementary restraint in the event of a crash, and automatic warnings for the driver of conditions such as doors ajar and brake system failure. Recent efforts at intelligent vehicles and highways (so-called intelligent transportation systems) are aimed largely at safety concerns as well as expediting travel time. Yet again, as with emission controls, the rise of record keeping systems would prove to have important effects on passenger safety.

Entertainment, Conviviality, and Control
Some aspects of driving were considered entertainment from the start. Sporting models were produced and racing became popular very early. The first automobile radio was installed in a 1919 custom Cunningham town car, and in 1930 the Galvin Manufacturing Corporation developed the first mass production car radio and sold it under the name "Motorola." From that time the automobile evolved as a platform for entertainment while under way. In the early 1960s Earl Muntz began aggressively marketing the continuous loop four-track magnetic tape cartridge as an aftermarket product for cars. In 1965 Edwin Lear, who had played a pioneering role in car radio, introduced an eight-track continuous loop magnetic tape cartridge system with superior price-performance, and the format was quickly adopted by recording companies. The era of custom sound had begun, and the technology rapidly evolved to include FM stereo radio, compact cassette tape decks, compact audio CD decks, and MP3 players.

Custom sound systems were popular in the aftermarket throughout the 1970s and 1980s, and the OEMs, recognizing the profit potential, began to incorporate sophisticated systems in their models, even listing the brand names of the systems as a key selling point. This explicit incorporation of equipment from up-market audio companies signaled a direct connection between the traditional home entertainment market and the automotive environment. This occurred during a period when even more people were leaving the cities for the suburbs, thus increasing their commuting time—time now available for listening to music or the radio. The compact cassette had allowed people to record music at home for playback in their cars, and the audio CD and digital MP3 have accelerated the trend. The automobile has become an extension of the home-based entertainment zone, and that trend is continuing with flat-screen LCD displays and DVD players that allow passengers to watch recorded material in the car.

Another dimension of automobile-based entertainment is voice communication. The first radiotelephones were installed in autos in the 1940s but were used mainly for business. Casual use emerged with analog cellular telephony in the mid-1980s and accelerated after the introduction of digital cellular in the early 1990s. The first cellular phones were, in fact, explicitly designed for automobile use. The high power demands of the analog transceivers and the poor battery technology of the time made so-called mobile cellular radios cumbersome. The advent of the lightweight digital handheld telephone freed the user from connection to the automobile, but the phones could be used while driving through a "hands free" system (increased risk of accident from driver distraction has caused some jurisdictions to outlaw telephone use without a hands free system).

The growing functionality of digital cellular communications allows the telephone to be incorporated in the automobile platform to communicate the status and location of the car for potential service providers and also to control different functions within the car. For example, the OnStar[4] service, which is available for many GM-manufactured vehicles, provides roadside assistance for finding directions, unlocking doors if locked out, remote vehicle diagnostics, personal calls, and other needs, including stolen vehicle tracking or summoning emergency assistance. The success of OnStar as conceived is debatable, but it represents the first instance of the automobile as an *addressable device in a wireless communications network*.

Telephony is only the beginning of wider network possibilities. Traditional telephony, which is now largely digital, might be moving toward a new paradigm called "voice over Internet protocol" (VOIP) that uses digital packet-switching to replace the traditional circuit-switched system. With VOIP a cellular phone is basically an Internet terminal, capable of carrying not only voice but any other service possible over the Internet. In addition, location-finding technologies such as the Global Positioning System enable cellular telephone devices to provide pinpoint geographic location. Whether embedded in an auto's electronics or plugged in by the operator inserting a cell phone into a slot, digital cellular telephony can become a tool by which every vehicle becomes part of a large network.[5] Some attempts are also on the way to view cars as active routers in such networks, sending and receiving packets and routing them to their final destinations.

Information technology lies at the heart of the likelihood that every new automobile will become an addressable node in a global communications network. By combining technologies such as geographic location, automobile performance monitoring, operator behavior, and time monitoring, it is easy to imagine the evolution of services such as real-time dynamic insurance pricing or dynamic time-based and use-based pricing of roads and other infrastructural services. The vehicle would keep track of where and when it is driven as well as all stops and other actions taken.[6] An insurance company would charge premiums as a joint function of the place and time and the way the vehicle is being operated. The potential for surveillance with such technology is obvious, and the full behavioral aspects of such a scenario are not yet clear. Still, it is likely that many people will trade at least some information about their whereabouts for benefits such as dramatically lower insurance premiums or special offers delivered to them in their automobile for products or services available in the immediate vicinity. Such technology also has important implications for managing other kinds of automobile-related risk, including monitoring the whereabouts of minors or tracking individuals under court-ordered restraint.

Production and Distribution Systems

Expediting and Coordinating Production and Distribution

Much recent research on IT in the automobile industry has focused on production and distribution of new vehicles and remarketing of used vehicles as part of the e-business transformation (see, e.g., summaries in Helper and MacDuffie 2001). This is appropriate given the important strides made in coordinating manufacturing and distribution networks. It is also useful to recognize the long history of organizational innovation in the industry. The development of production engineering and large-scale vertical integration of production at Ford early in the century was a watershed in the history of enterprise. Walter Chrysler's innovations in nonvertical integration through exploiting the independent parts supplier industry created an OEM that bought rather than made a large fraction of its products. Alfred Sloan in the 1920s and 1930s developed the so-called M-form organization, the multidivisional strategy that maintained uniform production of core components by dedicated divisions but decentralized design, motors, and marketing to the brands across the spectrum of price points. The M-form organization became the paradigm for the remainder of the twentieth century. Even American Motors, an amalgamation of companies that were losing market share, contributed innovations in the form of extensive outsourcing of content and supply chain management in the 1960s and 1970s. Important contributions were made by the OEMs of Sweden, France, Germany, Italy, Britain, and Japan as well. Of particular importance to IT's role in transformation is the Japanese development of techniques for quality assurance and lean manufacturing. The hallmarks of these developments are the evolution of a particular supplier-OEM relationship called the "voice" model and the use of an innovative information management process called the kanban system for coordinating production. The supplier-OEM tradition arose after WW II when Japan was struggling to rebuild its shattered industries with a depleted workforce and a dramatically changed social structure imposed by the occupation. A key part of the rebuilding was the establishment of strong bonds between labor and management that included long-term job security in exchange for labor willingness to work across boundaries between trades and to join management in problem solving. This cooperative strategy extended to suppliers and OEMs; the OEMs agreed to stay with key suppliers in good times and bad, while suppliers agreed to help the OEMs with improvements in design, production, and efficiency. The kanban system was a formalized feedback system using paper cards that would accompany part and component production and at key points in the processes be passed back up through the supply chain to streamline the flow. The system enabled the just-in-time inventory control strategy that eventually caught on among the global OEMs. The Japanese auto industry established itself as the producer of high quality, reliable cars that could be sold at the lower price end of export markets in the United States and Europe. Japanese technical innovation also demonstrated that it was possible to do some things that the more established automobile industries did not believe possible, as with the Honda CVCC engine.

Meanwhile, the U.S. industry was pursuing strategies that were more dependent on exploiting information technologies, especially digital computers and process control devices in manufacturing. The important breakthrough was numerical computer-controlled (CNC) tools that could be programmed to machine specific shapes with great precision and rapid throughput. The ability to program robotic tools to execute complex tasks on a repetitive basis allowed them to spread into body painting, glass handling, and other tasks that carried potential for causing problems in worker safety and health. The technology also eventually extended to the assembly process, with entire assembly lines choreographed by computers.[7]

The development of computer-aided design tools that can produce "programs" for CNC tools based on higher level geometric specifications soon became a part of the manufacturing technology. Initially, CAD/CAM tools were used to automate manual drafting, but as mathematical modeling was incorporated they began to be used for integrated solid modeling, wherein defined geometries could be tested through finite element analysis and other techniques before being fabricated. A designer could design a part, input the parameters of expected use, specify materials and processes for manufacturing, and execute a set of virtual tests to determine whether the part would meet performance expectations. This dramatically reduced the amount of experimental engineering required to create reliable and cost-effective components and has been extended to whole modules and to vehicles as a whole (such as crash testing). Geometric models are now widely used in the early phases of product design for simulation and to obtain feedback from potential customers. These systems offer a suite of tools for integrated product management across the whole value chain and life cycle and are used in computer-based modeling of the production process to facilitate materials requirement planning (MRP).[8]

MRP, the other OEMs' answer to the Japanese kanban strategy, emerged from the need to get better control over inventory and inbound and outbound logistics, as well as to coordinate manufacturing costs. The early MRP systems were expanded to incorporate a broad array of manufacturing resources and eventually to embrace the enterprise as a whole through enterprise resource planning (ERP), such as SAP's R3 and MySap software suites[9] and similar software platforms. German manufacturers, especially Daimler, have been leading adopters of this technology. The next steps in this area are more standardized interfaces and processes for managing data through industry-wide product specification standards and standardized business processes such as EbXML.[10]

These developments have not yet produced an impact as strong as the Japanese revolution in quality assurance and lean manufacturing, but their accumulated effects have been important. Automobile production worldwide is far more efficient than ever, as can be seen in declining car prices and improved quality, reliability, and durability. This is partly due to major improvements in materials technology—plastics, coatings, lubricants, and so on—but ultimately is due mainly to advancing knowledge about how to build what the market needs efficiently and effectively. In contrast to just two decades ago, it is now difficult to buy a new automobile of poor quality.

An important area of IT application during the past decade has been supply chain intermediation. Individual OEMs have constructed their own IT-based supply chain coordination mechanisms, all variants of the value-added network applications that draw upon electronic data interchange (EDI) standards.[11] Although EDI specifications offer general industry-wide standards for exchange of documents in the supply chain (e.g., ANSI X12 and its subgroup ASC X12H, or U.N. EDIFACT), each OEM has adapted these standards to its specific purposes. These systems link suppliers to OEMs and facilitate the documentation of order entry and fulfillment, inventory information, and payments. During the past five years this closed hub-and-spoke model has been expanded in two ways. First, many of the large players have adopted the concept of on-line supplier market portals that match supplier capabilities and OEM needs through auctions and other coordination mechanisms. The major example of this is Covisint,[12] a Michigan company created by a number of U.S. OEMs and their major suppliers to create a marketplace for auctions in which demand and supply can be matched throughout the industry and to develop tools whereby design and production collaboration between suppliers and OEMs can take place. It was hoped that Covisint, launched with great fanfare in December 2000 and riding the crest of the dot.com wave, would revolutionize the supply chain, driving down prices and improving coordination among suppliers and OEMs, but it has failed to achieve its goals of becoming a market clearing mechanism for high value-added parts and transacting $300 billion in business. There have been many successful auctions through Covisint, but most have been for commodity parts; there is little evidence that it has changed the operation of the supply chain, and there are doubts about whether it can support the full array of relationships between OEMs and suppliers now common in the industry.[13]

Another important development in supply chain intermediation has been the emergence of e-collaboration systems or supplier collaboration portals. They offer a unified channel for suppliers to interact on product design information, quality statistics, order cycle fulfillment, demand estimation and planning, and logistics. The main value of these systems[14] has been dramatic cost reductions compared to VAN-based solutions, easy global reach of technologies enabling global coordination, and integration of supply chain information in one accessible platform. These systems are in their early phases of adoption, but they already have changed supplier relationships in specific situations.

Marketing and distribution have been influenced by IT in two basic ways. The first is by providing purchase-relevant information, although such services have been around in print form for a long time. The Kelley Blue Book of prices was first published in 1926, and automobile magazines containing new car reviews are as old as the industry. Extending these publications to the Web was not difficult.[15] The more important contribution is that users can compare features and prices of specific models and find a large amount of information on any given model quickly. The general view among marketing and distribution experts is that such services have dramatically reduced the traditional advantage of dealers over buyers by shifting the information asymmetries to the benefit of buyers.

The second form of IT-enabled marketing and distribution has been the reference provider that attracts potential buyers with the promise of attractive pricing and matches the buyer with a dealer willing to sell at the price the intermediary has negotiated. There was much speculation about whether the OEMs could begin selling directly to customers using the Web, thus starting disintermediation, but most U.S. states forbid the direct sale of automobiles from OEMs to customers, and in any case, the dealers provide important services that OEMs would have difficulty providing currently. The services that have emerged in cyberspace have demonstrated that the right kinds of services available on the Web can provide an important consolidation function, bringing potential buyers into an investigation of particular brands and models, collecting data from the customers, and passing that information on to participating dealers in the customer's area.

Some of these services, such as Auto-by-Tel.com[16] or CarsDirect.com,[17] place possible customers in touch with dealers and also provide meta-markets for other car-related services such as insurance, car loans, and so on. Others, such as FordDirect.com,[18] are a joint effort of OEMs and dealers to steer customers to particular products and to offer possibilities to configure and specify cars to order. And still others, such as Auto-Nation.com,[19] are basically the Web storefronts of large dealer chains of both new and used cars. The effect of these services on consolidating distribution channels seems to be less significant thus far than the effect of the prepurchase information services. Yet, considerable evidence suggests that customers are increasingly turning to the Internet for key parts of their purchasing activities.

After-Sales Markets: Manufacturer-Customer Relationship Construction and Maintenance

There has always been a peculiar relationship between automobile manufacturers and customers. Few mass-produced products have inspired the same brand recognition and loyalty that automobiles do. The cultural and social presence of the automobile even extends to national identity, with considerable anxiety expressed when an iconic marque goes out of business (e.g., American Motors) or is purchased by a foreign OEM (e.g., BMW's acquisition of Rolls-Royce). This kind of relationship might be important for the manufacturer in its efforts to persuade purchasers to be repeat customers, but it does not attach to specific vehicles owned by specific customers, especially after warranty expiration, nor to used car resale. With one exception, the only residual relationship between the manufacturer and customers is the replacement parts market. The exception is financing.

The Commercial Investment Trust Company partnered with Studebaker to create the first automobile financing program in the United States in 1915. Other OEMs developed their own financing subsidiaries, some of which are among the largest financial companies in the world. Financing new purchases connected the OEMs to their customers in important ways, incorporating into their records systems employment, income, and other information that would never have been collected in a straight sales transaction. Although there was relatively little exchange of this information between the manufacturing and marketing arms and the financial subsidiaries, the latter were creating within the OEMs the skills in large-scale records systems that would prove important later. The collateral for the loan was the vehicle itself, and the company's rights of ownership had to be recorded to enable recovery in the event of loan default. Thus, the OEMs began to develop interorganizational information systems involving government regulatory agencies.[20]

Financing has also evolved in subtle ways. The rise of leasing as an alternative to purchase became popular in the 1980s and is now stabilized at the level of 25% of all car "trades." The reasons for this are complex, and their analysis is beyond the scope of this chapter. Some are institutional in origin and relate to taxation rules that make it more economical to lease in certain situations. Some of them are competitive and deal with the need to increase customer loyalty and to better understand customer behaviors. The fortunes of OEM-based leasing have waxed and waned for various reasons, not the least being their lack of expertise on running large leasing programs. The important fact of leasing is that it has demonstrated an operationally viable alternative to outright ownership that might be a harbinger of things to come.

With a lease, the automobile user establishes and maintains a very different relationship with the dealer and the OEM than in the past. In the traditional sales model, the buyer could disregard the OEM once the warranty was ended. Similarly, when the primary source of dealer profit was new car sales, the only incentive the dealer had to engage the buyer after purchase was to cultivate future purchases. With the market decreasingly a matter of new car sales, dealers as well as OEMs are trying to garner a larger share of what used to be the aftermarket. Leasing contributes to that trend, because when a lease expires the user looks for a replacement vehicle while the used vehicle returns to inventory. Even if leased autos account for only a fraction of total new vehicle transactions, the OEMs and dealers must develop complete protocols for handling leases, building the capacity to treat all transactions as though they were leases even though not all are.

This becomes significant when added to the fact that the relationships among OEMs, dealers, and buyers have changed dramatically. The main causes of this are the increased service life of automobiles together with a major change in warranty structures. New vehicle warranties were introduced very early to induce customers and to redress variance in materials and build quality, but it was in the interest of the OEM to make the warranty as restrictive as possible. For many decades the standard U.S. new car warranty was 12 months or 12,000 miles. Now it is difficult to find warranties shorter than 36 months or 36,000 miles, and competition has made warranties a major factor in marketing. Hyundai's recent introduction of a 10-year, 100,000-mile power train warranty far exceeds any previous standard warranty, but warranties of five or more years are not uncommon. The "power train" clause is important, because it denotes that different aspects of an automobile might be handled differently. Consumables such as fuel, lubricants, tires, batteries, and light bulbs were almost never covered by new car warranties. Similarly, crash damage was covered not by warranty but by insurance. The warranty declared only what the OEM determined to be its responsibility. The increasing standard warranty is an explicit recognition that the OEM is accepting more responsibility for the vehicle after sale.

In fact, U.S. OEMs have been forced to accept a large increase in responsibility for the after-sale lives of their automobiles.The federally mandated warranty on emissions-related components in effect required manufacturers to assume liability for mechanical system failures of a large part of the vehicle's value for a long time into the future. Because every manufacturer had to meet the mandate, there was no comparative advantage in simply abiding by the law, but there was potential marketing advantage by extending the warranty significantly to cover major aspects of the vehicle that tend not to fail, thereby making it seem that the company was voluntarily accepting responsibility for its products in a most laudable fashion. An OEM that failed to offer the new, higher standard warranty risked appearing as though it was not confident in the quality of its product. In fact, the new Hyundai warranty was created explicitly to overcome the company's image as a builder of inexpensive and not particularly reliable cars. Conversely, prestige marques were able to stay with somewhat shorter standard warranties because the consumer knew those companies' reputations for quality and reliability. The rise in the standard warranty was accompanied by the advent of an "extended warranty," for a significant additional fee. In just two decades, the effective warranties on the majority of new cars sold in the United States exceeded seven years and 70,000 miles.

Extended warranties are, in essence, automobile health insurance contracts, and the financial paper on them is usually held by a financial firm rather than an OEM. The financial firms' return is generated the same way as with health insurance: through actuarial analysis of the likely benefits to be paid out and establishment of premiums that will cover those costs and return a profit. The relationships among the user, dealer, and OEM shift dramatically under this model. In the past, once a vehicle was out of warranty, the owner had to pay the dealer or an independent service agency for repairs. Dealers did not necessarily relish working on older models because they wanted to avoid the arguments with owners over high service costs and they did not want to deal with the supply chain problems of servicing older vehicles. The average dealer would thus fill this service with a 50/50 mix of warranty and "customer pay" work despite the fact that the latter category dwarfs the former. The independent service market was there for older vehicles, and the dealers were happy to see the older vehicles go there. OEMs cared mostly about reimbursing dealers for warranty service. The situation today is very different.

Improved customer information on new automobile characteristics and pricing make it more difficult for dealers to make large profits on new car sales. At the same time, the extended warranty structure has brought a boon in profitability of service. The vehicle owner has, in effect, "prepaid" the repair costs of and does not need to haggle over the price of service. The OEM or the financial company that pays for the repair is interested only in meeting the legitimate claims against the warranty. The dealer is in an excellent position to conduct service in a way that produces profits. In many cases, dealers make substantial profits from selling extended warranties themselves and thus can offer lucrative sweeteners such as forgiveness of deductibles if the purchaser brings the vehicle to the dealer for all service needs. In this way, new car dealers and OEMs have become much more tightly coupled to the automobile user base.

The other way in which the relationship between OEMs and customers has changed is through passenger safety. The 1972 Pinto incident established the precedent that the OEM carried liability for product design as well as for quality. This was a vital distinction, with tremendous implications. The concept of OEM responsibility for build quality was already well established. The OEM would take care of problems directly attributable to the manufacturing process, and it was assumed that normal operations would reveal such problems. Even when a defect tied to manufacturing resulted in a significant loss, the liability was linked only to that particular instance, and there was no assumption that the manufacturer had any way of knowing that the event might occur. Design liability, on the other hand, creates responsibility for the entire class of losses that might occur as a result of a defect and creates the potential for much more serious liability if it can be demonstrated that the manufacturer should have known, or even worse did know, about the defect. Since the Pinto judgment, statute and case law have refined the parameters of design liability, with far-reaching implications.

It seemed that the only big implication of the Pinto case was that manufacturers have to be more careful in design. Of course, this is true, and safety design has become a major concern as well as a marketing claim, backed up by government-sanctioned crash tests. The deeper story, however, lies in residual liability when good faith design efforts are insufficient. No complex artifact like an automobile can be designed to be perfect, and every vehicle on the road is "unsafe" in some sense. Tens of thousands of deaths and millions of injuries each year attest to this. The question is, what liability does a manufacturer have for a defect discovered after the product is built and distributed? As the law and practice have developed, the answer is, until the problem has been fixed in all good faith. This is the origin of product safety recalls, which are a major factor in the U.S. automobile industry. Manufacturers become aware of safety-related defects and issue recalls to owners of the vehicles. The recalls typically state the nature of the problem and what might go wrong under certain conditions and instruct the owner to bring the vehicle to an authorized service center (usually a dealer) for repair at the manufacturer's expense. Note that in many cases a design defect if safety related will be repaired at the manufacturer's expense even though the vehicle is out of warranty. Of course, not every possible safety concern for every vehicle on the road results in a recall. But the ubiquity of the recall practice has radically changed the relationship between manufacturer and customer.

Returning to the perspective of IT, automobile-related record keeping did not evolve for executing product recalls but for the purposes described earlier: property registration, operator regulation, tax collection, financing, and insurance. Nevertheless, the systems created for those purposes enabled the recalls seen today. Without extensive record keeping it would not be possible to make a "good faith effort" to fix every vehicle still in operation within the jurisdiction by notifying everyone who owns such a vehicle, regardless of how many times ownership has changed hands since the original sale.

The combination of the records systems, long-term warranty structures, a product design liability, and performance mandates on emissions have fundamentally altered the relationship between the OEMs and automobile users, expanding it to include owners who might never have purchased a new car. Moreover, the situation now encompasses a constellation of players—OEMs, users, dealers, financing companies, insurance companies, and government agencies. In the years to come this will also include parts manufacturers as technologies such as radio frequency identifiers (RFIDs) are used to identify major components of the car so that inspections can be made more accurately and reliability and safety information can be transmitted nearly immediately upwards in the value chain.

Conclusion: IT and Enterprise Transformation in the Car Industry

IT has been profoundly important in the transformation of the automobile industry, but the process has been far more subtle and complicated than the rhetoric about the Internet and the Web suggests. Transformation has been a slow, infrastructural, accretionary process that has produced powerful cumulative effects. The industry made significant use of IT long before the era of digital computers or even electronic unit record equipment. In fact, the close association of information capability and the automobile industry began not long after the invention of Hollerith card sorting technology. The automobile industry coevolved with modern IT and in myriad ways incorporated that technology as it grew, so the full effects are difficult to spot because they are infrastructural and invisible.

The only visible change of importance for the industry due to the Internet and the Web has been the rise in information access for purchasers, which has eroded much of the traditional advantage in information asymmetry held by dealers. There is no sign of radical disintermediation and probably will not be as long as distribution regulations remain unchanged. The other major experiment to transform the industry by changing the supply chain (e.g., Covisint) has proved to be less significant than predicted. In time, radically increased information sharing and distribution through e-collaboration platforms and improved effectiveness of the supply chain through integrated software platforms and deployment of intelligent RFID tags could have major impacts. CAD/CAM systems are being used as product life cycle management tools and integrated into enterprise resource planning systems. Their significance lies in their ability to speed up the cycles between design and manufacturing and make product design increasingly collaborative and information intensive.

The most important impacts of IT in the transformation of the industry have been in the enhanced product characteristics of the vehicles and in the evolution and diffusion of records systems. The former has seen the automobile evolve into a platform of integrated computing technologies so extensive that a contemporary model has more than 60 processors that collectively have as much computing power as a desktop PC. The latter has transformed the relationship between OEMs, dealers, automobile users, and other actors and enabled sustained institutional regulation and new market-based services. These kinds of changes are deeply infrastructural and typically visible only on breakdown (Star and Ruhleder 1996). Such infrastructure is so familiar that it escapes notice, but it is the most important factor in any long-term assessment of sociotechnical change (Edwards 2001).

It is difficult to predict the future, but one can argue that the industry will be transformed from its old role as a product industry into what is essentially a service industry. This does not mean that there would be no "product"; rather, there could be a shift in what is considered essential to being competitive within the industry. The industry of the future could be focused on providing personal rapid transit and related services (entertainment, safety, risk mitigation, information support) for individuals and households, and within this model the product itself could play a minor role. Under such a scenario one can easily imagine a shift from individual ownership toward a rental or service provisioning model.[21]

The likelihood of a shift from product to service becomes more salient when considering that OEMs have been forced to internalize negative externalities generated by automobile use. Air pollution and passenger safety hazards have become their responsibility, and the European Union recently adopted rules making them responsible for final disposal of the automobiles they manufacture, bringing their responsibility full circle, from initial manufacture to final disposal. The time may come when they will be unable to deflect liability arising at any point in the life cycle. This is important in its own right but takes on greater importance given the dramatic growth in the population of used vehicles. Thus, the question arises, Why would an OEM that cannot shed the liability of its products and that seeks to capture more value from those products throughout their life cycle ever sell the products in the first place? It would seem more sensible to build, field, and dispose of the products through the life cycle and collect rents on use and ancillary services along the way.

Many arguments can be raised against this scenario—intransigent customer behavior, cultural preference for ownership, regulatory interventions such as antitrust—but there is no reason *in principle* why the industry might not shift toward a rental model. Moreover, IT is making such a shift increasingly feasible. The serious impediments that exist arise mainly from the industry's lack of experience with the idea, the inertia of existing ways of doing business, and the predictable opposition of incumbent interests that benefit from the current arrangements. Such impediments might prove insurmountable, and others could be added. Our purpose is to illustrate the potential magnitude of the transformation now under way.

Two issues deserve more detailed discussion. One is the growing vehicle service life in light of design liability and warranty obligations. No machine lasts forever, so the problem is bounded, but there is a big difference between 10 and 25 years. This becomes more salient given discussions under way regarding the so-called modular vehicle platform concept (Helper and MacDuffie 2001; Holweg and Pil 2001). In this design, an expensive platform contains the major motive and control systems of the vehicle, while interchangeable body components allow the vehicle to be reconfigured over time. The idea has emerged in part to cope with the challenge of deploying vehicles propelled by hydrogen fuel cells, which will be expensive to produce and will last a long time. The customer would obtain the platform on a long-term contract and reconfigure the vehicle's other parts as necessary.[22] The required records systems already enable such long-term strategies, and IT can greatly assist in the design, manufacture, and operation of such vehicles. Such a scheme need not involve customer ownership at all; it lends itself readily to the rental model. It is not important for our purposes whether the shift to long-lived platforms comes to pass. The point is that the transformation scenario of moving from product to service is robust across a number of feasible futures.

A different but somewhat related question is the likely rise in cross-border aftermarkets. For the past half century the main focus of the industry has been new vehicle manufacture and sale. The international dimensions of this revolve around economies of production and distribution, compliance with regulatory standards and content expectations, and provision of adequate service infrastructure. The traffic in used automobiles across borders has been an altogether different matter. Used vehicles move across country boundaries from "higher" to "lower" in socioeconomic terms. Vehicles that no longer pass the strict German safety inspections are sold across borders into countries that have lower regulatory constraints. The European Union's normalization of regulations and lowering of trade barriers across borders should accelerate transborder sales of automobiles. A similar phenomenon is under way in North America as trade barriers fall. The trend could be accelerated by growth in information services that facilitate such sales, increased interoperability of government records systems, and the ongoing globalization of the OEMs and large suppliers. The opening of truly international markets in used cars would bring changes not heretofore seen in the industry. How such a shift would alter the likelihood of a shift from product to service is difficult to determine, but it is not difficult to imagine multinational corporations developing offerings to provide a growing portion of the world's population access to personal rapid transit using fleets of vehicles owned from beginning to end by OEMs.

Our goal has been the explication of the mechanics of transformation in the automobile industry as a result of information capability brought by new information technology. Critics can refute the historical account, and time will tell whether our predictions will hold up. These stories are not particularly important in and of themselves. Their main role is to serve as vehicles to demonstrate a strategy of analysis that embraces much broader definitions of industry and much broader definitions of IT than transformation accounts typically embrace. This strategy is fraught with difficulties and risks, particularly related to the complexity of the stories that emerge. Nevertheless, such strategy is warranted given the extraordinary importance of IT in the transformation of human enterprise.

Acknowledgments

This research is part of the Global Electronic Commerce Project. It is supported in part by the National Science Foundation. The views herein are those of the authors and should not necessarily be ascribed to the National Science Foundation. We acknowledge the help of Glenn Mercer, John Kish, Andrew Wyckoff, Ken Kraemer, Nick Berente, Joe LaMantia, and Diane Bailey, and the assistance of Vladislav Fomin and Sean McGann of our research team during this project.

Notes

1. The automobile industry is one of the most extensively researched sectors of enterprise. The brief citations herein do not do justice to the great wealth of work on which our understanding of the industry depends.
2. Ford's Model T was the product of the world's first vertically integrated mass production system that, in 17 years, reduced the vehicle assembly time by a factor of 24 and the selling price by a factor of 3.
3. The technology also improved engine performance dramatically, assisting in the effort to remove the octane-enhancing agent tetraethyl lead from gasoline, a move necessitated by concerns about lead poisoning in the atmosphere.
4. See http://www.onstar.com/us_english/jsp/index.jsp. The discussion of OnStar is for illustration only. Similar solutions are also under way by other OEMs (for Ford see http://www.databaseanswers.com/telematics_ford.htm; for Daimler-Chrysler see, e.g., http://java.sun.com/products/consumer-embedded/automotive/whitepapers/ITCruiser-Whitepaper.pdf) and by competing independent service providers such as ATX technologies (see http://www.atxtechnologies.com/index.asp).
5. An important design boundary that also defines important competitive boundaries is whether the phone address is the address of a person or the address of the car. OnStar is based on the latter premise, while telecommunication opera-

tors and some alliances between phone and car manufacturers seem to prefer the former business model as mobile phones (and addresses) are becoming truly personal through number portability.

6. This principle is also patented by Progressive Insurance ltd based on their experiences with the pilot Autograph system. See http://www.progressive.com/newsroom/2nd_patent.asp?IMAGE1.X=21\&IMAGE1.Y=4.

7. This in fact resulted in specific production-related IT standards and networking protocols of which the MAP protocol is the most significant.

8. See, e.g., http://www-1.ibm.com/solutions/plm/.

9. See http://www.sap.com/.

10. For details see http://www.ebxml.org/.

11. For more information see http://www.disa.org/.

12. See http://www.covisint.com/.

13. This refers mainly to the "exit" model, in which OEMs switch to the supplier that provides the best pricing and quality, and the "voice" model, in which OEMs work closely with suppliers in long-term partnerships to achieve appropriate quality and price.

14. See, e.g., http://www.e-ventus.com/.

15. See http://www.kbb.com/.

16. See http://www.autobytel.com/.

17. See http://www.carsdirect.com/home.

18. See http://www.forddirect.fordvehicles.com/default.jsp.

19. See http://www.autonation.com/Corporate/Default.asp?Page=Home.

20. It is worth noting that insurance requires even more extensive information about individuals than financing, including government-kept data on operator licensing and driving behavior (e.g., chargeable violations and accidents). A few OEMs experimented with offering insurance to purchasers of their automobiles, but the practice never caught on. Nevertheless, over time the records systems involving financing, insurance, property, and operators did coevolve and become intertwined and, to a growing degree, interoperable.

21. We do not have space here to analyze different options available to do this, such as share systems (already in place) and pay-per-mile systems. One counter argument is the emotional attachment of owners to specific cars (40% of owners name their vehicles).

22. This scenario presumes that technological progress will not render the platforms obsolete before their service life is over.

References

Abernathy, W., and K. Clark (1985) "Innovation: Mapping the Winds of Creative Destruction," *Research Policy* 14(1): 3–22.

Bonnier, W. T. (2001) "On Privacy: The Construction of 'Other' Interests." PhD Thesis, Dept. of Management Information Systems, University of Calgary.

Edwards, P. N. (2001) "Infrastructure and Modernity: Scales of Force, Time, and Social Organization in the History of Sociotechnical Systems." In *Technology and Modernity: The Empirical Turn*, ed. P. Brey, A. Feenberg, T. Misa, and A. Rip. Cambridge, Mass.: MIT Press.

Helper, S., and J. P. MacDuffie (2001) "E-Volving the Auto Industry: E-Business Effects on Consumer and Supplier Relationships." In BRIE-IGCC E-conomy Project, *Tracking a Transformation: E-Commerce and the Terms of Competition in Industries*. Brookings Institution Press, pp. 178–213.

Holweg, M., and F. Pil (2001) "Successful Build-to-Order Strategies: Start with the Customer," *Sloan Management Review*, Fall, 74–83.

Nader, R. (1965) *Unsafe at Any Speed*. New York: Pocket Books.

Star, S. L., and K. Ruhleder (1996) "Steps Toward an Ecology of Infrastructure: Design and Access for Large Information Spaces," *Information Systems Research* 7(1): 111–134.

12

The Role of Information Technology in Transforming the Personal Computer Industry

Kenneth L. Kraemer and Jason Dedrick

The computer industry rarely stands still, and the ability to respond to a changing environment is fundamental to survival. While change is normally incremental, more dramatic transformations have taken place on occasion as well, driven by new technologies such as the microprocessor, the personal computer, and the Internet, by corporate strategies such as IBM's decision to rely on outside suppliers for key components in its original PC, or by the success of new business models, such as Dell's direct sales/build-to-order approach. The emergence of the PC and IBM's strategy for entering the market led to a transformation of the industry structure from vertical integration to horizontal specialization and to the creation of a global production network for computer hardware. Dell's success is now transforming the industry from supply-driven to demand-driven production, literally reversing the polarity of the supply chain.

Major changes in firm and industry structure have been driven mainly by competitive pressures in the industry and strategic responses to those pressures. However, information technology has played a vital role in enabling many of these changes. Motivated in part by the desire to demonstrate the capabilities of their own products, computer companies have long been leaders in using IT to improve their own processes and to coordinate activities within the firm. In addition, they have developed IT networks linking suppliers, customers, and business partners to improve efficiency throughout the value chain. The Internet has been especially important in supporting the shift to demand-driven production by making it easier to coordinate information-intensive processes across a global production network.

The Changing Structure of the Personal Computer Industry

The personal computer is based on a modular architecture whose components, peripherals, and software can be designed independently and integrated into the final system using standard technical interfaces (Ulrich, 1995). This product architecture, which became the dominant design with the introduction of the IBM PC in 1981, led most PC makers to rely on outside suppliers of most components and peripherals. The result was a horizontally segmented industry structure in which most firms compete in one or two segments only, making components, subassemblies, systems, or peripherals, developing software, or providing sales, distribution, technical support, and other services (Grove, 1996). Today, all of the components needed to assemble a desktop PC are available from a vast network of suppliers and a finished system can be built with little more than rudimentary technical knowledge.[1]

By the mid-1990s, the PC industry had matured into a fairly well-established industry structure based on specialization of functions across the value chain (figure 12.1). PCs were assembled by the major vendors using standard assembly line production methods, with production volumes set to meet demand forecasts. Components and subassemblies were shipped by suppliers to meet production schedules. Finished systems were sent to distributors, which held inventory for sale to retailers and resellers, which also held inventory for sale to the final customer. All of this entailed high levels of inventory throughout the system and many transfers and touches of the product on the way to the customer. The biggest winner under this industry structure was Compaq Computer, which supplanted IBM as the number one PC vendor in the United States

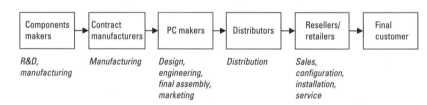

Figure 12.1
Indirect sales value chain.

and worldwide in 1994 by combining aggressive pricing with widespread distribution and a reputation for high quality.

Factors Disrupting the Industry Structure

Rapid Decline in Prices, Acceleration of Product Cycles

This apparently stable industry structure was disrupted by a series of changes in the late 1990s. One factor was the rapid decline in PC prices, which had long remained in the $2,500 range. By the end of the millennium, thanks to falling component costs and a shift of manufacturing to low-cost locations, the average selling price of a PC had been cut in half. While much of the price decline reflected lower components costs, price wars among PC makers had driven gross margins down as well. In addition, there was acceleration in the rate of product cycles, driven by competition in the microprocessor market, that led Intel to speed up its introduction of new processor generations. This led to faster depreciation of components and finished goods, putting a premium on minimizing inventory throughout the value chain (Curry and Kenney, 1999).

Direct Sales and Demand-Driven Production

A more fundamental destabilizing force in the industry was the success of the direct sales, build-to-order strategy exemplified by Dell and Gateway. Under this model, PC makers assemble systems as orders come in, usually allowing customers to choose from a set of configurations, and ship the product directly to the customer (figure 12.2). Direct sales bypasses distributors and retailers, taking out their profit margin. Meanwhile, business processes were fundamentally altered by the shift from supply-driven (build-to-forecast) to demand-driven (build-to-order) production. The direct sales/build-to-order model reduced inventory across the supply chain, giving it a significant cost advantage. It also allowed PC makers to achieve product differentiation through customization, in

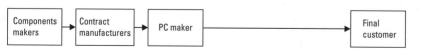

Figure 12.2
Direct sales value chain.

an industry whose products were otherwise almost impossible to tell apart (Kraemer et al., 2000; Dedrick and Kraemer, 2002).

The direct model, particularly as executed by Dell, achieved superior performance in measures such as net profit margin, return on equity, and inventory turnover and also enabled the PC vendor to develop a close relationship with the final customer. Indirect sellers such as Compaq, IBM, and HP worked with channel partners to develop hybrid direct delivery processes, leading to major improvements in efficiency across the industry. For instance, the industry average for inventory turnover rose from 12.9 turns per year in 1998 to 92 turns in 2001 (Kraemer et al., 2000; Hoovers Online, 2002).

Widespread Adoption of the Internet

The third factor in the transformation of the industry was the widespread adoption of the Internet by PC makers and their customers in the late 1990s. Dell and Gateway had been selling PCs directly to customers for over a decade by 1995, when the Internet first became available for commercial use, and had managed to capture a combined U.S. market share of just 10%. By 2000, that share had tripled to over 30%. While the Internet was not the only factor, it did play to the advantage of the direct vendors, which found it easy to offer online sales using the same infrastructure for product configuration and order fulfillment already in place to support telephone or catalog sales.[2] With no distributors or resellers to contend with, there was no problem with channel conflict when the direct vendors began to sell online. The Internet also allowed computer makers to offer a variety of services such as customized extranets to large customers and online support to smaller ones.

Changes in Firm and Industry Structure

The result of these foregoing forces has been a significant transformation of the overall industry structure and of supply chains.

Outsourcing of Production

The shift from indirect to direct selling and supply-driven to demand-driven production in particular has been accompanied by changes in the way that PCs are produced and by whom—namely, the large-scale outsourcing of production to contract manufacturers, both at home and abroad. From the beginning of the industry, most PC makers relied on outside suppliers of components (the major exception being Japanese vendors who sourced many components internally). Motherboard and notebook manufacturing were initially done in-house by the major PC vendors such as IBM, Compaq, and Apple. Other PC makers, such as Dell and Gateway, outsourced motherboard and notebook manufacturing from the start.

The companies that originally specialized in board assembly included contract manufacturers (CMs) such as SCI, Solectron, and Flextronics, as well as Taiwanese firms such as Asustek, GVC, and FIC. Other Taiwanese firms such as Quanta, Arima, and Inventec specialized in notebook production. Over the years, both groups have extended their capabilities and expanded their activities well beyond simple board assembly. The major CMs have expanded globally, invested in advanced manufacturing equipment, and also added capabilities such as new product introduction, parts procurement, production planning, logistics, and after-sales services (Sturgeon, 2002).

Taiwanese manufacturers developed strong design skills focused on the PC and became what are now called "original design manufacturers" (ODMs). Taiwanese ODMs often develop their own designs for motherboards and base notebook models, and PC vendors select from those designs to create their products (Dedrick and Kraemer, 1998). In other cases, PC vendors work with ODMs to design new models. Some ODMs also offer final configuration and after-sales services in major markets (S. H. Chen, 2001).

PC makers have outsourced much of their final assembly to CMs and ODMs, which operate increasingly in locations with low-cost labor such as China and Eastern Europe.[3] However, the move from build-to-forecast (BTF) to build-to-order (BTO) production means that rather than having long runs of the same product, firms must assemble PCs to fill individual orders, making final assembly a more complex, information-intensive process. Most PC makers have chosen to do BTO production internally, partly because they have developed sophisticated order fulfillment applications to integrate order entry, manufacturing, financial, and logistics functions. Some CMs also have these capabilities, and there may be a shift to more outsourcing of assembly. For instance, in early

2002, IBM sold its desktop assembly plants in the United States and Scotland to Sanmina-SCI (Dyrness, 2002), while HP sold its only PC assembly plant in Europe, located in France, to Sanmina-SCI (Ristelhueber, 2002). Outsourcing depends on effective communication and coordination with suppliers. As product cycles have shortened and time-to-market has become more critical, effective coordination has required increasing use of IT within and among firms. PC makers have upgraded their internal IT capabilities while also building electronic linkages with suppliers (Kraemer et al., 2000; Dedrick and Kraemer, 2002). In addition, major contract manufacturers have made investments in advanced IT systems that enable them to manage complex production processes, optimize capacity utilization across multiple plants, and manage inventory. Historically, Taiwanese suppliers have been slow to develop their own IT capabilities, but in recent years the ODMs have invested in enterprise information systems, adopted EDI, and are now investing in e-commerce technologies in conjunction with their customers and suppliers as part of a Taiwanese government program (T. J. Chen, 2001).

Flexible Supply Chains

With the success of direct sales, greater use of outsourcing, and increasing capabilities on the part of contract manufacturers, distributors, and other supply chain members, the PC industry has shifted to a more flexible industry structure, as seen in figure 12.3. Depending on the product, manufacturing may be done by the PC maker or a CM or ODM, while final assembly might be done by a CM, a PC maker, or even a distributor such as Ingram Micro. Some products are sold direct, some through one intermediary, and some through two. Customer services are provided by PC makers, resellers, and specialized service providers.

The capabilities of the CMs, distributors, and other specialists such as logistics companies give PC makers the flexibility to design different supply chains for individual products and markets. To illustrate, in the European market, Compaq outsources production of standard desktops to two Taiwanese manufacturers in the Czech Republic. Compaq is involved only in information flows and does not take physical possession of those products (figure 12.4). On the other hand, it produces configure-to-order PCs at its own assembly plant in Scotland and ships them

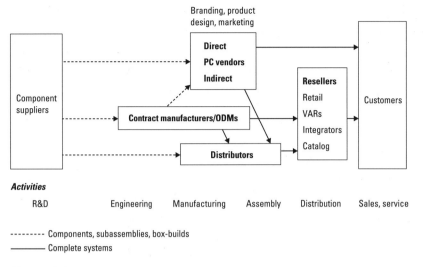

Figure 12.3
PC industry structure, 2002. Adapted from Kenney and Curry 2001.

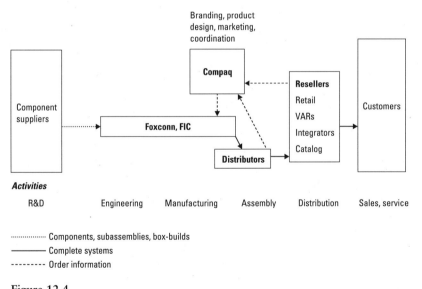

Figure 12.4
Compaq Europe supply chain for standard desktop PCs. Source: News reports and company interviews.

directly to customers (figure 12.5). Compaq uses a somewhat different organization of its supply chain for these products in the United States, with more reliance on channel partners. However, in order to offer custom configuration, it bought Inacom's final configuration facilities, bringing that process in-house after trying to work with outside distributors before.[4]

Another example of flexible supply chains is IBM's use of different delivery systems for desktop and notebook PCs. IBM has outsourced desktop assembly in the United States and Europe to Sanmina-SCI, which provides build-to-order configuration and fulfillment of third-party options orders. This moves IBM facilities off its books and takes advantage of Sanmina-SCI's lower cost structure and manufacturing capabilities. For notebook PCs, however, IBM has concentrated its production in its own joint venture facility, IBM-Great Wall, in Shenzhen, China.[5] The difference is that IBM's ThinkPad notebooks are a flagship product in the corporate market, and one in which IBM's design skills are an important differentiator. Since notebook manufacturing and design require much closer integration than in the case of desktops (due to issues

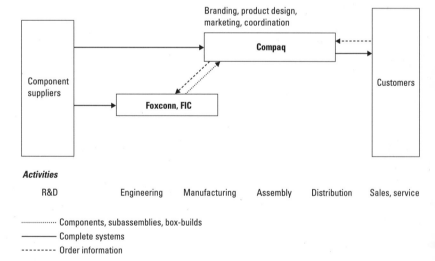

Figure 12.5
Compaq Europe supply chain for build-to-order desktop PCs. Source: News reports and company interviews.

such as miniaturization, battery life, heat dispersion, and the need for ruggedness), IBM finds advantages in coordinating the two internally. Also, notebooks are light enough to ship economically by air, so it is possible to produce for the global market from one low-cost location and still meet delivery requirements.

Other PC makers use still different supply chains for different products. Dell assembles desktops in its own facilities, taking advantage of its long experience and skills in BTO assembly, while it outsources notebook production to Taiwanese ODMs, which have skills in notebook design and manufacturing that Dell lacks. Apple outsources most of its notebook and desktop production to Taiwanese suppliers such as Hon Hai and Quanta and performs final assembly in-house only for its high-end BTO products.

Thus, the transformation that has occurred has provided greater flexibility within the industry as a whole, and also greater flexibility within individual firms, enabling PC makers to tailor their supply chains to product lines and even individual products. Such flexibility involves tremendous complexity and requires tight coordination of the various supply chains, which has been possible to manage largely though the innovative use of IT.

The Use of IT, Internet, and E-Commerce in the PC Industry

The PC industry has invested in a variety of IT systems and applications over the past decade, including internal IT applications by PC makers and their suppliers, customers, and business partners and external networks and applications that link those firms together. In particular, there has been a large-scale adoption of the Internet and an expansion of e-commerce as means of integrating the value chain. These investments have been associated with the major changes in business processes described above, such as outsourcing, demand-driven production, and direct sales.

Internal IT Systems

PC makers have reorganized the factory floor, replacing standard "progressive" assembly lines with cell production, in which a small team builds individual PCs to order. They have adopted just-in-time inventory systems and in some cases moved suppliers into the assembly plant to support build-to-order production. They also have linked sales, assembly, and service functions together so that production reacts quickly to demand, salespeople have information to push products that are available in inventory,[6] and tech support has complete information on each unit that is shipped.

These changes in business processes have been enabled by the introduction of manufacturing planning systems, factory floor applications, order management systems, and enterprise resource planning systems, as well as middleware to link the various applications together (Kraemer et al., 2000; Dedrick and Kraemer, 2002). Such systems have enabled firms to improve operational efficiency, reduce inventory, and better coordinate sales, manufacturing, procurement, and customer service. They have given managers better information to make decisions and have provided the necessary infrastructure to support online sales. Dell has gone the farthest in linking internal information systems and processes via a mix of custom and purchased applications, while IBM and Apple have implemented SAP enterprise software as a key part of revamping their fulfillment processes.

Similar investments have been made by major value chain partners such as contract manufacturers and distributors, enabling them to offer a wider range of services. CMs such as Flextronics and Solectron have systems that optimize use of plant capacity and manage the production process from procurement through manufacturing, delivery, and customer support. Sanmina-SCI's order management system can take orders from customers such as IBM and HP and manage the entire order fulfillment process. Distributors such as Ingram Micro and TechData can provide order fulfillment for PC companies and online retailers using the distributors' own product databases and order processing systems. These information systems can be utilized to serve multiple customers and thus reap economies of scale for the CMs or distributors making the IT investment.[7]

The robustness of internal IT systems is one reason that the most complex processes such as build-to-order production and custom configuration are usually handled by the PC vendor or by a single partner. Internal IT systems can coordinate a variety of functions, such as checking technical specifications, financing options and availability of components, managing complex production schedules, downloading software, tagging products, and transferring relevant information to sales and service personnel. Until more robust standards and sophisticated applications are available on the Internet to manage such processes across company boundaries, it is likely that these processes will remain integrated within one firm or a close partnership.

External Networks

The PC industry was an early adopter of the Internet, using the web to sell its products, provide customer service, and communicate with suppliers and business partners. The most aggressive was Dell, which began selling online in 1996 and by 2000 claimed that half of its sales volume was web-enabled in some way (Kraemer and Dedrick, 2001). It also offered a variety of online services, many of which were tailored for large corporate and institutional customers. For instance, Dell's Premier Page portals[8] are customized for corporate customers and include capabilities for online procurement, order tracking, asset management, software upgrading, and technical support. Large customers can have Premier Pages customized further to link to their own internal procurement systems, allowing orders to be sent directly from the customer's information systems to Dell's order management system.

Direct vendor Gateway likewise was early to offer online sales and service (Dedrick et al., 2001). Indirect vendors such as IBM, Compaq, and HP were more cautious about selling online due to concerns over channel conflict but have increasingly promoted direct sales and offered online services to improve efficiency and customer service. Apple has offered online sales and support and has gone further, offering various online services to consumers, including the popular iTunes music service.

The direct, build-to-order system was not created on the Internet, but it is well-suited to the Internet, given the thousands of possible product configurations and the need to match procurement and production to constantly shifting demand. This requires an integrated order management system able to provide information to external partners to manage production planning, procurement, payment, order tracking, and technical support. In the past, these interfirm transactions were usually handled by simple faxes and phone calls and in some cases by electronic technologies such as EDI. In recent years, web-based applications and extranets have been used to communicate with value chain partners.

The PC industry uses the Internet for a variety of functions, including product configuration, sales transactions, information exchange, and customer service. Pure Internet sales account for only about 10% of PC sales in the United States (IDC, 2001), while Internet-enabled sales account for a larger share.[9] A large percentage of business-to-business sales probably are carried out on the Internet or via EDI. For instance, Intel has shifted the majority of its sales to the web, while Dell, HP, IBM, and others conduct much of their procurement online.

In spite of the large investments in IT and the industry's reputation as a leading user of the Internet, the PC industry value chain is still linked by an uncoordinated mix of information systems, ranging from EDI to web-based applications, e-mail, faxes, phone calls, and in-person meetings. There are few common standards across the industry. EDI is expensive and limited in capabilities and therefore used only with major suppliers or customers. Creating closer links between separate firms' internal IT systems often requires costly customization. Outside the United States, the situation is worse, especially in Asia, where the largest share of manufacturing takes place. This lack of standardization is the main driver of RosettaNet, an industry effort to set XML-based standards for exchanging information on the Internet. Major distributors, CMs, component makers such as Intel, and some PC makers are leading the drive to these industry-wide standards.

IT and e-commerce are most often used to automate and standardize transactions between existing business partners. There have been attempts to create public and private B2B exchanges by companies such as ECNet, eConnections, E2open, and Viacore. Viacore is currently setting up private hubs for Cisco, HP, Arrow Electronics, and others based on RosettaNet Partner Interface Processes (PIPs) (Spiegel, 2002). Some of these companies have struggled, however; for instance, eConnections went out of business in 2002, and E2open shifted from trying to build exchanges to selling software that integrates customers' various supply chain management applications. ECNet, based in Singapore, pulled out of the U.S. market after failing there and now concentrates on Southeast Asia.

Even Dell had a brief fling in running an online marketplace. In September 2000, Dell unveiled the Dell Marketplace in partnership with software company Ariba to give small businesses a single site for purchasing office supplies and other goods and services from different suppliers. However, in February 2001, Dell pulled the plug on the Dell Marketplace, saying that customer interest was not sufficient to justify continued operation (Kraemer and Dedrick, 2001). In general, the difficulties faced by public trading exchanges have been related to governance issues and to the unwillingness of participants to share information with suppliers, customers, or competitors because of their implications for market power, both of which have limited participation. In contrast, in private networks market power and governance is in the hands of the network orchestrator who has worked out acceptable relationships among the participants.

Impacts of IT, the Internet, and E-Commerce on the PC Industry
What impacts have internal IT and interfirm networks had on the PC industry? We find that the Internet and other forms of e-commerce, combined with internal IT applications, have enabled many of the recent changes in the structure of the industry.

• Information technology has been a critical enabler of the shift from supply-driven to demand-driven production by allowing the timely movement of large amounts of complex data up and down the value chain. The internal IT systems of PC makers transmit necessary information among internal units and with external partners to fulfill their functions. The internal IT systems of CMs and distributors are important resources that enable these firms to expand their capabilities and geographic scope. As a result, they can offer a wider range of services, and PC makers can outsource entire processes to them.
• Use of the Internet enables greater speed and flexibility in the value chain, which is especially important in demand-driven processes. The Internet provides a common infrastructure and set of standards to all firms, without the need for investment in expensive proprietary network infrastructure. As a result, at least some forms of information can be exchanged quite easily, such as product and price information, sales and production forecasts, inventory information, and technical documentation. This information is sufficient to support a wide range of transactions, given the standardization of most products and some processes.

• The Internet and EDI support the industry's modular production and distribution network by standardizing information sharing and allow for greater flexibility in designing the value chain. Direct shipment from a contract manufacturer to the customer is simpler and faster than having the product change hands several times but requires complex information flows to trigger and record physical actions and financial transactions.

• The direct sales model is highly compatible with the Internet. The cost advantages of the direct model would likely have caused disruption with or without the Internet, but the Internet accelerated the shift in market share toward direct vendors and the urgency of indirect vendors to react. This disintermediation has not been nearly as pronounced outside the U.S. market, however, due to factors such as lack of trust in online transactions, customer desire to see and touch the product before buying, and the preference to have a service relationship with a local merchant.

• IT investments can lead to higher returns on assets or overall investment as transaction volumes increase. The investment in an order management system is mostly a fixed cost, with the marginal costs of handling additional transactions being very low. By contrast, each time a physical good is handled, there is a marginal cost. If information can be used to reduce the number of times a product is handled, or to reduce the need for inventory, the cost savings can be significant.

The driving forces behind the changes in the PC industry have been a combination of the direct sales model, falling prices throughout the industry, and faster product cycles increasing the importance of inventory costs. The use of IT, the Internet, and e-commerce has enabled many of the changes in the industry and has helped shape the new forms of industry organization that have emerged. While IT has enabled many changes in the PC industry and had positive impacts on performance, the adoption of the technology has been slowed by various factors. These include the limited capabilities of some members of the value network and the discomfort many companies feel about sharing information with trading partners.

While major PC companies and their immediate suppliers are large companies with relatively sophisticated IT capabilities, there are thousands of second-tier suppliers as well as distributors, resellers, and service providers whose capabilities are much more limited. Major improvements in supply chain management might be realized if these firms were linked electronically, since PC makers and others would have a more complete picture of supply and demand. However, smaller firms often have limited internal IT systems and are accustomed to interacting with partners via phone, fax, and maybe e-mail. There are some efforts to bring such suppliers into the industry's electronic networks, such as Taiwan's Plan A and Plan B e-commerce projects, but so far these plans have taken time to reach down to smaller second- and third-tier companies (T. J. Chen, 2001).

Attempts to encourage greater information sharing across the value chain have been stymied by the fear that sharing information will put firms at a disadvantage relative to their competitors, suppliers, or customers. For instance, components suppliers don't want PC makers to know how much inventory they have on hand, since the PC makers could squeeze them on price if they have extra inventory to move. Attempts to create industry standards for information sharing such as RosettaNet suffer from the same types of problems. First, a large number of participants must agree on standards that might make their own information more transparent. Also, an industry leader such as Dell is not anxious to join a standards effort that might reduce some of the competitive advantage it has achieved by developing its own private networks with suppliers and customers.

So, while the PC industry is relatively advanced in IT use compared to other industries, it remains a highly decentralized industry linked by a mix of IT and communications networks, overlaying a dense network of interpersonal relationships and tacit knowledge. Coordination of business processes across the value network involves many idiosyncratic relationships between firms, rather than a consistent standardized approach. While the transformation taking place in the industry has had some dramatic results, e.g., on inventory turnover, the process is driven largely by the activities of individual PC makers and their immediate partners. This trend is likely to be reinforced by the adoption of private e-commerce exchanges by firms, such as HP, which are adopting industry standards but applying them in a proprietary setting.

Conclusions

The PC industry has been transformed significantly since the mid-1990s. Facing intense price competition, faster product cycles, and pressure from Dell's rapid market share gains, the entire industry has moved to adopt new structures and processes. The result is greater use of the direct sales channel, demand-driven production, and modular production networks. Within these networks, firms are flexible in designing value chains for different products and markets, with each firm selecting a different mix that takes into account its own capabilities and strategies.

While this transformation has been driven by competitive pressures, the use of IT, the Internet, and e-commerce has been a key element in this transformation. They have enabled and supported the shift from supply-driven to demand-driven production and the formation of different value chains to most effectively support demand-driven production processes. They also enabled changes in the structure of the industry's global production network, making it possible to coordinate design, production, and logistics on a regional or global basis. As a result, PC makers have been able to locate these activities where costs are low and key skills are available, or close to major markets.

The transformation of the industry, while significant, has not been comprehensive. Some large distributors have disappeared, but this is more of a consolidation than a disintermediation, since distributors such as Ingram Micro, Tech Data, and Arrow Electronics still play critical roles in the value chain. The direct model has not gotten much traction outside the United States and a few other markets, so Dell's industry leadership is not global.

In terms of industry structure and processes, the major transformation has been the shift from supply-driven to demand-driven production and the creation of more flexible, information-intensive value chains to support this complex process. This change has led to dramatic reductions in inventory, better use of assets, and leaner operations throughout the industry. Beyond this, outsourcing and globalization have been going on for decades but have accelerated in recent years in the face of intensified price competition.

In the end, this price competition has driven the margins out of the industry so thoroughly that only Dell has consistently been profitable in PC production, while others have dropped out (Packard-Bell), merged (HP and Compaq), or scaled back their scope (IBM and Apple). The biggest winners have been Microsoft and Intel, whose near-monopoly positions have allowed them to capture nearly all of the profits generated by the industry in recent years.

Implications for Managers

The changes taking place in IT and organization structure have implications for managers in the computer industry and in other industries with similar characteristics such as rapid product cycles and product obsolescence and the need to use global production networks. These are illustrated by the automotive, apparel, and electronics industries currently, but we expect that as more industries are impacted by the pressures of globalization, these changes will have implications for other industries as well. They suggest that IT-enabled structural change is a source of competitive advantage for firms that apply IT to coordinate their own value chains and take advantage of the capabilities of the modular production and distribution network. The sources of competitive advantage in the new IT-enabled organization structure are the substitution of information for inventory, better matching of supply and demand, and the ability to tap into external economies in the global production network. Preliminary evidence for such competitive advantage is provided by comparison of Dell's performance with that of the PC industry (table 12.1).

The Dell case illustrates that while external economies can be accessed by any firm, the demand-driven organization is best positioned to take advantage of these economies because it can use real-time information moving up and down the value chain to drive the production network

Table 12.1
Performance indicators: Dell versus industry averages, 2002

	Dell	PC industry
Net profit margin (%)	6.1	3.5
Return on assets (%)	14.4	6.9
Return on equity (%)	44.6	16.5
Inventory turnover (per year)	109	91

Source: Hoovers Online 2002.

in response to demand and, when necessary, to manage demand in response to product availability. Information systems carry the signals that coordinate the whole process and therefore contribute to the firm's competitive advantage. Shifting from supply-driven to demand-driven production is not a simple matter, and managers need to understand the degree to which business processes need to be redesigned, the amount of organizational change required, and the demands on the corporate IT function. However, firms competing against an efficient demand-driven opponent will likely not survive without making the transformation, as Dell's competitors are finding out in the PC market.

Policy Issues

The changes taking place in the computer industry value network raise important questions for national policymakers concerned with jobs, trade, and technological competitiveness in their countries. Using IT, the Internet, and e-commerce, computer makers have been able to move production to low-cost locations while still retaining the high levels of coordination across the supply chain. Partly as a result, the U.S. balance of trade in computer hardware fell from a small surplus in 1991 to a deficit of nearly $29 billion in 1998, and hardware production jobs in the United States have declined (Dedrick and Kraemer, 2002).

While U.S. consumers enjoy the benefits of cheap computers, and U.S. companies rely on offshore production to survive in a highly competitive market, the loss of U.S. jobs is an issue for policymakers. Interestingly, even relatively low-cost locations such as Taiwan, Mexico, and Malaysia are losing employment to even cheaper locations such as China. These changes are not primarily driven by the use of IT or the Internet, but the process is clearly facilitated by those technologies. Additionally, whereas in the past such transformations were limited to production work, there is now a growing trend toward U.S. companies shifting knowledge work such as engineering design, software programming, and customer support services offshore as well. They are using IT, including telecommunications, to support outsourcing to places such as India, China, and the Philippines, with a potential loss of knowledge jobs in the United States. As a result, policymakers need to be aware of these global transformations and their potential impacts on jobs and economic activity.

Acknowledgments

This research has been supported by grants from the Sloan Foundation and the U.S. National Science Foundation (CISE/IIS/DST).

Notes

1. This is less true for laptop or notebook PCs, which require more sophisticated design and manufacturing skills to achieve the required size, weight, durability, and energy management.

2. This is true outside the computer industry as well. Mail order retailers such as Lands' End have likewise used their order fulfillment infrastructure to support web sales and have been highly successful. In fact, about 75 percent of all online retail sales in the United States are accounted for by retailers who have no physical stores, including catalog and other direct sellers.

3. For instance, in 1995, Apple employed around 1,800 people in manufacturing in Cork, Ireland, which served the European, Middle Eastern, and African markets. In 2000, Apple employed only about 400 people in manufacturing, having outsourced most of its manufacturing to Taiwanese firms located in the Czech Republic and Taiwan (interview with Tommy O'Connell and Martin Collins, Apple Computer, Cork, Ireland, November 2000).

4. Figures 12.4 and 12.5 refer to Compaq's operations as of 2001, prior to its acquisition by HP. Anecdotal evidence suggests that the combined company's PC business has utilized the BTO and supply chain capabilities that Compaq had already developed.

5. Interviews with IBM PC division executives, August 7, 2003.

6. For instance, if 60 GB hard drives are in short supply, they might offer a discount on 80 GB drives that are plentiful.

7. In reality, a certain amount of customization is often required to link the information systems of business partners, but the underlying applications and databases can be used to support multiple partners.

8. Premier Pages have been relabeled Premier Dell.com since 2001.

9. IDC defines pure Internet sales as those for which the order was placed and payment made online. Other sales, referred to as Internet-enabled sales, may involve use of the web for information gathering, configuration, and order placement but do not include payment online.

References

Chen, Tain-Jy (2001) "Globalization and e-commerce: Growth and impacts in Taiwan." Center for Research on Information Technology and Organizations, University of California, Irvine. http://www.crito.uci.edu/GIT/publications/pdf/taiwanGEC.pdf

Chen, Shin-Horng (2001) "Global production networks and information technology: The case of Taiwan." Chung-Hua Institution for Economic Research, Taipei.

Curry, James, and Martin Kenney (1999) "Beating the clock: Corporate responses to rapid change in the PC industry." *California Management Review* 42(1): 8–36.

Dedrick, Jason, and Kenneth L. Kraemer (1998) *Asia's Computer Challenge: Threat or Opportunity for the United States and the World?* New York: Oxford University Press.

Dedrick, Jason, and Kenneth L. Kraemer (2002) "Globalization of the personal computer industry: trends and implications." Center for Research on Information Technology and Organizations, University of California, Irvine.

Dedrick, Jason, Kenneth L. Kraemer, and Bryan MacQuarrie (2001) "Gateway Computer: using e-commerce to move beyond the box and to move more boxes." Center for Research on Information Technology and Organizations, University of California, Irvine.

Dyrness, Christina (2002) "IBM to get out of PCs." *News & Observer* [Raleigh, NC], January 9, D1.

Grove, Andrew S. (1996) *Only the Paranoid Survive: How to Exploit the Crisis Points that Challenge Every Company and Career.* New York: Currency Doubleday.

Hoovers Online (2002) Various company reports, including Dell, Gateway, Microsoft, Intel, IBM, Hewlett-Packard, Compaq. http://www.hoovers.com/

IDC (2001) "U.S. PC channel sales: Fourth-quarter 2000 and yearend review." International Data Corporation.

Kenney, Martin, and James Curry (2001) "The Internet and the personal computer value chain." In *Tracking a Transformation: E-Commerce and the Terms of Competition in Industries*, The BRIE-IGCC E-conomy Project. Brookings Institution Press, Washington, D.C.

Kraemer, Kenneth L., and Jason Dedrick (2001) "Dell Computer: Using e-commerce to support the virtual company." Center for Research on Information Technology and Organizations, University of California, Irvine. Working paper.

Kraemer, Kenneth L., Jason Dedrick, and Sandra Yamashiro (2000) "Dell Computer: Refining and extending the business model with IT." *The Information Society* 16: 5–21.

Ristelhueber, Robert (2002) "HP to sell PC plant in France to hungry Sanmina-SCI," *Electronic Buyers' News*, January 21, 3.

Spiegel, Rob (2002) "Arrow builds e-business hub: Viacore creates Arrow Connects! with BusinessTone connectivity tool." *Electronic News*, September 23, 22.

Sturgeon, Timothy J. (2002) "Modular production networks: A new American model of industrial organization." *Industrial and Corporate Change* 11(3): 452–496.

Ulrich, Karl (1995) "The role of product architecture in the manufacturing firm." *Research Policy* 24: 419–440.

13

IT and the Changing Social Division of Labor: The Case of Electronics Contract Manufacturing

Boy Lüthje

The impact of information technology on business, the economy, and society cannot be examined without an analysis of the profound changes in the productive structure of global capitalism. In the electronics industry, a new model of outsourced manufacturing has emerged as the centrepiece of globalized production networks: *contract manufacturing (CM)* or *electronics manufacturing services (EMSs)*. This form of network-based mass production is closely linked to the disintegration of the value chain and the emergence of the "Wintelist" (Borrus and Zysman, 1997) model of competition and the rise of "fabless" product design companies in key sectors of the IT industry. In contrast to the general perception of the "informational economy" (Carnoy et al., 1993; Castells, 1996) as service- or science-based, the rise of the CM model demonstrates that manufacturing still matters in the "new economy" (Cohen and Zysman, 1987). This development also highlights the interaction of new information networks with the restructuring of production, work, and the global division of labor in technologically advanced industries.

In this chapter, we take a closer look at the restructuring of production and commodity chains in the assembly of IT hardware (such as computers, internet switching, and telecommunications hardware) and the development of information networks, the Internet in particular. Based on recent debates among political economists, industrial sociologists, and geographers, the changes in the productive system of the IT industry will be analyzed as longer term shifts in the "social division of labor" (Sayer and Walker, 1992; see also Fröbel et al., 1977, Henderson, 1989, Gereffi, 1999) in a core sector of advanced capitalism. We start with (1) a general explanation of the contract manufacturing model and (2) a brief discussion of the developing division of labor between brand-name firms and their contractors, including the organization of work on the shop floor and the global production networks of the industry. Following this, we discuss (3) the related development of new forms of e-commerce in electronics and (4) the impact of the current recession in the IT industry on the development of IT-based production networks. Our conclusion (5) will also briefly sketch some future research issues.

Contract Manufacturing in the Wintelist IT Industry

Contract manufacturing is one of the fastest growing segments in the IT industry. Growth rates have been averaging 20% to 25% per year during the 1990s; the current recession has interrupted the growth but most likely has not ended it. According to industry consultants Technology Forecasters, the global market volume in the year 2000 was $88 billion. The leading players of the industry, most of them former small subassembly companies, were hardly known a decade ago. In the year 2001, the biggest firm had annual revenues of more than $15 billion. Market concentration has been developing rapidly, with five companies of North American origin (Solectron, Flextronics, SCI, Celestica, and Jabil Circuits) emerging as the key players. The names of these companies are unfamiliar even to many insiders since CM providers do not post their brand name on any product. The *Los Angeles Times*, therefore, called the EMS industry a system of "stealth manufacturing."

Contract manufacturing integrates a wide array of productive functions pertaining to circuit board and hardware assembly, as well as product engineering at the board and systems level, component design, process engineering, parts procurement, product fulfillment, logistics and distribution, and after-sales services and repair or sometimes installation services. From the standpoint of the labor process, these functions can be grouped around the design and assembly of printed circuit boards and related components, the final assembly of systems (called *box-build*), and logistics and inventory-related work (Lüthje et al., 2002). Contract manufacturers are serving a growing range of product markets from personal computers and servers, to Internet routers and switching gear, communications equipment (especially mobile phones), consumer products such as computer games or television sets, and industrial and automotive electronics, as well as space and aircraft electronics. The wide range of manufacturing services and products that contract manufacturers provide distinguishes them from traditional subcontractors, or "board stuffers," in the electronics industry. Whereas the latter firms perform labor-intensive assembly processes strictly controlled by the brand-name owner (also called OEM), CM companies develop and manage complex production processes, often cross-national in scope (Sturgeon, 1997 and 1999). Contract manufacturing is also different from more sophisticated sub-supply arrangements in the IT industry, particularly from *original design manufacturing (ODM)*. As opposed to contract manufacturing, ODM companies own the design of the product that is supplied to OEMs and sold under their brand names (as typical for computer monitors or for notebook computers supplied by Taiwanese manufacturers to flagship companies like HP, Compaq, or Dell).

Contract manufacturing is closely related to the new forms of specialization in the IT sector, characterized by the generalization of vertical disintegration and the commodification of an increasing array of IT products previously offered as part of larger computer and communications systems (Ernst and O'Connor, 1992). The term for the ruling duopoly in the PC industry, Wintelism, has become an analytical concept for the generic forms of corporate organization and market control in the vertically specialized computer industry (Borrus and Zysman, 1997; Borrus, 2000). The leading industry players are focusing on the engineering and design of key products in highly specialized market segments. Their mission is the definition of new product markets through the development of breakthrough technologies and their rapid commercialization, creating control and economies of scale in the respective market segments. The "PC revolution" of the 1980s, in which merchant producers like Apple or Compaq together with Intel and Microsoft in the microprocessor and software fields became global industry forces, and the subsequent emergence of the networking equipment industry led by Cisco epitomize this development (Lüthje, 2001). A most significant element of this shift is the fact that an increasing number of vertically integrated OEMs have been embracing the rules of Wintelism (Lüthje et al., 2002).

Changing the Social Division of Labor

Wintelism, as a mode of competition and market control, and contract manufacturing, as a form of manufacturing, are highly complementary. With regard to the production process, the profound changes in the intrasectoral division of labor are characterized by the following "stylized facts":

(1) The once tightly integrated value chain has become commodified, i.e., most IT products are complex commodities, assembled from traded parts and components supplied by various industry segments. The control of the time cycle of new technologies and products has become the chief problem of manufacturing organization in the industry.

(2) As market control has shifted away from assemblers towards product definition companies (Borrus and Zysman, 1997), product innovation is increasingly decoupled from manufacturing.

(3) In contrast to Fordist and also "Toyotist" industry models, there are no "focal corporations" (Sauer and Döhl, 1994) that coordinate the value chain through their own manufacturing operations. The "supplier pyramid" governed by large-scale final assemblers (as in the auto or TV industries) is replaced by networks of interacting industry segments. Hierarchy is defined by the flagships' ability to control technology development in key market segments.

(4) The acceleration of technology and product development has produced enormous instability across the value chain. Rapid expansion through the creation of new product markets is accompanied by old-style cycles of overproduction and surplus capacities—a situation that is at the core of the current slump in the high-tech industry.

New Relationships between Brand-Name Companies and Manufacturers

The hallmark of the contract manufacturing industry is a new type of relationship between brand-name firms (OEMs) and their contractors in manufacturing, resulting from vertical specialization in the most advanced sectors of the computer and telecommunications industries. The brief history of the CM industry during the 1990s reflects the trend of vertical disintegration (for an in-depth history of the early stages see Sturgeon, 1997 and 1999). The birth of contract manufacturing was marked by IBM's entry into the PC market in 1981 when Big Blue contracted the assembly of its motherboards to a no-name manufacturing company, SCI of Huntsville, Alabama. In Silicon Valley, some of the vertically specialized newcomers in the computer and network equipment industries, Sun and Cisco in particular, teamed up with specialized contractors like Solectron (a former solar energy company) or Flextronics, who subsequently became the leading players of the new industry. The relationships between vertically integrated OEMs in the United States and Europe rapidly developed during the second half of the 1990s. This happened mainly through the acquisition of entire plants through contract manufacturers, such as IBM's card assembly business in North Carolina and Texas, or Texas Instruments' Customer Manufacturing Division, which also included sales of related plants in Europe and Asia. In 1997, Swedish telecommunications manufacturer Ericsson was the first European OEM to sell off entire production units, followed by Europe's largest electronics producer, Siemens, who sold an important server manufacturing facility in Germany in 1999 and several other PC and mobile-phone plants in 2000. The current slump in the IT industry has been producing a new round of outsourcing deals, this time led by Alcatel of France and, once again, Ericsson (Lüthje et al., 2002).

The rapid expansion has brought about a highly differentiated spectrum of outsourcing relationships, emerging from various corporate strategies and traditions as well as from nation- and region-specific manufacturing practices (see table 13.1). For "fabless" technology definition companies such as Cisco, 3Com, or Microsoft (for its X-Box game console), contract manufacturers perform full-scale system manufacturing, which may include every aspect of PCB assembly, system integration, and testing. Vertically integrated OEM companies maintain similar production relationships through their outsourced plants, often in competition with their own remaining facilities. Vertically specialized mass producers in the computer industry, such as Dell, Compaq, or HP's Computer Systems Division, who still consider final assembly as an important interface with the customer, use contract manufacturers for the large-scale manufacturing of printed circuit boards or preassembled product kits. In addition, such companies outsource systems assembly in key foreign markets, mostly to medium-sized local contract manufacturers (as practiced, for instance, by Compaq in Germany or in China). It should be noted that the major OEMs from Asia—Japanese *keiretsu* and Korean *chaebol* in particular—have been relatively reluctant to use contract manufacturing in their core region. Sales of major assembly operations to CM companies until very recently mostly occurred in foreign markets outside Asia.

Vertical Reintegration among Contract Manufacturers

As table 13.1 demonstrates, the generalization of contract manufacturing is producing growing diversity and thereby increasing heterogeneity in the shape of production networks. The trend towards large-scale manufacturing cooperations of high diversity fosters *vertical reintegration* on the part of contract manufacturers. Major CMs have acquired specialized design and manufacturing capabilities in components and software as well as in supply chain management and logistics. A company such as Solectron owns sophisticated technology subsidiaries in the field of ASIC and chipset design. Flextronics has built a global business unit in the design and manufacturing of printed circuit boards; the company is also developing a telecommunications networks servicing unit, following acquisitions in this field from Ericsson. The rationale behind this vertical integration is traditional economies of scope, a trend also reflected by the transition of fully integrated manufacturing units from OEMs, which also include classical manufacturing support functions like tool-

Table 13.1
Types of OEM–CM integration

Fabless company—minimal final assembly and testing (Cisco, Sun, etc.)	Full-scale manufacturing and supply chain management (engineering, logistics)
Full-scale outsourcing of product lines and/or plants (IBM, TI, Siemens ICM, etc.)	Full-scale manufacturing and supply-chain management, plant conversion
Large-scale final assemblers with high volume outsourcing of key components (Dell, Compaq, HP CSD, etc.)	Mass production of key components (dedicated lines)
Customized final assembly in key markets (Compaq, Dell, HP PCD in Europe and Asia)	Final assembly (box-build) (includes local CM partners)
Still open: keiretsu and chaebol strategies	e.g., Sony/Solectron, Acer/Solectron, Mitsubishi/Solectron

Source: Lüthje et al. 2002.

and-die making. With regard to the IT sector as a whole it can be said that vertical specialization "at the top" (among OEMs) is matched by vertical reintegration at the level of standardized manufacturing processes.

The Shop Floor: New Patterns of Work Organization in Manufacturing

The specific role of contract manufacturers as "global supply chain facilitators" as well as the problems of integrating an increasingly complex division of labor is also engendering profound changes in the labor process and in shop-floor management practices. Contract manufacturing is producing a pattern of "flexibilized" manufacturing work with some common characteristics across the industry and its different locations. The defining elements of this form of work do *not* result from basic innovations in manufacturing technology, although large CM firms can be considered leaders in the use of advanced PCB assembly equipment and IT-based supply chain management. Basic work procedures—automated and manual PCB assembly, systems assembly, and warehouse and logistics jobs—are standard and well known throughout the electronics industry. The CM workplace is not very different from traditional electronics manufacturing operations; the predominance of state-of-the-art manufacturing environments in mid-sized to large plants sets working conditions apart from traditional board stuffing shops.

The unique characteristics of manufacturing work in the CM industry rather result from the nature of integration into the global value chains of the IT industry. Some basic characteristics of CM work can be summarized as follows:

• *"Work without a product"*: Because CM plants do not manufacture their "own" products, quality management and workplace control have to be refocused on customer orientation. Manufacturing has to be organized as "service work."

• *Relatively low wages with high variable proportions*: Because most CM plants are located in low-cost areas, manufacturing wages and benefits are rather modest, and bonus-oriented pay systems (including stock ownership and options) have to ensure customer orientation.

• *Labor flexibility*: The constant and very rapid change in production volumes is managed by an extensive use of various kinds of flexible employment.

• *Quality management based on restricted teamwork*: In most plants there is an ideology of "team orientation" but no formal structure of work groups, etc., as known from team concepts in other industries.

• *A heavy reliance on women and minority workers*: As in most areas of electronics manufacturing, the majority of the manufacturing workforce is female. In the United States, in California in particular, the workforce is mainly recruited from ethnic minorities in disadvantaged labor market positions.

CM companies pursue strict standardization of the labor process to ensure uniformity of work procedures on a global scale. Common processes are developed as a distinguishing feature of the CM model, designed to offer a uniform interface for OEMs seeking global one-stop shopping for manufacturing services. Solectron, a two-time winner of the prestigious Malcolm Baldridge Award for state-of-the-art quality manufacturing, is using the Baldridge certification criteria as a tool to trim the practices in every plant worldwide along a set of company-wide common processes. For similar purposes, Flextronics has a materials management concept developed by consultants under the name demand flow technology (DFT) that includes uniform work prescriptions for every manufacturing workplace worldwide.

Under such policies, however, we can observe highly divergent manufacturing practices across plants and regions. As we have discussed at length elsewhere (Lüthje et al., 2002), in the United States, we clearly see a low-wage/high-flexibility model, the hallmark of which is the sometimes extremely high proportion of temporary manufacturing workers in CM plants. In German and Swedish plants there is a higher degree of work integration, more sophisticated automation practices, and also a stronger role for unions and legal employee representations (such as works councils in Germany). Union wage standards are widely accepted, even in nonunion plants, although there is an increasing tendency toward concessionary bargaining and a surprisingly widespread use of temporary labor.

These differences, not surprisingly, reflect the general environment of industrial relations in the United States and the respective European countries. The coexistence of different work practices under strategies of globalized quality management reflects the limits to standardization and centralization of management control. This is reinforced by the continuous acquisition of manufacturing assets from OEMs; the growing variety of outsourcing relationships is reflected by a growing diversity of technologies and work practices on the shop floor. All of this points to the well-known fact that manufacturing know-how cannot easily be transferred across different regions and nations since it is rooted in specific local traditions of work, education, and technological learning. The CM industry is a particularly striking example here, since uniformity in working procedures—sometimes characterized as a McDonald's approach to manufacturing—is exposed as a defining element of the business model.

Transnational Production Networks

Through their continuing acquisitions CM companies act as transnational network builders, assembling a variety of plants with different manufacturing practices in specific national and global markets. Contract manufacturing, therefore, can be characterized as a mode of integrating, coordinating, and regulating diverging economic, social, and cultural conditions in global production systems (Lüthje et al., 2002). In 1996, the leading contract manufacturer, Solectron, had about 10 locations worldwide; in the year 2000 there were almost 50. In a distinctive way, CM companies strive to build a presence in every region in the triad of the capitalist world economy, combining operations in the lead economies with mass manufacturing in developing countries of the respective regions. For North America, Mexico has emerged as the prime low-cost location; for Asia, it's Malaysia and China (the latter already hosting the largest number of CM plants around the world), and for Europe, Hungary, Poland, Czechia, and Romania.

As in other segments of the IT industry, globalized just-in-time production is transforming older international divisions of labor based on the transfer of manual assembly processes with simple technologies to the Third World (cf. Fröbel et al., 1977). In contract manufacturing, technologies and processes in developed and in developing countries are rather similar. "Full-package production" (Gereffi, 1999) in low-cost locations is supported by the global standardization of work procedures pursued by major CM firms. A certain hierarchy between locations, however, is defined by three elements that have been typical for the development of most transnational CM networks during the 1990s:

(1) The lead position that product introduction centers (PICs) in developed countries play in prototyping and in the ramp-up of new product lines towards volume manufacturing. This implies inequality in the distribution of engineering capacities and the access of plants in low-cost locations to advanced engineering know-how within the global production system. It also implies a greater relative significance of skilled labor in strategic plants in developed economies.

(2) The location of specialized products with high diversity in manufacturing requirements and low volumes (low volume/high mix) in developed countries vs. standardized mass production (high volume/low mix) in low-cost locations.

(3) The concentration of specialized units in the design and manufacturing of critical components in developed countries.

This international division of labor is exemplified by the development in Eastern Europe. The buildup of CM capacities took off around 1997 and peaked in 2000/2001 before the current slump in the IT industry. The rapid shift of Western European OEMs towards fully outsourced manufacturing models was the driving force behind this development. In a very short period of time the leading contract manufacturers built plants, mostly in *greenfield* locations (see table 13.2). Most of these plants, which by mid-2002 together employed roughly about 18,000 workers, perform high-volume manufacturing, especially of consumer products such as PCs, printers, or mobile phones, for the entire European market. Printed circuit board assembly, basic hardware assembly, and final product configuration are the main functions of theses operations. The geographical proximity to the major Western European markets permits product delivery by truck to the distribution centers of customers and retailers within a time frame of 24–48 hours.

The production sites in Eastern Europe are integrated into the global chains of production and parts procurement of OEM and CM companies. This is particularly true with regard to the supply of electronic components, such as microchips, preassembled motherboards, raw printed circuit boards, and electronic displays, and passive components such as resistors or coils, which are mostly sourced from Asia. Components procurement is usually managed not by the local plants but on a centralized base, governed by large-scale contracts between CM companies and OEM customers.

Table 13.2
Manufacturing sites of the five leading CMs in Eastern Europe (09/2001)

Company	Hungary	Czechia	Poland	Other	Products
Solectron				Romania	PCBA
Solectron	Budapest				System assembly
Flextronics	Zalaegerzeg				PCBA, system assembly
Flextronics	Sárvár				PCBA, system assembly Industrial park with suppliers
Flextronics	Tab				PCBA, system assembly
Flextronics	Nyíregyháza				PCBA, system assembly
Flextronics		Brno			PCBA, system assembly
Flextronics			Gdansk		PCBA, system assembly Industrial park (under construction)
SCI/Sanmina	Tatabanya				PCBA
SCI/Sanmina	Pecs				Enclosures
Celestica		Rajecko			PCBA, system assembly
Celestica		Kladno			PCBA, system assembly
Jabil	Tiszaujvaros				PCBA

Note: PCBA = printed circuit board assembly. Source: Company information (Internet).

Within this geographic division of labor we see highly complicated chains of production emerging for individual products. The case of mobile phone production for a major European OEM, which we studied recently, may help to illustrate this point. The OEM has shifted the entire manufacturing of this product to a major U.S. CM. The CM is managing the basic manufacturing, product fulfillment (final assembly, software application, and testing), and the logistics for the European market. To this end, the CM is operating a huge fulfillment center in Hungary with a capacity of seven million handsets per year. This center is receiving the manufactured handsets without software ("dummies") from a factory of the CM in China. The Chinese operation is handling the assembly of the printed circuit boards and of the handsets, including supporting functions like the manufacturing of plastic enclosures. The highly complex printed circuit boards are sourced from the CM's PCB manufacturing facility in China, with some engineering support coming from the CM's operation in Germany. The dummy handsets are shipped to the Hungarian fulfillment center by aircraft, where the operating software is applied according to the orders from telecom operators or retail chains. This task, again, is of high organizational complexity, as each handset has to be equipped with software in the language for the country of destination and the specific requirements of each operator and retailer. After being packed in boxes, the handsets are finally delivered just-in-time to 15 different European countries, a process run by a Dutch trucking firm under a long-term contract with the CM.

Holding the Pieces Together: Information Networks, E-Commerce, and the Manufacturing Value Chain

The changes in the social division of labor in IT hardware manufacturing engender enormous challenges for the organization of manufacturing value chains. Information networks play a prominent role in the integration of labor processes in the highly flexibilized and globalized environment of the CM industry. Following our analysis above, the basic challenges can be summarized as

• Integration of highly diverse manufacturing cultures and conditions of production between OEM and CM and across a networks of plants with sometimes very different organization and work practices;

• Management of an increasingly complex organization within contract manufacturers, as related to the vertical reintegration of the production system of individual companies;

• Organization and control of a service-oriented manufacturing environment; and

• Integration of global production processes along very different conditions of production in various regions and countries.

Contract manufacturers have been at the forefront of developing new IT-based models of supply chain management. Apart from more traditional Electronic Data Interchange–based systems of data exchange, leading contract manufacturers were at the forefront in the implementation of Internet-based forms of supply chain and manufacturing management. However, the real organizational changes associated with the Internet have to be assessed in the context of the existing practices of supply chain and shop floor management. Here, our observations are pointing to a highly diverse picture.

Internet-Based Factory Models

Only a few companies have developed Internet-based integration of manufacturing and enterprise resource planning systems. Full-scale "virtual factory" relationships seem most advanced between fabless OEMs and CMs (see table 13.1). Cisco's manufacturing organization can be considered a leading-edge model. The assembly of Cisco routers, switches, etc. is integrated into an order and resource planning system that is entirely based on Internet standards. Major contract manufacturers are part of this arrangement. They also manage delivery and repair services, offering a seamless interface to the Cisco customer. Major CMs are operating manufacturing plants fully dedicated to Cisco product lines. The virtual integration of these plants into the Cisco organization has to be secured by a sophisticated control system for manufacturing data on the part of Cisco and through a high degree of personal interaction between engineers of both companies, supported by the physical proximity of the respective operations in Silicon Valley and other high-tech manufacturing centers around the world (Roberts, 2000).

However, plants that are operating within an Internet-based virtual factory framework do not differ significantly in their work organization from more conventional ones. Given the high degree of control over manufacturing data that existing IT networks offer to OEMs, tighter control of the shop floor through Internet-based data networking does not seem very likely. As opposed to traditional subcontracting arrangements, OEMs clearly tend to leave the management of the labor process to their manufacturing partners. The most important impact of the Internet on shop floor conditions, therefore, may probably be indirect: the expansion of e-commerce-related configured-to-order manufacturing is likely to increase the pressure to flexibilize work and employment (Lüthje et al., 2002).

As can be observed in many consumer goods industries, e-commerce-based direct sales strategies impose enormous organizational challenges on traditional assembly, warehousing, and logistics work. For instance, PC world market leader Dell's Internet-based ordering system (Dell and Magretta, 1998) dramatically increases flexibility requirements in final assembly. Because each computer product has to be configured to customer order, manual assembly work is making a remarkable comeback in computer plants. Dell's assembly operations in the United States almost entirely rely on manual labor with relatively low formal skill requirements. In Europe, some of the most successful indigenous contract manufacturers as well as high-end OEMs like Hewlett-Packard are using similar practices. One contract assembler for consumer PCs in Germany is operating almost entirely on the basis of manual labor (drawn from a local labor pool in an electronics industry center in East Germany with an average unemployment rate of 17%; Lüthje et al., 2002).

The role of Internet-based virtual factory models, however, should not be overestimated. Most OEM-CM cooperations are functioning on the basis of IT networks that are not specifically operating on Internet technology. Our example from mobile phone manufacturing cited above may illustrate this. On the part of the OEM, the whole process is controlled through the OEM's ERP system, which has a design and supply chain management system specifically developed for mobile handset manufacturing. The CM's ERP software has been adapted to work with the OEM's system, allowing comprehensive monitoring of order, quality, and delivery data for each individual handset in every stage of the manufacturing and logistics chain. For this purpose, each handset has a bar-code label, allowing built-to-order manufacturing from the basic assembly processes through the final stages of product fulfillment and delivery. Given the enormous costs as well as the reliability requirements in such a system, a migration to Internet-based networking architecture is likely to happen more slowly, perhaps as part of a general remodeling of the OEMs ERP system.

The longer term impact of the Internet, however, may emerge from the role of contract manufacturers as "global supply chain facilitators." As we discussed, the specific organizational know-how of transnational EMS firms is in the integration and coordination of different work practices and production cultures within worldwide production systems. Internet-based manufacturing promises to facilitate this function because it requires the definition of uniform interfaces between manufacturing procedures in different plants and locations. The multiple options of customizing IT networks within a uniform architecture makes the Internet ideal to deal with different types of customer relations, production cultures, and political and social regulations. In this perspective, probably the strongest driver for Internet-enabled network architectures is economies of scale derived from flexibility in standardization.

Procurement and Electronic Marketplaces

Of special importance here is the procurement of electronics components and parts. Contract manufacturers have developed sophisticated know-how in managing this portion of the value chain and exert considerable buying power. Parts and components are either purchased by contract manufacturers on behalf of their customers or by OEMs themselves. Contract manufacturers also have relationships with global electronics parts distributors such as Arrows or Avnet. Local sourcing of parts and components, as already mentioned, remains relatively limited, usually restricted to nonstrategic items such as cables, sheet metals, or plastic parts. The calculation of prices, volumes, and availability of parts is essential to the CM industry. The problem is complicated by the cyclical nature of most component markets.

Electronics parts markets sprang up in the year 2000 at the peak of the Internet boom. The two most important ones were The High Tech Exchange, which encompassed major computer and chip manufacturers (among them HP, Compaq, Hitachi, Samsung, NEC, and, from the CM side, Solectron; *Electronic News*, San Jose, May 8, 2000), and

E2open.com with strong participation from major players in the telecommunications and networking fields (Ericsson, Hitachi, LG Electronics, Matsushita, Motorola, Nokia, Nortel, Philips, Seagate, Toshiba and, again, Solectron). Supported by technology from IBM, i2, and Ariba and with financing from major investment banks such as Morgan Stanley, E2open boasted a combined parts purchasing volume of $200 billion per year (*New York Times*, May 30, 2000). The concept of both projects is mirroring e-market initiatives in other industries, such as Covisint in the auto industry or Chemplorer in the chemical industry, which include major manufacturers, e-commerce software companies, and telecom network operators.

Integrating Purchasing and Design

As part of the development of e-marketplaces, contract manufacturers have been actively supporting new application service providers (ASPs) in the electronics design field. Their software concepts go beyond mere parts trading. Design ASPs develop database systems that integrate parts purchasing with the design process and product introduction at the assembly line level. Startup companies like Silicon Valley–based Spin Circuit are promoting data-exchange systems for the design of printed circuit boards and hardware, bringing together OEMs, parts producers, distributors, and contract manufacturers. These design gateways promise seamless interfaces between product designers, manufacturing engineers, and parts suppliers who all will become part of a single Internet-based exchange system. Design engineers may even be able to change their product layouts according to the cost and availability of parts tracked in online databases. Some contract manufacturers are heavily supporting startups in this field because early participation seems to offer the opportunity to control crucial nodes in global Internet-based manufacturing networks.

The future impact of the integration of component design and trade on supplier networks and on manufacturing and engineering work is still difficult to assess. The suggestion seems plausible, however, that global electronics parts and component markets will foster the de-localization of sourcing relationships, which is already characteristic for the contract manufacturing industry. On the labor side, we may expect substantial rationalization of engineering work, an increased separation of product and process engineering, and a diminished role for personalized cooperation between product and manufacturing engineers within local industry networks. In qualitative terms, engineering work may become still more oriented towards nontechnical, commercial factors such as cost and parts availability. It may also mean increased competition for engineers in developed countries from outsourcing of engineering work to low-cost regions. However, such a development will mainly reinforce existing trends, since contract manufacturers are actively making use of local engineering talent, which in some locations, especially in Eastern Europe, is widely available.

Managing the Crisis: Supply-Chain Management and Global Overcapacities

The developments described above mark profound changes in the production system of the IT industry, which seemed hardly conceivable a few years ago. However, the dynamics of vertical specialization and reintegration under the auspices of Wintelism is also creating new potentials of crisis and massive economic and social risks (Lüthje et al., 2002). As the current recession in the IT industry illustrates dramatically, contract manufacturing is at the center of the restructuring of key industry segments engendered by enormous overcapacities in computer, data networking, and telecommunications equipment. Contract manufacturers seem to be among the prominent victims of the crisis.

The unexpected stagnation of growth rates, which became visible in 2001 (see above), indicates a massive rupture in the ultra-rapid growth pattern of the 1990s. However, the slowdown seems to be only temporary in nature, since the recession is also accelerating the outsourcing of manufacturing on the part of brand-name firms. Recent developments such as IBM's sale of most of its PC manufacturing capacities to Sanmina-SCI (*Wall Street Journal*, 01/11/2002), the acquisition of Hewlett-Packard's PC factory in France by the same contract manufacturer (*Financial Times*, London, 01/19/2002), or the transfer of large-scale manufacturing assets from Lucent to Solectron, including major plants such as the one in North Andover, Massachusetts, seem to justify the expectation of further growth.

Against this background, the role of contract manufacturing in the regulation of industrial overcapacities becomes particularly visible. As in most other industries, overcapacities in IT systems manufacturing are structural in nature and global in scope. In IT, this development has become particularly pronounced as the massive influx of speculative capital has enforced technological progress and market cycles. Latent overcapacities have been built up in almost every segment of the industry, particularly driven by the acquisition of plants in industrialized countries and by the expansion of mass manufacturing capacities in low-cost locations in Asia, Mexico, and Eastern Europe (Lüthje, 2001). Since 2001, the major contract manufacturers have suffered heavy losses and declining revenues. The leading companies had to take massive restructuring charges for plant closures, overpriced acquisitions, and inventory excess (*Financial Times*, London, 06/21/2002). At the same time, large-scale job-reduction programs and plant closures were carried through. Layoffs mostly affected volume-production sites in low-cost areas in the United States, such as Texas, the Carolinas, or Georgia, in similar areas in Europe (Scotland and Ireland in particular), and even more so in the newly established manufacturing complexes in low-wage countries such as Mexico, Malaysia, or Hungary (*Financial Times*, London, 10/26/2002). Loans of some very large CMs were downgraded to junk bond status.

One particular source of financial losses was the huge inventories of electronic components that contract manufacturers held on behalf of their customers. These components were either purchased directly by the manufacturers and built into the assembled products or were owned by brand-name firms or parts distributors and assigned for use and inventory management to the manufacturing service companies. The complicated arrangements in this field produced what a well-known consulting firm aptly called a "supply chain disaster." In many cases, the ownership of excess inventory was indeterminable. Both sides, contract manufacturers and OEM customers, tried to leverage their respective customers or manufacturing partners to take financial responsibility for significant amounts of excess parts and components (*Electronic News*, San Jose, 04/09 2001). Few details became public, but market data indicate that contract manufacturers had to bear the major portion of the excess. According to iSuppli, a well-known consulting firm, at the peak of the crisis, during the third quarter of 2001, contract manufacturers held about 49% of the global inventory excess in semiconductors, estimated at $5.9 billion globally (*Electronic Business Asia*, Hong Kong, 03/2002; *Electronic News*, San Jose, 03/06/2002). At every end of the production chain, OEMs and CMs have reacted with massive restructuring of their supply and purchasing operations. This process is put forward within the context of highly publicized mergers in key industry sectors, intended to reduce global overcapacities, as in the case of the recent takeover of Compaq by Hewlett-Packard. Major OEM companies have been centralizing their relationships with contract manufacturers across various corporate departments and product lines. In order to counterbalance the growing bargaining power of large contract manufacturers, some first-tier OEMs have started to establish key accounts for their relationships with contract manufacturers, resulting in the selection of a small number of preferred CMs for worldwide operations.

Contract manufacturers on their part are further centralizing supply chain management. Solectron recently announced a sweeping reorganization of its global purchasing organization, designed to overcome inefficiencies caused by the rapid growth of recent years. According to company sources, responsibility for purchasing decisions will be shifted away from individual plants or customer specific teams to a company-wide organization, as already practiced by other competitors. The target is a radical cut in the number of key suppliers and concentration at the global and the regional level (with purchasing organizations for North America, Europe, and Asia). About 250 suppliers are targeted to make up about 80% of Solectron's purchasing spending of roughly $14 billion annually, down from 550 in recent years. The supply chain organization is developing company-wide schemes for supplier evaluation, bidding, and purchasing procedures. This standardization is considered crucial for the broad implementation of web-based supply chain tools, with the intention to lift the proportion of web-based purchasing procedures from 35% to 75% of the overall purchasing volume (Carbone, 2002).

The slowdown in the IT industry and the related centralization of supply chains has also impacted most business-to-business (B2B) projects. According to a study by A.T. Kearny in 2001, out of the 17 public exchanges in the electronics industry announced during the year 2000, few have reached strategically relevant trading volumes. Almost 80% of the companies that have committed to an exchange have yet to transact any volume outside of pilot activities (*Electronic News*, San Jose, 08/06/01). According to this and other accounts, the reasons for this development at least in part are structural. In an industry dominated by large trading partners with close, direct relationships and high visibility, the benefits of bringing together multiple buyers and sellers are not apparent. One consequence has been that major OEMs such as Dell and CMs such as Celestica have refocused their B2B strategies on the development of private exchanges with direct connections to suppliers and customers (*Electronic News*, San Jose, 09/03/01).

At the same time, the recentralization of supply chains inside and between major OEMs and contract manufacturers is likely to further complicate the development of industry-wide supply chain standards (as promoted by industry consortiums such as Rosettanet). Inside the industry, there seems to exist a widespread consensus that a sweeping overhaul of the supply chain organization is needed; however, there is no consensus on the questions of how this should be achieved and who should do it (Spiegel, 2002). The major problem seems to be that contract manufacturers and OEMs are connected very closely but the "downstream" integration of business processes towards parts suppliers, distributors, and smaller components producers is lacking. Supply chain integration at this level seems to be much more complex—not only for the greater diversity and number of participants but also because the global production networks of the vertically specialized IT industry include scores of small and mid-sized companies in newly industrializing countries, especially in Asia. So far, only very few initiatives have been launched to integrate this element of the supply chain into what is called the "seamless" web of global production (for a discussion of this question and the related problem of knowledge transfer in global production networks see Ernst, 2001, and the recent special issue of *Industry and Innovation* edited by Ernst and Kim, 2002).

Some Preliminary Conclusions

In this chapter we have traced the relationship between new models of mass production in the context of vertical specialization and information technology networks. Our findings support the observation that information technology is not a driver of organizational change *per se* but part of a complex shift in the social division of labor, which ultimately is related to the demise of vertically integrated mass manufacturing as was prevalent in the era of Fordism (Lüthje, 2001). In this context, information technology and Internet-based models of supply chain management in many ways do facilitate and support vertical specialization.

However, our analysis also points to the double-sided relationship between "fragmentation and centralization" (Ernst and O'Connor, 1992) in the IT industry, i.e., vertical specialization and the global reconsolidation of production chains. Contract manufacturers have emerged as major actors of recentralization. In a certain sense, vertical specialization "at the top" of IT value chains (i.e., the trend towards fabless product definition companies) is matched by vertical reintegration "at the bottom" (i.e., of manufacturing assets). In the context of the Wintelist IT industry, CMs act as integrators of vast manufacturing networks but do not have the capacity to control product development and thereby market cycles.

Many of the instabilities and economic risks of the CM model are rooted in this specific constellation. The current trend of recentralizing supply chain management is one strategic answer. As we have discussed, this trend is also likely to favor further centralization of e-commerce networks in electronics component trading. There is more research needed to support this thesis. Beyond the question of centralization vs. decentralization this set of problems points to the very fundamental question of how the architecture of e-marketplaces is constructed by complex networks of corporate actors and their interaction in global market places (see Bar, 2001). This research would also have to address the problems of regulation and coordination inherent to systems of network-based mass production (such as Wintelism in the IT industry), which has not been examined systematically in the international academic debate so far. With regard to e-commerce networks, one central question could be how standardization along the supply chain could be supported by strong public standards and institutions.

Acknowledgments

This chapter summarizes results from current research funded by Deutsche Forschungsgemeinschaft (DFG) under the title "Neue Produktionsmodelle und internationale Arbeitsteilung in der Elektronikindustrie. IT-Kontraktfertigung im pazifischen Raum und Mittel- und Osteuropa." Most of the research for this chapter was carried out in close collaboration with Martina Sproll and Wilhelm Schumm; both deserve thanks for their comments and suggestions. The conceptual framework of the following analysis, with emphasis on the situation in Asia, was developed in an article for a special issue of *Industry & Innovation* coordinated by Dieter Ernst and Linsu Kim. Our special thanks extend to the numerous individuals in the contract manufacturing industry who spent their time with us on interviews and plant visits.

References

Bar, F. (2001) The Construction of Marketplace Architecture. The BRIE-IGCC E-Conomy Project: *Tracking a Transformation. E-Commerce and the Terms of Competition in Industries.* Washington, D.C.: Brookings Institution Press, pp. 27–50.

Borrus, M. (2000) The resurgence of U.S. electronics. Asian production networks and the rise of Wintelism. In *International Production Networks in Asia. Rivalry or Riches?*, ed. M. Borrus, D. Ernst, and S. Haggard. London: Routledge, pp. 57–79.

Borrus, M., and J. Zysman (1997) Globalization with borders: The rise of Wintelism as the future of global competition, *Industry and Innovation* 4 (2): 141–166.

Castells, M. (1996) *The Rise of the Network Society.* Oxford: Blackwell.

Carbone, J. 2002. Solectron prepares for the upturn. *Purchasing.* October 24.

Carnoy, M., M. Castells, S. Cohen, and F. H. Cardoso (1993) *The Global Economy in the Information Age: Reflections on Our Changing World.* University Park: University of Pennsylvania Press.

Cohen, S. F., and J. Zysman (1987) *Manufacturing Matters: The Myth of the Post-Industrial Economy.* New York: Basic Books.

Dell, M., and J. Magretta (1998) The Power of Virtual Integration: An Interview with Dell Computer's Michael Dell. *Harvard Business Review*, pp. 73–84.

Ernst, D. (2001) *The New Mobility of Knowledge: Digital Information Systems and Global Flagship Networks.* Honolulu, Hi.: East-West Center Working Papers, Economics Series, No. 30.

Ernst, D., and D. O'Connor (1992) *Competing in the Electronics Industry: The Experience of Newly Industrializing Economies.* Paris: OECD.

Ernst, D., and L. Kim, eds. (2002) Special Issue: Global Production Networks, *Industry and Innovation* 9(3) (December).

Fröbel, F., J. Heinrichs, and O. Kreye (1977) *Die neue internationale Arbeitsteilung.* Reinbek: Rowohlt.

Gereffi, G. (1999) International trade and industrial upgrading in the apparel commodity chain. *Journal of International Economics* 48: 37–70.

Henderson, J. (1989) *The Globalisation of High-Technology Production.* London: Routledge.

Lüthje, B. (2001) *Standort Silicon Valley: Ökonomie und Politik der vernetzten Massenproduktion.* Frankfurt/New York: Campus.

Lüthje, B., W. Schumm, and M. Sproll (2002) *Contract Manufacturing: Transnationale Produktion und Industriearbeit im IT-Sektor.* Frankfurt/New York: Campus.

Roberts, B. (2000) Ready, fire, aim. E-business might be risky, but doing nothing is riskier. *Electronic Business*, July.

Sauer, D., and V. Döhl (1994) Arbeit an der Kette—Systemische Rationalisierung unternehmensübergreifender Produktion. *Soziale Welt* 45, no. 2: 197–215.

Sayer, A., and R. Walker (1992) *The New Social Economy: Reworking the Division of Labor.* Cambridge, Mass./Oxford: Blackwell.

Spiegel, R. (2002) Taking on the supply-chain mess. *Electronic News* 48(1) (January 1): 23 ff.

Sturgeon, T. (1997) *Turnkey production networks: A new American model of industrial organization?* BRIE Working Paper 92A, Berkeley, Calif.

Sturgeon, T. (1999) *Turn-Key Production Networks: Industry Organization, Economic Development, and the Globalization of Electronics Contract Manufacturing.* Ph.D. dissertation, University of California Berkeley.

V
Community and Society, Home and Place

14

Public Volunteer Work on the Internet

Lee Sproull and Sara Kiesler

There are many examples of public voluntary work that are enabled by the Internet, from improving the abilities of low performing middle school students and the health of sufferers of chronic pain to providing high quality technical support to software users and producing high quality scientific image analysis. Such activities represent socially desirable goals with potentially important economic implications and require organizing the behavior of many people to achieve them. Projects directed toward these endeavors have all been successfully accomplished over the Internet by volunteers who are often strangers to one another. These net-based voluntary communities that produce socially beneficial outcomes represent an IT-enabled organization of important social activity. This chapter examines the still early days of this exciting new potential for "transforming" how public voluntary cooperative activity is organized.

Organizing Voluntary Activity Offline and Online

Many of today's models for organizing voluntary activity were invented at the beginning of the twentiethth century, a time when work, commerce, and leisure were all locally organized. These models focused on local needs, depended upon recruiting local volunteers, and organized volunteer participation through local face-to-face meetings. Even when the organizations were national in scope, the headquarters organization was small and all of the action was at the local level. Examples include youth scouting, parent–teacher associations (PTAs), the Red Cross, the League of Women Voters, service clubs, and civic associations. Self-help and self-improvement organizations such as Alcoholics Anonymous were also developed under these same local models. People voluntarily met face to face on a regularly scheduled basis to provide cognitive and emotional support to one another. Even today, civic, social, and self-help organizations typically organize people's voluntary participation in two-hour to three-hour blocks of time at a prespecified time and a particular place. However, people today are less motivated or less able to show up once a week at a scheduled time in a particular place for a two-hour to three-hour meeting. Thus, many of these organizations are in decline today (Putnam, 2000). Contrast this place-bound and time-bound model for organizing participation with that of any time/any place electronic voluntary communities. In electronic voluntary communities, people can participate in small units of time, at any time, and from any place (with technology and net access).

People have been voluntarily cooperating over the net since its earliest days, sharing resources and helping one another solve problems, mostly with a technical emphasis. Indeed, the technical operation and improvement of the net itself have been supported by voluntary cooperative work through the net-based Internet Engineering Task Force (undated). Beginning in the early 1990s, when access to the net began to become more widespread, both the number of people using the net and their social diversity increased, and this continues today. At the same time, the number and diversity of net-based voluntary cooperative communities began to grow, and this also continues today.[1] It is too soon to tell if they are "transforming" how public voluntary cooperative activity is organized. Their numbers and scope are small relative to business transformations. Their metrics for success are not so easily defined or tracked. Nevertheless, there is sufficient evidence to examine their potential.

It is impossible to provide accurate estimates of the number of net-based voluntary cooperative work groups and members. More than 54 million Americans have participated in one or more public online groups of all types but only a fraction of these are cooperative work groups (Horrigan, 2001).[2] A reasonable estimate is probably that between 10 and 15 million people worldwide participate in hundreds of thousands of net-based public volunteer cooperative communities.[3] These communities organize the productive electronic contributions of many people, most of them strangers to one another, to produce socially beneficial outcomes. This characterization does not preclude the possibility that volunteers may also meet face to face; however, the primary organizing modality is electronic. Many of these are self-organizing communities with no external sponsorship. Some, particularly in medical, educational, and scientific domains, may be sponsored by nonprofit organizations. Some, usually in technical domains, may be sponsored by for-profit organizations. But in all cases what makes them vital is the net-based contributions of their public volunteer members.

By now, the basic dynamics of voluntary electronic communities are familiar. Anyone with net access can participate at any time from any place. Someone posts a seed question, a proposal, or other relevant contribution such as a piece of code to a group and others reply with answers, comments, or improvements. People can read or send one or more messages at their convenience. If they have discretionary network access at work, they can participate during the day during the interstices between work activities. Or they can do so from home. They can fit their contributions into their own time schedules.[4] Each message itself consumes a rather small unit of time and attention and represents a voluntary microcontribution to the community.[5] This kind of microcontribution is the basic building block of the electronic community. Some people may devote hours a week to online voluntary activity, but they can do so in small units of time at their own convenience.

Many people emphasize the convenience of any time/any place participation. Just as important are the mechanisms for aggregating and organizing those microcontributions into larger units for efficiency and social effectiveness. Both technical and social mechanisms are necessary to organize the smallest units of contribution into larger units that are useful to participants. People can indicate that their contribution is a response to a previous one and can display messages as "threads"—a seed message and all responses to it. Threads organize message microcontributions so that everyone can see their constituent parts, making it easy for potential contributors and beneficiaries to see what has already been contributed. Cooperative communities may have tens or hundreds of active threads, which require a level of aggregation beyond the self-organizing thread. Here a human designer may suggest or impose a topic map or topic architecture in order to group contributions into more general topic categories. A topic common across most electronic communities is the FAQ (frequently asked questions). Additionally, the focus of a group usually suggests the structure of its topic map. Health support communities usually have topics for symptoms, medications, and side-effects, negotiating the healthcare system, and managing relationships with family and friends. Civic communities may organize contributions around civic functions such as the garden club or public library. Software development communities may organize contributions around such items as new features, bug reports, patches, or documentation. Statistical aggregation mechanisms organize contributions of ratings and votes. Custom software aggregates and organizes contributions to scientific projects. In all cases the high order organizing is done by human beings.

The next section describes examples of net-based voluntary cooperative communities to illustrate their dynamics and scope and suggests something about their potential social impact. In every case, volunteers participate at their own convenience and benefit from the aggregated and organized contributions of others.

Case Examples

Volunteer Technical Support

From the earliest days of the net, people have cooperated to exchange technical tips and solutions to problems in voluntary user groups.[6] Today, in more than 50,000 net-based groups people voluntarily cooperate to expand their knowledge about technical topics and solve technical problems for one another.[7] Volunteer user groups are recognized as providing high quality support. They have won industry awards and vendors have added them to their corporate strategy for technical support (Foster, 1999; Morris, 2001). (See Moon, 2004, for an extensive analysis of these groups.)

Volunteer Health Support

In Internet health support groups, people voluntarily provide informational and emotional help to one another (Boberg et al., 1995; Brennan, Moore, and Smyth, 1995; Davison, Pennebaker and Dickerson, 2000; Galegher, Sproull, and Kiesler, 1998; Mickelson, 1997; Mackenna & Bargh, 1998; Sharf, 1997; Winzelberg, 1997). Most Internet health support groups are self-organized and are populated primarily by non-professionals—people with health problems or concerns and their families and friends. They have no entrance criteria except for online access, and they all rely on the voluntary microcontributions of their members. There are more than 600 health support groups on the Internet, many of which are accessed through web pages or bulletin boards, Internet mailing lists, and chat sites (www.psychcentral.com). These groups offer the possibility of encountering many different perspectives on a problem, of finding people with similar experiences and pain, and, at the same time, of communicating in comparative comfort and psychological safety.[8]

More than 6.5 million Americans have participated in an online health support group, which represents about 10 percent of Americans who have gone online in search of health information (Fox, 2002). Frequent health information seekers are more likely to have joined an online support group in comparison with occasional health information seekers—13 percent of those who look for health information several times a month or more have done so, compared with 6 percent of those who look every few months or less. Those who are most in need are more likely to take advantage of online support groups. Ten percent of health information seekers in fair or poor health consulted an online support group the last time they searched for health information, compared with just 1 percent of those in excellent health. And 14 percent of those in fair or poor health have participated in an online support group, compared with 5 percent of those in excellent health (Fox, 2002).

Members of voluntary health support communities may derive physical and psychological health benefits from their participation in addition to information and social benefits. Although the evidentiary base for these benefits is quite small, it comes from carefully designed studies that use either random assignment or statistical procedures to control for other factors that could influence health status. Benefits for active participants include shorter hospital stays (Gray et al., 2000), decrease in pain and disability (Lorig et al., 2002), greater support seeking (Mickelson, 1997), decrease in social isolation (Galegher, Sproull, and Kiesler, 1998), and increase in self-efficacy and psychological well-being (Cummings, Sproull, and Kiesler, 2002; Mackenna and Bargh, 1998).

Software Development Groups

The Internet supports a large and growing voluntary cooperative software development community. People from all over the world use the net to voluntarily contribute code, documentation, and technical support to open-source software projects just because they want to. No one tells people what to work on; people work on areas that they know and care about (Raymond, 1999, p. 32). Unlike proprietary software, open-source software is governed by licenses that encourage sharing and improving code. The largest repository of open-source projects reports about 55,000 hosted projects and 500,000 registered users (www.SourceForge.net).

The "modern" open-source software development community began in 1991 when a 21-year old Finnish student posted a program on the net that he had written and invited others to use it and contribute their own code to it (Torvalds, 1991).[9] That invitation, by Linus Torvalds, was the genesis of Linux, a PC operating system that had more than 27 percent of the server operating-systems market by 2000 (Price Waterhouse Coopers, 2001, p. 241). Its high quality is signaled by the fact that it has received industry awards in multiple years for best operating system of the year (www.InfoWorld.com). Linux was entirely, and still is primarily, a voluntary software development effort consisting of thousands of developers and other contributors distributed over 90 countries on 5 continents (Shankland, 1998; Raymond, 1999). Although Linux may be the most widely known open-source software development project, others are also notable. Apache, the most widely used server software on publicly accessible websites, used in 66.5 percent of sites in 2002, is open source (Netcraft, 2002). The Internet backbone relies in some measure on open-source software (e.g., Apache, sendmail, bind) (The Mitre Corporation, 2002). Various surveys indicate that open-source developers spend an average of 5 to 8 hours a week on their projects. More than 80 percent of open-source developers are not paid for their efforts (Gosh, Glott, Krieger, and Robles, 2002; Boston Consulting Group, 2002).

Individual contributions are aggregated and organized through both technical and social mechanisms. For example, the Linux-kernel mailing list plays a large role in organizing the voluntary contributions of kernel programmers. Feature freezes, code freezes, and new releases are announced on this list. Bug reports are submitted to this list. Programmers who want their code to be included in the kernel submit it to this list. Other programmers can then download it, test it within their own environment, suggest changes back to the author, or endorse it. From June 1995 to April 2000, about 13,000 contributors posted almost 175,000 messages to the Linux-kernel list. The mailing list has a topic structure that reflects the modular structure of the Linux kernel architecture. Contributions are also organized in part by volunteer module maintainers. In 2000, there were 147 designated module maintainers who review Linux-kernel mailing list submissions relevant to their modules, build them into larger patches, and submit the larger patches back to the list and to Torvalds directly. Over the years Torvalds and the community have come to know and trust the technical competence of these maintainers, most of whom still "work on Linux for free and in their spare time" (Linux-kernel mailing list FAQ, available at www.tux.org/lkml).

Volunteer Scientific and Scholarly Work
In contrast with other types of net-based voluntary cooperative work groups, voluntary contributions to scientific projects are not self-organized. Instead, a scientific agency or investigator designs ways that individual microcontributions can be organized and aggregated in the service of a scientific or technical goal and then invites volunteers to make those microcontributions. For example, from November 2000 to September 2001, NASA Ames Research Center invited net-based volunteers to identify and mark craters on images of Mars. As they described it, this was an experiment to see if "public volunteers (clickworkers), many working for a few minutes here and there and others choosing to work longer, can do some routine science analysis that would normally be done by a scientist or graduate student working for months on end" (http://clickworkers.arc.nasa.gov/top). More than 85,000 volunteers marked and classified craters.[10] The quality of their work was "virtually indistinguishable from [that] of a geologist with years of experience in identifying Mars craters" (http://clickworkers.arc.nasa.gov/documents/crater-marking.pdf).

Voluntary distributed projects that ask people to contribute unused cycles from their PCs may lead to important computational results.[11] But projects that engage the hearts and minds of volunteers may have greater social utility. The goal of Project Gutenberg, for example, is to make available online much of the world's public domain literature. The Project relies upon volunteers to scan in OCR images of books, to proofread pages, and to manage the consolidation and digital archiving of resulting texts. In 2002 volunteers proofread more than half a million pages, resulting in more than 800 books averaging 300 pages archived for public access (http://texts01.archive.org/dp/).

Volunteer Mentoring and Tutoring
Net-based mentoring and tutoring projects use the net to support voluntary interaction among interested parties on topics of mutual interest. As is the case with all net-based voluntary cooperative communities, people participate at their own convenience. Aggregation mechanisms organize the contributions for efficiency and effectiveness. Many online mentoring projects have a career development focus organized so that professionals in a field can volunteer advice or suggestions to novices or students thinking about entering the field. These programs may formally "match" mentors and protégés or simply provide directories of available mentors and allow people to self-select their mentors. Much of the mentoring occurs through one-to-one email, although more successful programs usually organize larger electronic group discussions and coaching. Protégés may ask for specific advice (e.g., on how to prepare for a job interview) or for more general opinions (e.g., which are good courses to take). Mentors may offer a glimpse into their work (e.g., here's what I'm working on today) or on their lives (e.g., here's how my husband and I manage family obligations). Successful matches result in more than information exchange. Mentors and protégés report that the personal relationships they develop are satisfying and even inspirational (www.mentornet.net).

Some mentoring and tutoring programs are more explicitly focused on educational projects in which adults, often professionals, volunteer to collaborate electronically with individual students or entire school classes on specific projects. Online mentoring and tutoring programs are reported to have resulted in educational gains for their participants.[12] They are also reported to have resulted in increased positive career aspirations.[13]

Discussion

As more people make voluntary microcontributions, the social and legal conventions supporting voluntary information sharing will become more important. Both information ownership and liability are central. For instance, open-source software licensing currently offers putative assurance to those who contribute code that their contributions will remain accessible to benefit others. In technical and health support groups, people are assumed to own their own words but advice usually comes with cautions. Liability may be assumed to reside with the recipient, or the supplier, of bad code or bad advice. These are issues of great complexity that will continue to be negotiated over a long period of time.

Quality control is an important challenge for volunteer groups. If volunteers produce erroneous contributions or ones of low quality, the group experience can rapidly spiral downward. Most technical and health support groups rely on peer review to manage quality control. The earliest, and still most common, quality control mechanism is cautionary or challenging messages posted in response to a problematic one. Some websites now give members the opportunity to rate the contribution of others' messages. Software then aggregates and displays these ratings in an overall quality index for contributions or contributors. When contributions are more structured than text, software can use redundancy and statistical aggregation to improve estimates. For example, in the NASA crater-marking endeavor, each image was marked independently by 50 people and a composite was created from all the markings. In Project Gutenberg, each page is proofread independently by more than one person. In all cases, quality control relies upon the fact that contributions are visible to all members or to error-checking software.

Another important challenge is encouraging and sustaining long-term volunteers. Whereas some utility can be derived from one-time contributors, many groups exhibit a participation structure in which a minority of participants does a majority of the work. As in the offline world, the care and feeding of loyal volunteers is important. Some groups use positive feedback to demonstrate the value of peoples' contributions. When contributions are under software control, it is easy to provide people with a running total of their contributions, such as the number of craters marked or the number of pages proofread. Systems that let readers rate contribution quality also offer this kind of feedback. But there is much to be learned about what motivates electronic volunteers. One of the most interesting design challenges now is to envision new kinds of microcontributions and aggregation mechanisms. Another way to say this is, What kinds of social problems can people make contributions toward solving in small amounts of time from their computer keyboards?

Not all worthy endeavors can be accomplished through net-based volunteer activity. When physical resources must be collected, organized, or dispersed, (e.g., building houses, donating blood, serving in soup kitchens), people must work at the same time in the same place, but even in these situations there may be a role for net-based volunteers (e.g., Sproull and Patterson, 2004). We do not advocate replacing offline volunteer work with online volunteer work. Nevertheless, there may be many opportunities to involve new people or tackle new problems with net-based volunteer work. The organizer of one net-based volunteer project observed:

This is the Internet equivalent, I suppose, of a barn raising. People come together and volunteer their talents toward a common and laudable cause. And this type of volunteerism . . . is the real essence of the Internet. It is something that literally couldn't happen any other way or through any other medium. . . . Such collaboration simply wouldn't work without the Internet. When some engineer offers . . . two hours of labor per week, which is about the norm, the only way to get anything done is to eliminate meetings, eliminate travel, eliminate the effects of time zones, eliminate as much overhead and friction from the process as possible. And what's left over is the work, itself (Cringely, 2002).

To that we would add, what is also left is the personal pleasure of contributing and the social benefit of the contribution.

Notes

1. This growth also characterizes online entertainment and economic communities, which are not the focus of this chapter. See Sproull (2003) for an overview of different types of online communities.

2. Members of voluntary entertainment groups sometimes use their groups to organize charitable activities, but that is not their major focus (Cravens, 1999).

3. This estimate is derived from combining estimates from different domains. Some groups and members may be counted in more than one estimate; many groups and members have surely been overlooked. The estimates include more than 55,000 open source projects with more than half a million registered users (www.SourceForge.net); more than 50,000 voluntary technical support groups on the net (Moon 2004); more than 3 million people who have downloaded and run the seti@home client; somewhere between 600 and 3,000 health support groups (www.psychcentral.com; http://groups.yahoo.com); and more than 6 million Americans who have participated in a health support group (Fox, 2002).

4. In one survey of online volunteers, convenience and schedule flexibility were the two most commonly cited reasons for individuals choosing to volunteer online (http://www.serviceleader.org/vv/admin/summary.html).

5. Studies have reported a mean message length ranging from 8 to 30 lines of new text (Galegher, Sproull, and Kiesler, 1998; Sproull and Faraj, 1995; Wasko and Faraj, 2000; Winzelberg, 1997). Other studies report that people spend 10–20 minutes per session participating in their online cooperative work group (Boberg et al., 1995; Brennan, Moore, and Smyth, 1995; Lakhani and von Hippel, 2003).

6. Voluntary user support groups existed long before the net (Levy, 1984). But like other offline voluntary organizations, their membership was small and they relied on face-to-face meetings (Armer, 1980).

7. There are 1,000 Usenet newsgroups under the comp.* hierarchy, and many more user groups using mailing lists or web-based communication forums. CataList, the catalog of public listserv lists, reports 1,106 lists with the keywords helpdesk, software, hardware, technical, users, and user. Yahoo! Groups hosts at least 47,000 groups devoted to programming languages, software, desktop publishing, and networking and operating systems. (These estimates are based on searches conducted in August 2002).

8. Davison, Penebaker, and Dickerson (2000) found more frequent use of online health groups relative to the incidence of problem occurrence for health problems limiting physical mobility such as multiple sclerosis and for problems entailing immense diagnostic uncertainty such as chronic fatigue syndrome.

9. Cooperative code sharing has been prominent in the computer science research community since the 1960s (Levy, 1984).

10. Every crater was marked an average of 50 times and classified an average of 7 times; 37 percent of the contributions were made by one-time visitors.

11. More than 4.2 million people in 26 countries have downloaded and run the Seti@home client, resulting in 27.36 teraflops, 1.87×1021 floating point operations, the largest computation on record (http://setiathome.ssl.berkeley.edu). See http://www.aspenleaf.com/distributed/ for an inventory of distributed projects.

12. Middle school students paired with college student e-mentors significantly improved their scores on standardized tests of attitudes toward reading (Lesesne, 1997). The International Telementoring Project reports significant educational gains for student participants (Lewis, 2002).

13. More than 90 percent of female college students who participated in Mentornet intend to remain in an SMET field (http://www.mentornet.net/Documents/About/Results/Evaluation/). Female high school students who participated in telementoring are more likely to pursue opportunities related to academic and career success such as finding an internship or joining a sci/tech club. And 28 percent reported that communicating with their mentors had altered in a positive way their attitudes about women in science (http://www2.edc.org/CCT/admin/publications/report/telement_bomhsg98.pdf).

References

Armer, P. (1980) SHARE: A eulogy to cooperative effort. *Annals of the History of Computing* 2(2): 122–129.

Boberg, E., et al. (1995) Development, acceptance, and use patterns of a computer-based education and social support system for people living with AIDS/HIV infection. *Computers in Human Behavior* 11: 289–311.

Boston Consulting Group (2002) The Boston Consulting Group hacker survey. Available at www.bcg.com/opensource.

Brennan, P. F., S. Moore, and K. Smyth (1995) The effects of a special computer network on caregivers of persons with Alzheimer's disease. *Nursing Research* 44: 166–172.

Cravens, J. (1999) Fan-based online groups use the Internet to make a difference. Available at http://www.serviceleader.org/vv/culture/fans.html.

Cringely, R. X. (2002) Chase 2.0 Is that a supercomputer in your jammies? Available at http://www.pbs.org/cringely/pulpit/pulpit20020502.html

Cummings, J., L. Sproull, and S. Kiesler (2002) Beyond hearing: Where real world and online support meet. *Group Dynamics: Theory, Research, and Practice* 6: 78–88.

Davison, K. P., J. Pennebaker, and S. Dickerson (2000) Who talks? The social psychology of illness support groups. *American Psychologist* 55: 205–217.

Foster, E. (1999) Best technical support: It may not be the guy on the telephones any more. Available at: http://www.infoworld.com/articles/op/xml/99/11/29/991129opfoster.xml.

Fox, S. (2002) Vital decisions: How Internet users decide what information to trust when they or their loved ones are sick. Washington DC: Pew Internet and American Life Project. Available at http://www.pewinternet.org/reports/toc.asp?Report=59.

Galegher, J., L. Sproull, and S. Kiesler (1998) Legitimacy, authority, and community in electronic support groups. *Written Communication* 15: 493–530.

Gosh, R.A., R. Glott, B. Krieger, and G. Robles (2002) Free/libre and open source software: Survey and study. The Netherlands: University of Maastricht. Available at http://www.infonomics.nl/FLOSS/report/.

Gray, J. E., et al. (2000) Baby CareLink: Using the Internet and telemedicine to improve care for high-risk infants. *Pediatrics* 106: 1318–1324.

Horrigan, J. B. (2001) Online communities: Networks that nurture long-distance relationships and local ties. Washington, D.C.: Pew Internet and American Life Project. Available at http://www.pewinternet.org/reports/toc.asp?Report=47.

Internet Engineering Task Force (undated) A novice's guide to the Internet Engineering Task Force. Available at http://www.ietf.org/tao.html.

Lakhani, K., and E. von Hippel (2003) How open source software works: "Free" user-to-user assistance. *Research Policy* 32: 923–943.

Lesesne, T. S. (1997) *ALAN Review* 24(2): 31–35.

Levy, S. (1984) *Hackers: Heroes of the computer revolution.* Garden City, N.Y.: Anchor Press/Doubleday.

Lewis, C. (2002) Evaluation Results from Teacher Surveys. Ft. Collins, Co: R&D Center for the Advancement of Student Learning, Colorado State University. http://www.telementor.org/research/2002-ExecSummary.htm.

Lorig, K. R., et al. (2002) Can a back pain e-mail discussion group improve health status and lower health care costs? *Archives of Internal Medicine* 162, 792–796.

Mackenna, K. Y. A., and J. A. Bargh (1998) Coming out in the age of the Internet: Identity "de-marginalization" from virtual group participation. *Journal of Personality and Social Psychology* 75: 681–694.

Mickelson, K. D. (1997) Seeking social support: Parents in electronic support groups. In *Culture of the Internet,* ed. S. Kiesler. Mahwah, N.J.: Lawrence Erlbaum Associates, pp. 157–178.

The Mitre Corporation (2002) Use of free and open-source (FOSS) software in the U.S. Department of Defense. Available at www.egovos.org/pdf/dodfoss.pdf.

Moon, J. Y. (2004) Sustaining voluntary technical support on the net. NYU Stern School: PhD dissertation in process.

Morris, J. (2001) Doing it right. Customer support management. Retrieved May 13, 2002, http://customersupportmgmt.com/ar/customer_support_doing_right/index.htm.

Netcraft. (2002) Netcraft Web Server Survey. Available at www.netcraft.com/survey.

Price Waterhouse Coopers (2001) 2001/2003 Technology Forecast. Menlo Park, Calif.: Price Waterhouse Coopers.

Putnam, R. (2000) *Bowling Alone.* New York: Simon and Schuster.

Raymond, E. (1999) *The Cathedral & the Bazaar: Musings on Linux and Open Source by an Accidental Revolutionary.* Cambridge, Mass.: O'Reilly.

Shankland, S. (1998) Linux shipments up 212%. CNETNews.com. December 18.

Sharf, B. (1997) Communicating breast cancer on-line: Support and empowerment on the Internet. *Women & Health* 26: 65–84.

Sproull. L. (2003 in press) Online communities. In *The Internet Encyclopedia,* ed. Hossein Bidogi. New York: John Wiley.

Sproull, L., and J. Peterson (2004 in press) Making infocities livable. *Communications of the ACM.*

Sproull, L., and S. Faraj (1995) Atheism, sex and databases: The net as a social technology. In *Public Access to the Internet,* ed. B. Kahin and J. Keller. Cambridge: MIT Press, pp. 62–81.

Torvalds, L. (1991) Free minix-like kernel sources for 386-AT. Info-minix Usenet group.

Wasko, M. M., and S. Faraj (2000) "It is what one does": Why people participate and help others in electronic communities of practice. *Journal of Strategic Information Systems* 9: 155–173.

Winzelberg, A. (1997) The analysis of an electronic support group for individuals with eating disorders. *Computers in Human Behavior* 13: 393–407.

15

The Internet and Social Transformation: Reconfiguring Access

William H. Dutton

In business, the secretary was once the prime gatekeeper who screened and prioritized calls for an executive, but this role is becoming less common as e-mail is used increasingly. However, if the proliferation of communication options becomes overwhelming, people might well look again for gatekeepers to filter, prioritize, and select information for the user or consumer.

The Internet and Society

The rapid diffusion across the globe of the Internet, the World Wide Web, and new digital media has generated many forecasts of their potential to transform enterprise, governance, education, leisure, and society at large. This potential has evolved from the Internet's roots in computer and telecommunications developments of the latter half of the twentieth century but also represents a novel phenomenon linked to this ever-changing, intertwining web of people and technology. The Internet is not a technical artifact fixed in a particular point in time and isolated from the people who develop, use, and are affected by it. Instead, choices about the design, use, and governance of the Internet—digital choices—are continuously changing the communicative power of individuals, groups, organizations, and nations as it reshapes physical and electronic access to information, people, services, and technology in all areas of society.

This process of "shaping tele-access" (Dutton 1999) or "reconfiguring access" (Dutton et al. 2003) is entwined with the coevolution of a diverse range of institutional and organizational structures with the layers of network services, platforms, and digital applications based on the ensemble of technologies encompassed by the Internet. The "digital choices" made by individuals and groups involved in these separate but interrelated social processes can result in many different outcomes for the same technology.

This broad conception of the Internet shows that its technical capabilities are bound up with a wide range of personal, community, economic, business, organizational, political, legal, and other dynamics shaping—and being shaped by—this pervasive network of networks. It also indicates how an understanding of the relationship between the Internet and society needs to be based not just on projections derived from the technical capabilities but on an appreciation of who has the skills, equipment, know-how, and motivation essential to design, produce, use, consume, and govern the relevant technologies (Dutton 1996, 1999).

The "Net Effect": Why a New Perspective Is Needed

A growing number of quasi-experiments and surveys, both cross sectional and longitudinal, have attempted to gauge the impact of the Internet on community and society. Meta research and syntheses of this evidence tend to support the notion that, on balance, individuals and households are better connected, or at least not more isolated, as a consequence of using the Internet and other information and communication technologies (ICTs) in their everyday lives (Katz and Rice 2002; Wellman and Haythornthwaite 2002; and Wellman and Chen, this volume). At the same time, such investigations also provide support for the notion of a diversity of "impacts," in the sense that the Internet is associated sometimes with greater isolation, at others with more connectedness (or even with no appreciable connection to social relationships). Thus, they have not ended the sometimes heated debates between those who view ICTs from a dystopian or a utopian (or even a "technology is irrelevant") perspective.

However, looking for the "net effect" in findings across multiple surveys might not yield the right answer if we are looking in the wrong place. Generally, the "impact" perspective that has dominated social research on technology and society from its emergence in studies of computers in the 1950s has been too long term, fragmented, narrowly focused, and technologically deterministic to be able to provide direction to understanding the full scope and diversity of the role in transforming our lives of a phenomenon as multifaceted and pervasive as the Internet (see chapter 1, this volume).

For instance, the emphasis in the dominant social-impact perspective on long-term impacts, such as on community, fail to take account of the more immediate role of ICTs such as the Internet in reconfiguring access to people, services, and technologies as well as information—with profound implications ranging far beyond the community. Taking a long-term perspective can provide valuable insights, but it is only by looking closely at the more immediate role of the Internet in reshaping access that we will be able to grasp a fuller understanding of how individuals, households, and organizations can use the Internet in strategic ways to shape their communicative power.

A technologically deterministic view of the impacts of technology also misses the intrinsic social nature of the Internet as a web of interacting individuals and institutions, involving a wide variety of resources and know-how. Studies of particular technologies (such as microelectronics, the answering machine, the mobile phone, or the personal computer) in particular contexts (such as the office, factory, home, or newspaper publisher) have brought some enduring insights into many underlying issues, processes, and forces relevant to ongoing ICT innovation. But they cannot take account of the way the Internet encompasses and integrates dynamically a vast, continually changing array of different technologies and environments.

Conventional analytical perspectives frame "access" as an independent variable shaping the use and impacts of a technology (figure 15.1). Most significantly, overall, this approach saw the impact of technologies on society as following some kind of inherent logic, which led to a view of access to a technology as leading to particular patterns of use and impacts that could be predicted on the basis of features of the technology. The alternative perspective—that of reconfiguring access—does not

Access to the Internet  **Uses and Impacts**

Figure 15.1
A conventional perspective on access to ICTs.

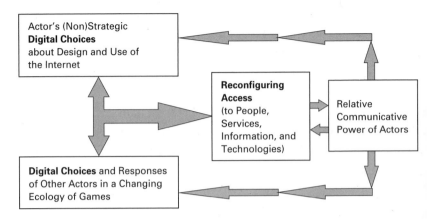

Figure 15.2
Reconfiguring access: a strategic perspective on digital choices. Adapted from Dutton et al. 2003.

see a technology like the Internet as creating automatic impacts. Instead, it regards access as an unpredictable outcome of the choices made by people and institutions about the design and use (or non-use) of the technology (figure 15.2). It also recognizes that the making of such digital choices changes the relative communicative power of different actors involved in a multiplicity of games. The interplay of actors in these games determines the ultimate social, economic, business, educational, and other outcomes tied to ICTs.

The Role of the Internet in Reconfiguring Access

The concept of reconfiguring access provides a framework for understanding the social implications of the Internet that builds on past social research insights. It acknowledges the significance of technical choices while incorporating the unpredictable impact of the Internet in coevolutionary interactions involving the choices made by multiple actors influenced by social, institutional, economic, technological, and other related factors. There is much evidence, even from traditional impact studies, to support the view that complex social processes determine outcomes of technology-based innovation in ways leading to different outcomes in different contexts (Dutton 1996, 1999; Dutton et al. 2003).

The reconfiguring access perspective challenges the cliché that "information is power," because it ignores the degree to which social and economic capital enable access to the expertise, knowledge, and data that is the basis of creating knowledge. The key is not information *per se*, but the ability to control access to information. Access is also shaped not only by technologies but also by institutional arrangements, public policy, geographical proximity (such as in gaining tacit knowledge from direct observation), and other social factors.

The role of ICTs in shaping access to information might be called "information politics" (Danziger et al. 1982: 133–135; Garnham 1999). However, information is only one element in what can be viewed as a far more general politics of access. Table 15.1 summarizes how social and technical choices about the Internet can reconfigure electronic and physical access to all four key interrelated resources (Dutton 1999): people (locally and globally), services (including the ability to render obsolete or create a new business or industry), and technologies (including equipment, know-how, and techniques)—as well as information (making some people information rich and others comparatively information poor).

There are many other ways in which ICTs can reduce, screen, reinforce, or alter access, such as shaping information content and flow by accident or design. ICTs also change patterns of interaction between people, information, communities, and organizations. The Internet can reduce travel, save time, and extend the geography of human community as a substitute for face-to-face communication or permit communication among people who might never have an opportunity to meet face to face. It may also replace valuable human contact with a much less rewarding form of communication, fostering social isolation.

How Digital Choices Reconfigure Access

The traditional social-impacts perspective provided a technologically deterministic view of the world. But the growing awareness that the outcomes of changes tied to technological innovations are determined to a great extent by the choices of people and institutions does not mean that technology then becomes seen as being of no special significance. The nature of the technology still matters in many critical ways. The choices

Table 15.1
How the digital choices about (non)use of the Internet reconfigure access

Internet provides access to	Kind of Internet activities	Examples
People Reconfigures how you interact with people, whom you communicate with, whom you know, where and when you interact with them	Intercreativity between individuals and within groups; other one-to-one, one-to-many, many-to-many communication	Always-on messaging and e-mailing; collaborative online working; online lectures to virtual classrooms; video conferencing and streaming; children producing digital content; developing a multimedia presentation between many people; online game playing; Internet-based interpersonal interactions, from chat rooms to e-democracy consultations
Services Influences what you can do online, when you can do it, and how much it costs to do it; where and when you buy other products and services; who pays what to whom —and how it is paid	Conducting electronic transactions and obtaining electronic services from distant or nearby sources	Fast online delivery of multimedia products and services, to any location, involving large amounts of data, e.g., downloading music and video; digital art collections; access by doctors to X rays at remote locations; e-shopping, e-banking, and other e-business interactions
Information Affects how and what you read, hear, see—and know	Retrieving, analyzing, and transmitting images, video, sounds, statistics, etc.	Online news streaming; listening to or watching archived or live radio and TV programs; exchanging large amounts of multimedia research or statistical data; Web searches for a huge variety of information sources

Table 15.1
(continued)

Internet provides access to	Kind of Internet activities	Examples
Technologies Shapes how and when you access the Internet and other ICTs	Producing and using broadband know-how, equipment, and techniques to shape access to, and use and consumption of, the Internet and other ICTs	Broadband telecommunications infrastructures; wireless network connections; Internet infrastructures; new digital multimedia; browsers to find information in Web searches; network security; antivirus, antispam and child-protection software

Source: Adapted from Dutton et al. 2003 (table 2).

made in the production of ICTs are as important in understanding the reconfiguring of access as are the choices made by users and consumers of ICT-based products and services. The design, development, implementation, and use of ICTs can reconfigure access by: changing cost structures; expanding or contracting the proximity of access; restructuring the architecture of networks; creating or eliminating gatekeepers; redistributing advantages between senders and receivers; or enabling more or less user control (Dutton 1999).

Changing Cost Structures

ICT innovations such as the microprocessor and e-mail have clearly contributed substantially to lowering the costs of some products and services. For instance, the costs of transporting and delivering purely digital services and products have been reduced to the often negligible cost of an Internet link that is already covered by an always-on broadband connection. However, the costs of producing and promoting digital content, such as a major motion picture, music by a highly paid artist, or a large information database, remain very high. This has meant that the free downloading of "bootleg" music, videos, and other professionally-produced creative content has posed a serious challenge to the cost structures in media industries. On the other hand, large companies with dominant control of telecommunications infrastructure, mass media, Web search engines, or other ICT infrastructures can design the technology in ways that advantage them in relation to their would-be competitors.

Innovations can disrupt established cost structures, for instance as the rapid growth since 2001 of Wireless Fidelity (WiFi) technology could do for cost structures of broadband Internet access. WiFi is a wireless-based local area network that can offer "hot spot" local broadband access points to Internet infrastructures (Dutton et al. 2003). WiFi start-up costs are lower, and its installation more flexible, than other broadband options because it uses unlicensed radio spectrum and relatively low cost equipment with low power consumption.

Expanding and Contracting the Proximity of Access

Distance, time, and control were dimensions of access identified in the 1950s by one of the earliest "technology and society" scholars, Harold Innis (1972 [1950]). They remain relevant to contemporary discussions of the social and economic role of ICTs. Changes in the ease, speed, and costs of gaining access to people, services, information, and technologies wherever they are located can have dramatic implications for democratic and community processes and for the structure, size, location, and competitiveness of business and industry throughout every sector of the economy. Depending on the strategies of users, government, regulators, and content and service providers, this can be done in ways that support either greater democratic empowerment at local levels or more accumulation of centralized political power; more competition and diversity in the provision of services, or more oligarchic control.

In her work on the telephone, Suzanne Keller (1977: 282) noted: "One of the most interesting questions is the meaning of 'near' once human yardsticks are displaced by electronic ones." This is ever more apparent with the Internet, which enables you to keep in regular, informal touch with people in distant locations and delivers files from around the world to your desktop as if they were stored on your own PC. This capability keeps being enhanced by a stream of innovations, such as WiFi, that extend local broadband access points into homes, classrooms, offices, college dormitories, coffee bars, Internet cafés, hotels, airport lounges, trains, and many other locations.

Restructuring the Architecture of Networks

The architecture of a technical network often reflects the social and institutional forces shaping it. For instance, the design of a "visual telegraph" network in France in the 1790s was so suited to the structure of the French state that the electric telegraph was resisted by French authorities, despite its many technical advantages, because it introduced two-way dialogue into a domain that had been entirely conceived in terms of a "one-way monologue" (Attali and Stourdze 1977; Dutton 1999: 62–63). On the other hand, the telephone and Internet are examples of what Kenneth Laudon (1977: 16–17) called "interactive" or "citizen technology," based on regular "horizontal" information flows among individuals and organized groups whenever they want. This reinforces the notion of a more democratic distribution of communicative power. In contrast, "vertical" communication structures, such as those used by mass media, allow a small group to broadcast to millions, following a more centralized pattern of power distribution.

One of the most basic features of the Internet is its ease in supporting one-to-one, one-to-many, many-to-one, and many-to-many horizontal networks of communication. Broadcasting, on the other hand, is still better suited to support one-to-millions (or billions) communication in vertical networks. The idea that the Internet can support more democratic patterns of communication is based on this divergence of its underlying architecture from traditional broadcasting and telecommunication networks, such as terrestrial television.

Creating or Eliminating Gatekeepers

Technological change can also alter the role of gatekeepers in the dissemination of information. In situations where there is a scarcity of media sources and outlets, "gatekeepers" such as producers, editors, and publishers play a critical role in deciding who gets on television, what is news, and what is fit to print. But such gatekeepers can now be bypassed through desktop publishing operations and the use of the Internet and Web for one-to-many communications, as demonstrated by many Web-sites run by one or a few journalists.

An illuminating example of the diverse personal, business, and technological factors that contribute to the reconfiguring of access by the elimination of a gatekeeper is provided by the development of the first automatic telephone switch (see also Dutton 1999: 64–65). In the late nineteenth century, Almond Strowger was an undertaker in Kansas City. In the course of running his business, he became incensed by the delays, rudeness, and negligence of telephone operators. Rumor has it that he also became convinced that one operator, the wife of a competing undertaker, was diverting business to her husband by falsely reporting the Strowger's line as being "busy" when prospective customers rang him up. His determination to eliminate human operators from the network led him to invent and build the Strowger Automatic Telephone Exchange, which was demonstrated as the first of its kind on November 3, 1892. The system had important consequences beyond just gains in the technical efficiency of the service. Whether by accident or design, the anonymity of automatic switching contributed to the privacy and simplicity of the phone call and therefore made it a more appealing service (Dutton 1999: 64–65).

A notable example of how the Internet can be used to bypass traditional media gatekeepers to enhance citizens' direct access to information was provided in 2003 by the UK government's inquiry led by Lord Hutton into the circumstances surrounding the death of the chemical and biological weapons expert Dr. David Kelly. During the course of the Inquiry, documentary evidence was published on a Website (http://www.the-hutton-inquiry.org.uk) that became the most popular site in the United Kingdom. The documents included many e-mails between: ministers and their civil servants and advisors; TV executives, news editors and journalists; and even experts in, and close to, government intelligence services. Direct public access to such information is unusual in the United Kingdom, where freedom of information legislation is relatively limited compared to countries such as the United States.

In business, the secretary was once the prime gatekeeper who screened and prioritized calls for an executive, but this role is becoming less common as e-mail is used increasingly. However, if the proliferation of communication options becomes overwhelming, people might well look again for gatekeepers to filter, prioritize, and select information for the user or consumer.

Redistributing Power between Senders and Receivers
With electromechanical switching technology, telephone calls became anonymous. This shifted power to the person calling, since the called party would not know who is calling, and, therefore, be more inclined to answer a call in case it was important. With answering machines and call line identification, information about the identity of the caller has become increasingly available to the receiver. This shifted the advantage to the receiver, relative to that in circumstances of anonymity.

Although the Internet was created by a culture promoting openness and freedom of expression, it can take away the anonymity that made automatic telephone switching so attractive. For example, e-mail was originally designed in the 1960s to identify the person sending a message, as its agreed protocol standard required the header of every message to identify the person sending it, the recipient, the date sent, and the subject. This can be used by recipients to help prioritize and screen messages in order to better manage their communications but also creates a documented trail that can be followed to strip away anonymity. The use of this capability was illustrated by the open Web publication of private e-mails at the Hutton Inquiry, which led many commentators to suggest that the public availability of so many e-mails might make people generally reluctant to discuss sensitive matters via e-mail and prefer methods less likely to be recorded, such as face-to-face or telephone conversations. This further highlights the complex mixture of motivations underpinning the technological choices that shape access.

User Controls
The proliferation of global communications channels opened by the Internet and other ICTs has added an important new dimension to traditional concerns about who controls (or frees) the content and access to communications and information. The banning or restriction of traditional printed information or terrestrial broadcast channels could be implemented relatively easily by national governments. Now, people can have instantaneous access to multimedia sources in almost any country: news, books, TV and radio programs, pornography and high art, online games and esoteric virtual academic discussion groups. This is seen as a major advantage in open societies, but a number of authoritarian countries have taken action to filter Internet content, limit or ban the use of

particular browsers, or block satellite TV channels. However, constraining access via the Internet can be difficult because many users are skilled in exploiting the technology to find loopholes through any censorship strategy.

Concerns have also been expressed by users who want to try to regain some control over ICT content, such as the violence chip in TVs and computer "net nanny" software that can facilitate parental control over the kind of content their children are allowed to access at home. Yet, many parents look towards the content providers to protect their children, while many in the broadcasting and Internet communities regard the use of screening devices as censorship. That exemplifies the complex issues involved where the nature of technical capabilities intertwine with public policy objectives and a diverse ecology of games to challenge fundamental social and political notions of freedom, control, personal responsibility, and shared community values.

This feeling of loss of control over content has been heightened by the explosive growth of e-mail from both senders acceptable to the receiver as well as unsolicited spam e-mail and viruses. This can be addressed by spam filters, virus checkers, and other tools—but spam filters can also filter out e-mails wanted by the recipients, and computer "hackers" (and "spammers") continuously produce new viruses (or spam) that are a step ahead of current virus checkers (and spam filters). Legislation enabling spam to be made illegal is in place in some countries, but not in others.

Unrealistic user expectations about the technology's capabilities can also contribute to concerns about loss of control over content. The ability of intranets, firewalls, and other security systems to protect the confidentiality of e-mail can enhance users' trust in the Internet. However, the Hutton Inquiry showed that the Web could be used to reveal to the world private interpersonal e-mail deemed to be in the public interest, despite all the supposed safeguards.

All this again demonstrates that reconfiguring access involves many interacting and evolving dimensions and actors.

Social Factors Shaping Digital Choices

As shown in the above discussion, choices concerning ICTs, and their social implications, are not random or unstructured. For example, the

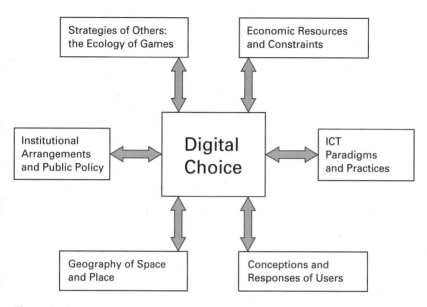

Figure 15.3
Social factors facilitating and constraining digital choice. Adapted from Dutton 1999: figure 12.1.

distribution of information "haves" and "have-nots" and other patterns of access are enabled and constrained by the social and economic contexts within which relevant actors at all levels make choices that have immediate and long-range cumulative consequences on reconfiguring access to people, services, information, and technologies. The main factors facilitating and constraining these digital choices can be categorized along the six dimensions summarized in figure 15.3 and discussed in the following subsections.

Economic Resources and Constraints
The size, wealth, and vitality of nations, companies, and other actors place major constraints on the development and use of ICTs in all arenas of activity. For example, the options open to a "high achiever, high-tech teleworker" differ from those of the "low-income, low-tech lone parent" (Silverstone 1996: 225–228). Likewise, the strategies of the dominant traditional telephone companies (telcos) will differ from those of new entrants to the communication industry (Mansell 1993). Such differences

are shaping, for example, the competitive battle to deploy high-performance broadband links. Telcos offering Digital Subscriber Line (DSL) broadband on their initially telephone-based networks and cable companies were the first main broadband providers, but an increasing range of other options are challenging for the position of the third main broadband "pipeline," particularly WiFi, satellites, and other wireless technologies (Dutton et al. 2003: 8–10).

Reconfiguring access can improve or undermine the economic vitality of a nation, business, household, or local community, thereby enhancing or exacerbating socioeconomic disparities. For example, the dominant telecommunications provider for the western Caribbean island of Jamaica was able to use its control over the only cable communications link from the island to restrict broadband supply capacity and favor its own vertically integrated Internet service provider (ISP) while driving others from the market (Dutton et al. 2003: 39–40). A senior official from the Jamaican Office of Utilities Regulation says this has placed serious constraints on Jamaica's ability to use broadband to help meet its need for substantial economic development (ibid.).

ICT Paradigms and Practices

ICT concepts and practices can become the foundation of powerful belief systems or "paradigms." These create a way of interpreting reality that is very different from that perceived by people whose thinking is embedded in another paradigm (Dutton 1996). For instance, visions generated by the notion of an information society, such as "virtual organizations," have been important shapers of social, political, and technological choices irrespective of their descriptive validity (Bloomfield et al. 1997). The very idea that we work in a virtual organization or live in an information society can influence public policy and the behavior of individuals, for instance, in the development of "virtual universities" using the Internet and other ICTs to reconfigure how, when, and where faculty teach, students learn, administrators manage, librarians work and interact with users, and so on (Dutton and Loader 2002).

At the same time, experience and knowledge about ICTs can influence or even create a paradigm change, for instance in the overall shift of work from manufacturing to services industry that has occurred in the "information age" (Freeman 1996; Bell 1999 [1973]). The growth of e-business and e-commerce in online shopping, banking, news media, and numerous other activities that have been enabled by the Internet and Web is having transformational effects within and between many sectors. Technologies therefore do make a difference, even if they do not determine social outcomes. For instance, ICTs can bias social choices by making some avenues more economically, culturally, or socially rational than others.

The ways in which ideas like the information society shape views about how the world works and, thereby, influence the decisions of individuals, firms, and governments is a major reason why alternative perspectives on the role of the Internet in society, such as the reconfiguring of access, are more than competing theories. They are also ideas that can shape decisions in everyday life, and in once-in-a-lifetime choices.

For instance, the underlying philosophy that shaped the emergence of the Internet from the 1970s—open access to information, communication, and collaboration between people—reflected the noncommercial and academic culture that infused the perceptions and motivations of those most influential in determining Internet capabilities (Reid 2000; Castells 2001). This culture also gave rise to the key trigger to the popularization of the Internet: the World Wide Web and its simple "click" user interaction. This was developed at the European Laboratory for Particle Physics (CERN) as a means of giving researchers quick and easy access to an online "hypertext library" (Reid 2000).

It was only in the 1990s that a strong push towards the e-commerce and e-business value of the Internet emerged. However, the culture shaping Internet technology had already created and embedded capabilities in Internet standards and infrastructure that were based on notions of trust and sharing. This proved to be at odds with many commercial and business interactions, where more priority is given to confidentiality, security, profitable pricing, protection of intellectual property rights (IPRs), and other factors. There has therefore been a move in Internet developments towards a focus more oriented to business needs, for instance in the establishment of greater online trust (Guerra et al. 2003) and security firewalls.

This has parallels with the movements that gave rise to the personal computer and more recently to WiFi. For example, the low cost and flexibility of setting up and using WiFi attracted many early adopters from a libertarian grassroots movement committed to developing "community wireless" capabilities that remain open to all. Business and industry has again subsequently caught up with the potential commercial scope of this technology, for example as shown by Intel's launch in January 2003 of its Centrino wireless mobile computing technology.

Conceptions and Responses of Users

The conceptions and responses of a wide variety of users, workers, consumers, managers, citizens, audiences, etc. also play an active role in shaping the implications of ICTs, often in ways very different from those expected by simply extrapolating from the perceived potential of the technology. Misconceptions of the user can therefore undermine the diffusion of ICTs (Woolgar 1996). Many innovative technological and market failures can be understood as a consequence of having a weak conception of the user, for example, in the notions of a "paperless" office that failed to take account of the many attractive features of paper for users.

The reconfiguring access approach also underlines how users' roles are not confined to being passive recipients of whatever is designed for them. Much motivation for the development of the Internet, personal computers, WiFi, and other technologies came from users with the skills and motivation to design and build their own systems—PCs were first known as "home brew" computers. The explosive growth in the use of text messaging on mobile cell phones was driven largely by teenagers discovering the value of this technique for themselves, when the providers of the systems had thought it would be used primarily by business for short messages. The grassroots communitarian movement that gave rise to the popularity of WiFi is another example of active user involvement in shaping technology-related outcomes.

Geography of Space and Place

One of the most prominent attributes of ICTs is the relative ease with which the new electronic media can overcome constraints of time and distance. This has been a key reason why e-mail has proved to be such a popular communication medium, even for only the occasional Internet user. But although the Internet can be accessed at any time from numerous locations anywhere in the world, there is evidence that ICTs might actually make geography matter more. Rather than undermining the importance of space and place, as argued by some, there is evidence that ICTs bring new significance to locations, for example, by opening up new options for locating call centers, technical support offices, and other facilities that can be delivered online (Cornford and Gillespie 1999).

At the same time, there is little evidence that the use of the Internet has led to a reduction in people having face-to-face meetings. For instance, the Internet can create virtual communities of interest that encourage people to travel to meet those they first contacted online— and with whom they might not have been in touch with at all without access to Internet-enabled interactions. This indicates that the Internet does not act primarily as a substitute for travel. It also shows how the Internet could reshape travel through the journeys to meet new online friends or business partners, as well as by giving people the ability to travel to where they need to be for face-to-face meetings while still remaining in close touch with work, family and friends through e-mail and other online interactions.

The Internet and other networks have made it possible for global enterprises to be managed efficiently through the flexible location and coordination of people, services, information, and technologies. The Internet has also given small local businesses the opportunity to tap into global markets through Websites and e-business links. Such flexibility has significant advantages for many enterprises, and for those individuals who are given the opportunity to participate in networked and online operations. But when a call center or other e-enabled facility is relocated to a different part of the world through e-enabled systems that take advantage of lower wages, costs, or other factors, some people might be configured out of access to that kind of employment in their own locality.

Institutional Arrangements and Public Policy

Choices about access are constrained by a variety of institutional arrangements and public policies because technical, social, policy, and organizational innovation are interdependent (Dutton, Blumler and Kraemer 1987; Freeman 1996). For instance, the design of an organization influences how the Internet is used within it, but the Internet also creates many new options for radically redesigning organizations and interactions between them, such as a "virtual organization" composed of separate private firms or public agencies employing ICTs to enable them to act as if they were part of the same real unit. Digital choices are strongly affected by institutional arrangements and policies in areas such as telecommunications regulation, standards, copyright, public-service broadcasting, and education. Nevertheless, public policy at local, national, and international levels can be responsive to technological change, as demonstrated in a variety of policy initiatives aimed at supporting advanced information infrastructures, including information superhighway in the 1990s and the expansion of broadband availability in the early 2000s.

Strategies of Others: The Ecology of Games
The struggle for communicative power to control and influence the design and use of the Internet and other ICTs generally takes place in a variety of different arenas at the same time. All actors are not involved in the same struggle: individuals and groups pursue different goals within their own domains in an "ecology of games" (Dutton 1999: 14–16).

A game is an arena of competition and cooperation structured by a set of rules and assumptions about how to act to achieve a particular set of objectives. An ecology of games is a larger system of action composed of two or more separate but interdependent games (Dutton 1992). Within each "game," players follow an established set of traditions, rules, and disciplines. All players have a role in shaping outcomes, playing different roles in different games, and often acting in many games at the same time. For instance, a specialist might pursue a technically elegant network design, while a top manager primarily seeks cost reductions. This places major constraints on the predictability of outcomes based on assessments of strategic aims, unless the varied goals of different actors in the wider ecology of games is well understood and orchestrated. Seeing outcomes of broadband Internet policy-making and actual use of the technology as resulting from such a continuous interplay between different stakeholders in different games is an important step towards enhancing an understanding of why the Internet is not on a predetermined path that will produce predictable results.

The complexity and significance of the ecology of games can be illustrated by looking at some of those related to the provision and use of broadband Internet capabilities (Dutton et al. 2003). The many players and interests with a stake in these games include business enterprises and their employees and customers; community groups, individual citizens, and their local and national governments; commercial content providers and consumers of that content; teachers, students, and educational administrators; broadband policymakers, regulators, and the companies and users affected by them; parents and their children; and scientists, researchers, and their sources of funding. There are many, many more.

The behavior and decisions of all actors affect those of other actors. For example, action by content providers to protect their copyright will constrain the choices consumers can make, while those consumers who find ways around copyright protection undermine revenues for content providers. WiFi providers give new choices for accessing broadband but might threaten the large investments in facilities of incumbent suppliers. Governments committed to a competitive market can help less powerful players in ways that limit the power and freedom to maneuver of large players, while global power-broker companies can decide, or threaten, to move operations to other areas as a lever in policy bargaining. And so on, with outcomes unfolding as the products of countless strategic and everyday decisions made by a myriad of players in many different games in different arenas. Table 15.2 illustrates a few of the main games shaping the implications of broadband Internet.

Broadband Internet could change the rules of some games, such as mergers of media companies to coordinate convergent ICTs. It can also open up new roles for those who have been communicatively empowered by broadband access, for example, when media consumers become media producers by setting up Websites offering news and discussion forums from a myriad of perspectives not seen in the mass media. This was highlighted in 2003 during the Iraq War by the daily Web log (blog) diary posted by the nonjournalist "Baghdad blogger" that became a source of news of everyday life under the bombing (Pax 2003). Schoolchildren making local history videos for use in history lessons as an alternative to professionally produced courseware is another example of how the Internet and other new media can place more control in the hands of users (de Sola Pool 1983).

Table 15.2
Illustrative games shaping the implications of broadband Internet

Game	Main players	Rules
Telecom regulation	Telecom firms, regulators, investors, consumers	Regulators umpire moves of competing firms, taking account of conflicting and complementary goals of players
Broadband pipeline supply	Telco, cable, wireless, and other broadband suppliers	Suppliers compete for market share and the position as the third main broadband pipeline (with DSL and cable)
Internet service provision	ISPs, WISPs, telecom and IT firms	ISPs use broadband applications and infrastructure access to win customers
Communitarian	Neighborhoods, community groups, Net enthusiasts	Individuals and groups seek free or low-cost, open access to the Internet, sometimes competing with commercial users or providers
Economic development	Governments, public agencies, investors	Players build ICT infrastructures to attract business, investment, and jobs to localities, nations, and regions
Developing country	Governments, nongovernmental organizations (NGOs), local activists, investors	Players seek to close social as well as economic divides in developing countries by the appropriate use of suitable ICT infrastructures
E-games	Pro/anti e-enablement players in government, business, education, etc.	Organizations put their vitality at stake through over/under investment in online infrastructures and applications
Implementation	Users, ICT product and service suppliers, consultants	Users struggle to implement and maintain broadband in order to reap the potential benefits
Consumer protection	Consumers, consumer groups, suppliers, regulators	Legislators and regulators respond to competing views of the consumer's interests in broadband Internet provision
New-media publishing	Media giants versus Internet entrepreneurs; media novices versus professionals	Established and emerging producers of Internet content compete to reach audiences
Copyright, IPR, digital rights management	Content providers versus consumers and ICT industries; regulators	Telecom, media industries, and users compete over interpretations of rights in access to information and services

Source: Dutton et al. 2003 (table 4).

Understanding the Intrinsic Social Nature of the Internet

This chapter has focused on how the Internet and other ICTs reconfigure access to people, services, information, and technology through an ecology of games between multiple players, as summarized in figure 15.2. Unpredictable outcomes are produced by the interaction of social and technical choices within the changing ecology of games that reconfigures access, but the analysis outlined in the chapter shows that actors can have an impact through the choices they make in different games. Those who understand the centrality of the way the Internet and related ICTs reconfigure access to local and global resources could be in a better position to decide whether, and how, to use the technologies to reap dividends to enhance their own situation and help to close social, economic, education, health, age, gender, and other divides.

The reconfiguration of access viewpoint contrasts with the traditional perspectives on the social and economic implications of extensive ICT use, which depicted access to ICTs as leading to uses that result in impacts determined largely by the technology. The central issue highlighted by the reconfiguration view is the design and use (or nonuse) of ICTs to reconfigure access strategically—opening up or closing off networks.

The relevance of this approach has been exemplified in this chapter by many examples of the use of broadband Internet and other ICTs and by drawing on enduring lessons from social research on earlier computer and telecommunications technologies. An adoption of this perspective would lead research on ICTs to:

1. Prioritize access, rather than information, as the prime focal point.
2. Move from a deterministic paradigm in which outcomes can be predicted to one anchored in an open society where the social, economic, and other consequences of any activity using ICTs is understood to be the outcome of an unfolding ecology of choices by multiple actors.
3. Understand better how changes in communicative power through digital choices that reconfigure access can be employed to address key questions about the role of the Internet and other ICTs in areas such as assessing or exacerbating the digital divide; creating a sense of greater community or more isolation; empowering citizens or the state, managers or employees; reducing or enhancing users' control over access; and restricting or expanding access.

4. Undertake investigations that explore how, and to what extent, technologies matter in shaping and reshaping access.

5. Focus on how ICT innovations are able to complement and enhance traditional ways of doing things, as well as offering a means of substituting old technologies, methods, and habits with new approaches using electronic services and products.

Reconfiguring access transforms a person's relationships with information, people, services, and technologies, but it might not represent a revolution in underlying societal institutions and processes. The Internet cannot in itself overturn entrenched and deeply rooted power bases and cultural and social influences, particularly since people tend to use ICTs in ways designed to reinforce their communicative power. But the shifts in communicative power enabled by the technology could open up possibilities for significant transformations in which many traditional ways of thinking, planning, and behaving can be rethought and reinvigorated. In these transformations, the interaction of social and technical choices within a changing ecology of games means that all actors—policymakers, practitioners, researchers, users, managers, clerks, teachers, students, and a multitude of others—can have an impact on outcomes.

References

Attali, J., and Y. Stourdze (1977) "The Birth of the Telephone and Economic Crisis: The Slow Death of Monologue in French Society." In *The Social Impact of the Telephone,* ed. I. de Sola Pool. Cambridge, Mass.: MIT Press, 97–111.

Bell, D. (1999 [1973]) *The Coming of Post-Industrial Society: A Venture in Social Forecasting.* New York: Basic Books.

Bloomfield, B. P., R. Coombs, D. Knights, and D. Littler, eds. (1997) *Information Technology and Organizations, Strategies, Networks, and Integration.* Oxford: Oxford University Press.

Castells, M. (2001) *The Internet Galaxy* (Oxford: Oxford University Press).

Cornford, J., and A. Gillespie (1999) "The Geography of Network Access." In *Information and Communication Technologies—Visions and Realities,* ed. W. H. Dutton. Oxford: Oxford University Press, 255–256.

Danziger, J. N., W. H. Dutton, R. Kling, and K. L. Kraemer (1982) *Computers and Politics.* New York: Columbia University Press.

de Sola Pool, I. (1983) *Technologies of Freedom.* Cambridge, Mass.: The Belknap Press of Harvard University Press.

Dutton, W. H. (1992) "The Ecology of Games Shaping Communication Policy." *Communication Theory* 2(4): 303–28.

Dutton, W. H., ed. (1996) *Information and Communication Technologies—Visions and Realities*. Oxford and New York: Oxford University Press.

Dutton, W. H. (1999) *Society on the Line: Information Politics in the Digital Age*. Oxford: Oxford University Press.

Dutton, W. H., and B. D. Loader, eds. (2002) *Digital Academe: The New Media and Institutions of Higher Education and Learning*. London: Routledge.

Dutton, W., J. Blumler, and K. Kraemer, eds. (1987) *Wired Cities: Shaping the Future of Communications*. New York: G. K. Hall.

Dutton, W. H., S. E. Gillett, L. W. McKnight, and M. Peltu (2003) *Broadband Internet: The Power to Reconfigure Access* (Oxford: Oxford Internet Institute). Available online at http://www.oii.ox.ac.uk

Freeman, C. (1996) "The Two-Edged Nature of Technical Change: Employment and Unemployment." In *Information and Communication Technologies—Visions and Realities*, ed. W. H. Dutton. Oxford: Oxford University Press, 19–36.

Garnham, N. (1999) "Information Politics: The Study of Communicative Power." In *Society on the Line: Information Politics in the Digital Age*, ed. W. H. Dutton. Oxford: Oxford University Press, 77–78.

Guerra, G. A., D. J. Zizzo, W. H. Dutton, and M. Peltu (2003) *Economics of Trust: Trust and the Information Economy*. DSTI/ICCP/IE/REG(2002)2, Paris: OECD.

Innis, H. (1972 [1950]) *Empire and Communications*. Toronto: University of Toronto Press.

Katz, J. E., and R. E. Rice (2002) *Social Consequences of the Internet*. Cambridge, Mass.: MIT Press.

Keller, S. (1977) "The Community Telephone." In *The Social Impact of the Telephone*, ed. I de Sola Pool. Cambridge, Mass.: MIT Press.

Laudon, K. (1977) *Communication Technology and Democratic Participation*. New York: Praeger.

Mansell, R. (1993) *The New Telecommunications: A Political Economy of Network Evolution*. London: Sage.

Pax, S. (2003) " 'I Became the Profane Pervert Arab Blogger.' " *Guardian*, 9 September, 2–3 (G2 Section).

Reid, R. H. (2000) *Architects of the Web*. New York: John Wiley & Sons.

Silverstone, R. (1996) "Future Imperfect: Information and Communication Technologies in Everyday Life." In *Information and Communication Technologies—Visions and Realities*, ed. W. H. Dutton. Oxford and New York: Oxford University Press, 217–231.

Wellman, B., and C. Haythornthwaite (2002) *The Internet in Everyday Life*. Oxford: Blackwell Publishing.

Woolgar, S. (1996) "Technologies as Cultural Artefacts." In *Information and Communication Technologies—Visions and Realities*, ed. W. H. Dutton. Oxford: Oxford University Press, 87–102.

16

Impersonal Sociotechnical Capital, ICTs, and Collective Action among Strangers

Paul Resnick

The notion of "capital" suggests a resource that can be accumulated and whose availability allows people to create value for themselves or others. People can do more when they have access to physical resources like buildings and tools, which are usually referred to as "physical capital." Money, or "financial capital," allows people to acquire many other kinds of resources. In the latter half of the twentieth century, economists began to think about education as "human capital." People who have more knowledge and skills can produce more, so it makes sense to think about spending on education as a form of investment rather than consumption (Schultz 1961).

Productive resources can reside not just in things and in people, but also in social relations among people (Coleman 1988; Putnam 1993). Following Coleman (1988), I define "social capital" as productive resources that inhere in social relations. In order to give the concept greater specificity, analysts sometimes restrict the definition to particular kinds of social relations, such as networks of interpersonal communication, trust, and intimacy, or widespread social trust, or norms of reciprocity.[1] Equating social capital with any particular pattern of social relations, however, would make it impossible to identify as social capital any new patterns of social relations.

Social capital can facilitate useful interactions among people. It helps people connect with information and other people. It helps them share and exchange resources. It helps them coordinate interdependent actions. Perhaps most importantly, social capital can help people overcome dilemmas of collective action. One collective action problem is procurement of public goods, where people might free ride, hoping that others will supply them.[2] A related problem is the overuse of common pool resources, where individuals might consume more than their fair share, creating a "tragedy of the commons" such as overgrazing a shared pasture (Hardin 1968). Another related dilemma of collective action is social mobilization, where everyone may realize that all will benefit if they all act, but individuals who act without others will fare badly, and no one acts out of fear that others will not join. Starting a labor strike (Kollock and O'Brien 1992) or starting the dancing at a party are examples of this problem.

Social capital can reside in many different kinds of social relations. Certainly, personal relationships can be productive resources, as suggested in the phrase, "it's not what you know, it's who you know." Granovetter (1973) and others have noted that "weak ties," those personal connections that involve less frequent interaction and less personal affection, are especially productive for some purposes, such as job finding, because they provide bridges to broader reservoirs of information and other information. This chapter, however, calls special attention to social resources that are not based on personal connections at all, what I call "impersonal social capital."

For example, "introducer systems" can generate personal ties when they are needed. In organizations, shared knowledge becomes a resource for shorthand in conversation, and established roles streamline decision-making and create legitimacy for decisions. In organizations and larger cultural groupings, shared values and a sense of collective identity make it easier to unite for a common purpose, and thus overcome problems of collective action. In markets, price signals identify mutually beneficial exchanges. Monitoring and sanctioning systems, whether emerging from individual action or organized by governments, can partially substitute for interpersonal trust in individual exchanges. Norms of generalized reciprocity, together with such monitoring and sanctioning systems, can help overcome problems of collective action. All of these are forms of impersonal social capital.

I use the term "sociotechnical capital" to refer to productive resources that inhere in patterns of social relations *that are maintained with the support of information and communication technologies* (ICTs). ICTs can be used to support personal relationships, as any teenage devotee of instant messaging can attest. But while ICTs can bridge time and distance, affective communication is more difficult through ICTs than face to face. ICTs are relatively more useful in supporting impersonal forms of social capital, where affective ties are not needed.

Elsewhere (Resnick 2002), I offer a systematic catalog of the building blocks that ICTs offer for establishing both personal and impersonal sociotechnical capital. This chapter highlights a few that are of particular relevance to impersonal social capital. It then explores potential transformations of important social enterprises, including news monitoring, electoral politics, and commuting.

Some ICT Supports for Building and Maintaining Social Capital

One useful capability of computers is that they can search, sort, and select on the basis of geographic coordinates. In the United States, for example, street addresses can be looked up in a database to determine latitude and longitude coordinates. Using geographic coordinates, searches can be conducted, say, to find all the items within a one-mile radius or to select the 10 nearest items. In the United Kingdom, the UpMyStreet.com web site invites users to enter the postal code where they live or want to make connections. The site then not only displays local organization listings (schools, cafes, etc.), but also shows the messages contributed by people who live closest to that postal code. The system can scale well as more users participate: with few users, readers might see messages from people 100 km distant; with more users, they would make connections in their neighborhood.

Taste matching is another useful capability. Word of mouth normally travels from person to person in a social network. Recommender systems (Resnick and Varian 1997) can supercharge this process, allowing recommendation sharing among people who may not know each other or be explicitly aware of each other's interests. Computers provide support for gathering feedback about information, products, or even people, either in the form of explicit ratings or traces of behavior such as clickstream or purchasing data. For example, amazon.com gathers explicit book reviews from readers and also mines purchasing behavior to generate bestseller lists and links from individual books to other related books.

Recommender systems can identify people who have similar tastes, simply by looking for similarities in their ratings, their clickstream behavior, or their purchasing data. Because the taste matching is automatic, people can get personalized recommendations from each other without ever establishing a personal relationship. Automated matching based on revealed tastes can also be used in a variety of situations for directly matching people for a variety of purposes, from dating to task group assignment to ride sharing.

A third function that ICTs can support is behavior monitoring through reputation systems. In large groups, it is hard for individuals to determine who to trust, and hard for the groups as a whole to encourage trustworthy behavior. This acts as an inhibitor to transactions that require risk-taking and to collective mobilization whenever there is an opportunity for free-riding. A reputation system gathers information about people's past behavior and makes it available to others. For example, at eBay and other auction sites, a buyer and seller can leave comments about each other after completing a transaction; these comments are visible to future buyers and sellers.

If people regularly provide honest feedback, and those with more positive feedback are treated better in the future, this can enable the maintenance of trust in a large online interaction environment (Resnick et al. 2000). Game-theoretic analysis suggests that this can be fairly effective, though not optimal, even if people remain anonymous and thus have the option of shedding bad reputations and starting over (Friedman and Resnick 2001). EBay is the most widely used site that relies on a reputation system: buyers and sellers leave comments about each other after transactions. Empirical analysis suggests that past reputations are somewhat informative in predicting future problems (Resnick and Zeckhauser 2002) and that buyers do reward sellers who have better reputations by paying higher prices for their goods (Resnick et al. 2002).

Areas Ripe for Transformation

Whenever a new way of accomplishing some function emerges, observers tend to notice first a substitution effect of the new for the old. Then there is what economists call an "income effect": an increase in how much the function is performed overall because that function has become less expensive relative to other goods in the economy. Finally, new structures emerge that rely on cheap ubiquitous availability of that function.

Malone and Crowston (1994) charted this progression in the realm of transportation (the horseless carriage; increased mobility; suburbs) and tried to anticipate it in the realm of coordination technologies. The same framework can help us understand and perhaps even foresee potential impacts of technologically-mediated forms of impersonal social capital.

The first wave of changes should be a substitution of impersonal sociotechnical capital for forms of impersonal social capital that do not rely on technology. For example, eBay's reputation system is a partial substitute for legal enforcement of fair trade (Bakos and Dellarocas 2003) and private, electronically-mediated dispute resolution services such as SquareTrade partially substitute for legal adjudication at substantially lower costs. Instead of markets, we see the emergence of barter systems with computers maintaining accounts so as to ensure equitable patterns of exchange over time (Cahn 2000). Instead of long-lived organizations with clear external boundaries and internal roles and authority relations, we see ad hoc groups using ICTs to organize their efforts in areas such as technology standard setting (Bradner 1996), software development (Raymond 1999) and even public service projects such as NetDay's efforts to wire schools for Internet connectivity (Haring 1996).

The second wave of changes, the income effect, should lead to more reliance on impersonal social capital of all kinds, technologically mediated or not. For example, while personal connections are still valuable, computerized matching services are gaining prominence both for job placement (e.g., monster.com) and for dating (e.g., match.com, friendster.com).[3] When choosing media, restaurants, and consumer products, people are turning increasingly to the "word of mouse" of strangers, on sites such as amazon.com, epinions.com, or movielens.org, not just the "word of mouth" of their friends.

The third wave of changes will be the emergence of new institutional structures and ways of life that would have been impractical without impersonal sociotechnical capital. These changes are harder to foresee, both because it is hard to imagine large-scale social transformations in general, and because it is difficult to evaluate whether a limited supply of impersonal social capital has been the key factor preventing these transformations until now. However, some hints of potential transformations are apparent in several arenas, as discussed in the following subsections.

News Monitoring and Opinion Formation

The news industry may be poised for a major transformation if more and more people begin to rely on advice from distant acquaintances and strangers to monitor the news and form opinions. The Internet has broken the mass media's near monopoly on publishing information about current events: there has been a proliferation of independent websites, and even publishing platforms such as indymedia.org. But most people still rely on the professionalized mass media to certify the accuracy of information and to indicate what is worth paying attention to.

Widespread use of recommender systems, however, has the potential to change this. For example, at Slashdot, a news and commentary site, there are often hundreds of comments about a story, but readers act as moderators by rating the comments up or down. By setting a threshold of 4 or 5 for which comments should be displayed, a reader can typically select less than a dozen comments about each story. Another site, kuro5hin.org, even lets the users choose which new stories appear on the front page. Similar technologies are being used to monitor what news stories are most linked to in weblogs, which often feature links to articles posted on other websites, with a bit of commentary.

The first generation of tools is also available now for tracking which news stories receive most attention from readers (e.g., http://www.nytimes.com/gst/pop_top.html) or links from weblog "bloggers" (e.g., http://www.technorati.com/cosmos/currentevents.html). More personalized versions of these are likely to emerge, using recommender system technology to identify stories read or linked to by others who share one's own tastes. If a large number of readers begin to depend on these distributed recommending processes, rather than picking a few sources and relying on their editorial staffs, the levers of influence for shaping public opinion will shift considerably.

Grassroots Politics

Electoral politics may also be poised for a major transformation. For instance, an income effect was certainly apparent in the lead-up to the 2004 U.S. presidential primaries, with more reliance on impersonal social capital and less on mass media, mass mailings, or endorsements from prominent individuals. Early in October 2003, months before the first presidential primary, more than 70,000 people reportedly attended simultaneous informal events for Democratic candidate Howard Dean. Venue selection for these was made by attendees themselves, coordinated through the website meetup.com. Candidates have shown greater commitment to energizing a large base of volunteers and small contributors than in any campaign in the previous few decades.

What will electoral politics look like if the current activity is the beginning of a longer term trend? We could see a return to grassroots political organizing for both presidential and local campaigns. Rather than an old-style ward organization, however, with favors traded among people who have established personal loyalties, we should expect to see a looser network, with information sharing and mobilization of coordinated action mediated by ICTs.

Moveon.org may be a harbinger of such a future. More than 2 million people have signed up through its website. Members participated in a discussion period and straw poll among Democratic presidential candidates. They receive email alerts soliciting donations to fund specific political advertisements, signatures on petitions, and calls to congress. There was even a distributed phone bank system to allow members to make phone calls to California voters prior to the gubernatorial recall election in 2003. Close friends or co-workers could easily be active members of moveon.org without knowing of the other's membership. Once people have joined, personal connections are not critical to any of the forms of participation.[4]

Semi-public Transportation

In the United States, there is tremendous unused transportation capacity in the form of unoccupied seats in private vehicles.[5] Not only would filling some of those seats reduce smog, congestion, and fuel consumption, but it also could create opportunities for increasing local social capital as people conversed in the car. The major barriers to ride sharing include coordination of routes and schedules, safety risks, social discomfort with sharing what are currently private spaces, and an imbalance of costs and benefits among the affected parties. Despite these barriers, ride sharing does occur. More than twice as many people in the United States share a ride to work in a private vehicle as use public transportation to get there (NHTS 2003), either informally or through formal carpools and vanpools.[6]

In a few cities around the United States, ad hoc ride sharing among strangers has emerged to enable drivers to use high-occupancy vehicle (HOV) lanes. For example, thousands of commuters from Virginia suburbs to Washington, DC, regularly use a completely informal system called "slugging" to fill cars in order to make use of HOV lanes. To solve the coordination-of-routes problem, conventions have evolved among drivers and riders for pickup and drop-off points. Often, pickup points are at or near public transportation stops, so that riders can fall back on public transport if there are not enough drivers that day. Commuter parking lots along highways, originally designed to support regular carpooling (Park 'n' Ride), are also popular pickup points. But sometimes restaurant parking lots are used, or indeed any place with space for cars to pull over that is near an HOV entrance. There are a limited set of destinations and their meaning is well understood (e.g., "Bob's" refers to the parking lot of a restaurant that used to be Bob's Big Boy, but no longer is). Generally, there is no signage: regular users just know where to go.

Conventions have also developed to address the problems of safety risks and social discomfort (LeBlanc 1999). Generally, riders or drivers line up and are matched in order of arrival, but either party can refuse the first rider or driver in line if they feel uncomfortable.[7] Riders normally arrange not to leave behind a lone female, allowing her to go ahead of the last male rider if necessary. To alleviate social discomfort, the illusion of private spaces is generally maintained. Riders are expected not to initiate any conversation, and need not respond to conversational overtures from the driver.

I am not aware of any news stories or web sites reporting any serious safety incidents such as rape, kidnapping, or murder. The system is not completely successful, however, in preventing unhappy matches. On the DC website slug-lines.com, the most common story is a rider's tale of an unpleasant ride: the driver didn't go to the promised destination, drove in an unsafe manner, or left something on the seat that soiled the rider's clothes. Posted stories involving the breakdown of the illusion of private spaces are generally happier ones: reunions of long-lost friends or finding that driver and rider's high school social networks had significant overlaps.

It is doubtful that transportation planners who wanted to encourage HOVs envisioned the instant matching that is occurring of riders and drivers who do not know each other. In fact, public officials sometimes discourage the practice. There are concerns about public safety, so much so that the Houston Metro Police Chief (Wall 2002) opined, "We think the No. 1 safety tip would be: Don't do it." The meeting points can create congestion problems and slow down public bus service. Some of the passengers are siphoned off from using (and paying for) public transportation, which hurts the viability of that enterprise. In fact, the term "slug" for people seeking shared rides apparently was first coined by bus drivers who would pull over to pick up bus riders only to be waived on, which they viewed as analogous to a bus rider using a fake coin, or slug, instead of paying a fare (LeBlanc 1999, p. 22).

Technology-based monitoring and reputation systems could reduce some of the trust problems inherent in ride sharing with strangers. A computer system could authenticate drivers and riders, and give them one-time codes to say to each other. Deviations from expected routes could trigger phone calls to confirm that nothing had gone wrong. Far more likely than actual violence is the kind of unpleasant ride described earlier. Here, a reputation system would be far more useful than posting horror stories on web sites. Passengers could refuse rides from drivers that other passengers rated as unsafe, and drivers could refuse to take passengers with a history of rude behavior.

Technology could also be used to coordinate the matching of rides to riders. For example, suppose riders and drivers had easy interfaces for entering their destinations. As a first step, drivers approaching a slug line might be able to identify quickly a waiting passenger wanting to go close to the driver's final destination, rather than the usual drop-off point. That would get the passenger closer, and avoid a drop-off stop that may be slightly out of the way for the driver. Now imagine that drivers and riders had a Global Positioning System (GPS) or other location-sensing devices that could monitor and transmit their locations to a central coordinating computer. That could enable pickups at points other than well-used slug lines, allowing more people to slug from home rather than first driving to a central pick-up point.

As housing and jobs sprawl over larger areas, and consumer tastes for convenience and privacy change, advocates of public transportation are fighting an uphill battle to maintain and improve systems. If enough people participate in ad hoc ride sharing, it could become more convenient and reliable than today's public transportation systems. During a transition period, public transportation will almost certainly be needed as a backup system around which ad hoc ride sharing can crystallize (many slugging routes in DC and elsewhere duplicate public transportation routes and use bus or subway stops as pickup and drop-off points). Eventually, however, impersonal sociotechnical capital may lead to a system of semi-public transportation in private vehicles that replaces public transportation entirely.

Conclusion

Much of the public discourse about social impacts of the Internet has focused on sociability. It is important to understand how mediated communication displaces, substitutes for, and complements face-to-face interactions, and how that impacts family and friendship. This chapter has argued, however, that larger structural transformations in society are likely to arise from new forms of organized interaction among strangers that ICTs can enable.

These transformations are not inevitable. For example, major engineering and incentive issues would need to be overcome to make a system of semi-public transportation a reality. Entrenched forces, such as car companies and public transit employee unions, might organize against ad hoc ride sharing (see Briggs 2002 for a fictional, humorous report of such opposition). A few well-placed scare stories and nuisance regulations might be sufficient to prevent a critical mass of adoption that would be necessary to make the system reliable.

Nor are these transformations inevitably good for society. For example, distributed news monitoring may be subject to even greater manipulation than are today's mass media. And such manipulations may be less easily detected and counteracted.

While large-scale societal changes are neither inevitable nor inevitably good, it is worthwhile to try to anticipate where they might occur. It is helpful to begin by examining the changing capabilities and cost structures for coordinating activity. With impersonal sociotechnical capital, connecting happens without personal connections and organizing without organizations. Over centuries, the process of modernization has included more and more coordinated activity among strangers, abetted by industrialization, urbanization, and the growth of government. ICTs are ushering in the next chapter in that process.

Notes

1. Social trust, sometimes called generalized trust, is, roughly, the belief that most people are trustworthy, so that others are assumed trustworthy until proven otherwise.

2. Economists define a public good as one that is *non-rival*, meaning that everyone can enjoy it without reducing the benefit it provides to others, and *non-excludable*, meaning that if procured, it is available to everyone, regardless of their contribution to its procurement. Examples include roads, public safety, and radio and TV broadcasts.

3. Friendster relies on a social network to generate a pool of candidate matches (friends of friends of friends, etc.). But, as in other matchmaking services, the selection among candidates is based on pictures and text in profiles, and users contact each other directly. This means there is little dependence on friends to introduce people or vouch for them.

4. This suggests, of course, that there are probably opportunities for moveon.org to be even more effective if it also tapped into more personal forms of social capital in organizing its political activity. For the purposes of this chapter, however, the remarkable thing to notice is how much grassroots political action can be organized without either personal connections or formal organizational structure.

5. Based on a 1991 survey, the mean occupancy for trips to work was 1.14 passengers and the mean occupancy for trips was 1.63.

6. Less than 5 percent used public transit to get to work. More than 90 percent used personal vehicles. The mean occupancy for "work" trips was 1.14. Thus, 11 percent of commuters were passengers in personal vehicles (and perhaps as many as 11 percent of drivers had passengers).

7. In two hours watching at one location, I never saw a vehicle or passenger passed over, so there may be a strong social norm against this.

References

Bakos, Yannis, and Chrystanthos Dellarocas. (2003) *Cooperation Without Enforcement? A Comparative Analysis of Litigation and Online Reputation as Quality Assurance Mechanisms.* Working Paper, Cambridge, Mass.: MIT Sloan. Available online at http://papers.ssrn.com/sol3/papers.cfm?abstract_id=393041.

Bradner, Scott. (1996) *RFC 2026: The Internet Standards Process—Revision 3.* Internet Engineering Task Force, Best Current Practice document. Available online at http://www.ietf.org/rfc/rfc2026.txt.

Briggs, Brian. (2002) *Ford Testifies to Stop Ride Sharing.* August 22. Available online at http://www.bbspot.com/News/2002/08/ride_sharing.html.

Cahn, Edgar S. (2000) *No More Throw-away People: The Co-production Imperative.* Washington, D.C.: Essential, p. 212.

Coleman, James S. (1988) "Social Capital in the Creation of Human Capital." *American Journal of Sociology* 94 (Supplement): S95–S120.

Friedman, Eric, and Paul Resnick. (2001) "The Social Cost of Cheap Pseudonyms." *Journal of Economics and Management Strategy* 10(2): 173–199.

Granovetter, Mark S. (1973) "The Strength of Weak Ties." *American Journal of Sociology* 78(6); 1360–1380.

Hardin, Garrett. (1968) "The Tragedy of the Commons." *Science* 162: 1243–1248.

Haring, Bruce. (1996) "NetDay96 to Turn California Students into Cybersurfers." *USA Today*, March 6, p. 7D.

Kollock, Peter, and Jodi O'Brien. (1992) "The Social Construction of Exchange." *Advances in Group Processes* 9: 89–112.

LeBlanc, David E. (1999) *Slugging: The Commuting Alternative for Washington, D.C.* East Point, Ga.: Forel.

Malone, Thomas W., and Kevin Crowston. (1994). "The Interdisciplinary Study of Coordination." *ACM Computing Surveys* 26(1): 87–119.

NHTS, *NHTS 2001 Highlights Report.* 2003, Washington, D.C.: Bureau of Transportation Statistics, U.S. Department of Transportation. Available online at http://www.bts.gov/products/national_household_travel_survey/highlights_of_the_2001/pdf/entire.pdf.

Putnam, Robert D. (1993) *Making Democracy Work: Civic Traditions in Modern Italy.* Princeton, N.J.: Princeton University Press, p. 258.

Raymond, Eric. (1999) *The Cathedral & the Bazaar: Musings on Linux and Open Source by an Accidental Revolutionary.* Cambridge, Mass.: O'Reilly.

Resnick, Paul. (2002) *Beyond Bowling Together: SocioTechnical Capital,* in *Human-Computer Interaction in the New Millennium,* ed. J. M. Carroll. Boston: Addison-Wesley, pp. 247–272.

Resnick, Paul, and Hal Varian. (1997) "Recommender Systems" (introduction to special section). *Communications of the ACM* 40(3): 56–58.

Resnick, Paul, and Richard Zeckhauser. (2002) "Trust Among Strangers in Internet Transactions: Empirical Analysis of eBay's Reputation System." In *The Economics of the Internet and E-Commerce,* ed. M. R. Baye. Amsterdam: Elsevier Science, pp. 127–157.

Resnick, Paul, Richard Zeckhauser, Eric Friedman, and Ko Kuwabara. (2000) "Reputation Systems: Facilitating Trust in Internet Interactions." *Communications of the ACM* 43(12): 45–48.

Resnick, Paul, Richard Zeckhauser, John Swanson, and Kate Lockwood. (2002) "The Value of Reputation on eBay: A Controlled Experiment." Working Paper. Available online at http://www.si.umich.edu/~presnick/papers/postcards/

Schultz, Theodore W. (1961) "Investment in Human Capital." *American Economic Review* 51(1): 1–17.

Wall, Lucas. (2002) "In Search of Slugs: Impatient Houston-Area Commuters Form Impromptu Car Pools." *Houston Chronicle*, December 2.

17

The Tech-Enabled Networked Home: An Analysis of Current Trends and Future Promise

Alladi Venkatesh

Introduction: The Significance of the Networked Home

This chapter extends our previous work (Venkatesh, Kruse, and Shin 2003) and examines the concept of the networked home as both a social institution and a technological construction. In particular, it addresses the transformational possibilities in the home owing to new technologies that have caught the attention of policymakers and the world of practice, including architects, home builders, community developers, digital technology producers, the entertainment industry, and other home-based service providers (Neibauer 1999, Bergman 2000, Harper 2000, Abel 2003, Bell, Heath, and Henry 2003, Turow and Kavanaugh 2003). While the concept of networks is not new to family studies, the new technologies of information and communication are requiring us to look at the home as an intersection point of sociology and technology (Kraut 2003, Robinson 2003, Wellman and Haythornthwaite 2003).

The concept of home networking has grown in prominence since the late 1990s (Magid 2000, Ruhling 2000, Business Week Online 2001, Miyake 2002, Frye 2003). With the emergence of mobile telephones and other personal communication technologies targeted for home use, home networking is receiving much attention because of the potential for dramatic shifts from current to new levels of practice. In this chapter, we address various issues concerning the networked home.

We define the networked home in terms of two major components: an internal household network, which consists primarily of network relationships with family, friends, and social circles; and an external network, which connects the home to outside agencies, such as schools, shopping centers, work/office, and other civic/community centers.

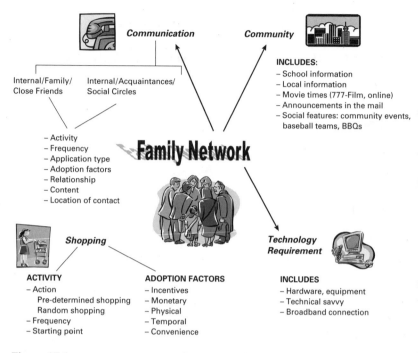

Figure 17.1
A visual representation of family networks.

Two major initiatives have contributed to a recognition of the increasing importance of the networked home concept: one concerned with technology and the other with community issues.

The technology initiative suggests that technologies available in today's fast-paced, electronic world can indeed connect people to people, people to machines, and machines to machines. Recent developments in communication technologies have been quite dramatic, especially in wireless/mobile telephones, satellite communication, and the Internet—all resulting in faster and more efficient communications globally. For a long time previously, the average citizen had only the residential phone as the primary technology for a communication network (Wellman and Tindall 1993)—which is still the case in many parts of the world.

The community initiative raises the question of how individuals and families can access these technologies. A related issue pertains to the transformational processes associated with the so-called "communication revolution," as these highly complex technologies make their way into the everyday lives of ordinary citizens. In the final analysis, however advanced the technology might be and whatever its desirable qualities are, its success will be measured in terms of community acceptance. These technology and community initiatives ultimately deal with the same questions: What are the current technological needs of families? What will their future preferences be? And what would motivate them to acquire new technologies as they are introduced into the social order? Thus, our focus in this chapter is on how to conceptualize the home as a user of network-related technologies and to ascertain what issues emerge in this context. Along the same lines, we ask how these new technologies can enhance the value of home networks and what trends are foreseeable in this regard. Several other related issues concerning standards, policy, and government regulation are relevant to this area, but are not discussed here.

Family Networks: Some Prevailing Approaches

Network approaches to the study of household or family behavior have a long tradition. The concepts and issues relating to social or community networks (Scott 1991, Wellman and Leighton 1979), or more specifically to family networks, have been well researched by scholars since the 1950s (e.g., Bott 1957, Milardo 1988). Elizabeth Bott's (1957) work remains the classic piece in this area. Many other scholars have followed her work with an elaboration of sociologically oriented themes. When we speak of networks in the family context in this chapter, we are referring primarily to social ties that emerge from these networks. To quote Szinovacz (1988, p. 7):

The themes of social affiliation and integration have been central to sociology and social psychology since their beginnings. We have learned that the specific characteristics of their social ties have important consequences for the individual and larger societal structures. Among these ties, interactions with and supports from relatives and friends have been and continue to be of primary importance. Whatever impact industrialization and urbanization may have had on nuclear families, they have not erased the social support functions.

The more recent developments in the area of family networks point to some new thinking in the structural analysis of social systems. These include patterned interconnections of family members with other families and social groups. The four types of networks widely discussed in the family sociology literature are kin networks, friendship networks, work/professional networks, and community networks (Grieco 1987, Milardo and Allan 1997, Roschelle 1997). The role of technology in fostering family networks began to appear in the literature around the turn of the century, with Wellman (1999, this volume) and his associates (Hampton 2003) taking the lead on a series of research studies concerning technology in the context of organizational and personal networks. The question that arises in the present context is whether technology is an enabling agent in fostering family networks or an active promoter of networking practices.

For a major part of the twentieth century, the radio and television (both one-way), plus the telephone (two-way), were the main media/communication technologies that found their way into the homes of ordinary citizens. They heralded, in a sense, the first communication revolution. More recently, we have seen a veritable rush of technologies entering the home, leading to the creation of new forms of networking possibilities (Miyake 2002). One of the key technologies since the late 1980s has been the home PC (Venkatesh 1996). Subsequently, the computer's networking value has increased dramatically, primarily owing to the Internet (Kraut et al. 1996). The rapid convergence of communication and information technologies has also contributed to networking possibilities in the home.

The rise of the Internet and other related technologies, with their attendant social consequences, suggests that traditional network approaches can be viewed in a new (technological) light. However, recent technological developments have caused us to identify the relevant category of inquiry as the "networked home," as an extension of the "networked family." This may not be just a semantic issue, because it brings to the forefront the fusion of technological networks and social networks. One can argue, therefore, that the emphasis on the networked home is inclusive of, and an extension to, the notion of the networked family. In order to add greater refinement to the concept of the networked home, we integrate the notion of "family" as a sociological group with "home" as a combination of physical, technological, and social spaces (Venkatesh and Mazumdar 1999, Lee 2000). The distinction between "family" and "home" is that we regard family as a social institution and home as a living space. In summary, while the idea of network itself is not new, technological advances have changed the character of family networks and introduced greater complexity and variety into home life (Kiesler 1997). Further, while networks can exist in the absence of technology, modern networks are highly technologically based.

Evolution of Home Networking Since the 1980s

Although computers were introduced into U.S. households in the early 1980s (Dutton, Kovaric, and Steinsfield 1985), it was only with the advent of the Internet and extensive e-mail connectivity in the mid to late 1990s that people began to use computers for home networking in greater measure. Since then, studies have reported a growing number of networking applications in the home. For example, Shih and Venkatesh (2004) identified 17 different uses of computers that were grouped into seven major categories (table 17.1): work at home, family

Table 17.1
List of activities covered by computer use

Activity Space	Activities
Work/employment related	1. Job related 2. E-mail (work related)
Family communication	3. E-mail (personal) 4. Writing letters/correspondence other than e-mail
Family recreation	5. Games/entertainment
Home management	6. Home management (recipes, family records) 7. Health information 8. Travel information/vacation planning 9. Financial management 10. Online banking
Home shopping	11. Shopping (frequently-purchased goods) 12. Shopping ("large-ticket" items) 13. Shopping (other)
Education/learning	14. School related
Information center	15. Reading news 16. Sports information 17. Community information

communications, family recreation, home management, home shopping, education/learning, and information access. Each of the categories gives rise to substantial networking opportunities for the family.

Work-at-Home Issues

During the first decade of home computing, prior to the emergence of the Internet in the mid-1990s, nearly 70 percent of computer use was devoted to job-related activities. That is, the computer was seen primarily as a work tool. This limited focus has broadened rather significantly since the 1990s. With the increasing domestication of the computer, it is no longer viewed as a mere work tool but has become a more versatile networking medium. Although work at home represents just one among many activities on the computer, it is still recognized as a major home-based activity. The literature on work at home is growing because it raises many technological and sociological questions. For instance, Salazar's (2000) study identified three sets of questions concerning computers and work at home:

• Why do people choose to work at home? How is work organized so that it can be done at home? What kind of environment is conducive to working at home?
• How do workers define boundaries between work and home? How is the work environment negotiated with other household members?
• What are the physical workspace and technological needs of those who work at home? How is the computer used to access materials?

These questions are particularly significant in the case of workers who are not self-employed and don't own their businesses, but are paid employees of an organization. Sociologically speaking, many people desire to work at home in order to be able to work without interruptions and be more productive and efficient, to have flexibility in their lives that permits them to be with their families who need their care and attention, and to create a sensible balance between work and family responsibilities. In terms of the technological factors that permit them to meet these objectives, home-based workers need to be able to transport work back and forth, to maintain regular contact with their colleagues in the work environment, and to perform as effectively as they would if they were physically present at the office. However, there are some barriers to accomplishing this. First, working in the home is not the same as working in the office. The social dynamics are totally different because there are no opportunities for personal give and take through face-to-face interaction with colleagues. Second, the physical remoteness is not fully overcome by the mere presence of technology, however advanced or sophisticated it might be.

In summary, while technology enables networking opportunities that permit people to work at home, it also limits social interaction in ways that can have both negative and positive consequences. Ultimately, the worker must weigh the trade-offs between working at home and working at the office in order to choose the best course.

Some Conceptual Issues

Three main conceptual schemes motivate our thinking in regard to the networked home.

1. *Network growth.* The relevant question here is what role technology plays in augmenting the communication networks, in reinforcing network relationships, and in facilitating the entry of new players into the networks.

2. *Network maintenance.* The issue here is how technology sustains social practices and priorities and how people function as active agents in the network.

3. *Negotiation and empowerment.* With the bevy of communication options at hand, how do people negotiate which options to use? Do they combine options to add value to their communications? How is technology changing the experiential aspects of communication, and how are people "empowered" by having many choices of communication?

We investigate these key areas by identifying the categories of communication from the perspectives of communication technology and of the actors engaged in the communication activity. Thus, the relevant questions are: How do people communicate? What are some important communication activities? How do communication activities maintain or change the family network?

Research Study of Social Networks and Communication in Families

In the remainder of the chapter, we report various conceptual issues resulting from a preliminary analysis of an ethnographic study of twenty-five families living in the Southern California area. This work was still in progress at the time of writing, and a more complete analysis of our study will be reported elsewhere in due course.

The Categories Analyzed

In our study, we focused on five modes of communication (e-mail, telephone, paper notes, cellular telephone, and instant messaging) and three forms of social networks:

1. *Family*. There are three major continuums to family communication: distant and local; strong and weak relationships; and strong and weak dialogue. A family member who is located distantly may be contacted once a month but may have a strong link to the person who initiates contact. There may also be family members with whom links are local but weak. It appears that the nature of the relationship drives which mode of communication is used to contact the other person, and the quality of the dialogue with that person.

2. *Friends*. Friends are like extended family, in the sense that they live outside of the household. One can have a more intimate relationship with close friends than with an extended family member, but with some other friends or acquaintances the bonds may be weaker. For children, especially teenagers, friends may be as important as family, if not more so.

3. *Community networks*. Through a variety of means provided by a technology-enabled network, the convenience and accessibility of community institutions are available to people more than ever before. Social clubs are formed online, creating the phenomena of cyber-neighbors. Principals of schools can now directly contact parents about their children's progress instead of arranging parent-teacher conferences. Of course, traditional means of community networking (e.g., phone call, letter through postal mail) are still in place, but the addition of Internet-based communication technologies is indeed changing the dynamics of the family network.

The Social Dimensions of Communication Technologies

The following are the main social dimensions of communication technologies we identified in the study:

1. *Strong relationship vs. weak relationship*. This continuum seeks to classify a relationship in terms of the bond between the actors in communication networks. Here is an excerpt from a 15-year-old female (Maria) from our sample:[1]

"For instance, my mother and I share a strong relationship, while my teacher and I share a weak relationship. It is also possible to have medium relationships. The type of relationship one has with another entity is not necessarily dependent on the spatial relationship (near or far) or the quality of the dialogue. I say about two sentences a week to my mother who lives in a different state, but I might send an e-mail daily to my teacher, who lives two miles away. But yet the relationship with my mother is still stronger."

2. *Distant vs. local.* This refers to the physical distance between the actors in the network.

3. *Engaged dialogue vs. snappy dialogue.* This relates to the length and involvement of a dialogue. An engaged dialogue is something that is usually of some length and carries a significant level of emotional and temporal investment. A long phone call is an engaged dialogue. On the other hand, a snappy dialogue is one characterized by its brevity, a certain sense of "to-the-pointness." Sending just jokes through e-mail is brief. Although seemingly "weak" in nature, snappy dialogues do not necessarily imply a lack of strong relationships. Sending pictures through e-mail, which is terse in its quickness, is not only a highly personal activity, but one that packs some emotional punch as well.

4. *Intimate vs. non-intimate.* Some conversations are intimate and some are not.

5. *Urgent vs. non-urgent.* Depending on the urgency of a situation, a person will decide which mode of communication is the best course of action.

6. *Socialization vs. informational.* Our respondents described the telephone as a device that exhibits a high degree of socialization. E-mails, on the other hand, are informative and get straight to the point.

7. *Formal vs. informal:* Formal communication refers to a correspondence that is structured, as in a typed document or letter. Informal correspondences are dialogic and unstructured.

8. *Vulnerable vs. protected.* Vulnerability refers to a qualitative assessment of communication situations that might render people susceptible to certain situations that they would rather avoid for fear of negative consequences.

9. *Faster vs. slower.* Refers to the speed of establishing contacts and communicating.

10. *Convenience vs. hassle.* This refers to whether the medium allows for an easy way to establish contact, or introduces some complexity and is perceived as a hassle.

Communication Technologies and Their Characteristics

We will systematically go through the five major communication technologies used in the family networks of our research participants—e-mail, telephone, paper notes, cellular phone, and instant messaging—and elaborate on the emerging trends and patterns that are unique to each of them.

E-mail

E-mail is one of the more popular communication mediums within a social network. It is a universal choice for sending messages to, and receiving messages from, family, friends, community members, and commercial agencies. Many of our respondents mention e-mail and the telephone in the same breath, as if they were either at the opposite ends of a complex negotiation process or just plain substitutes. In talking about e-mail, it was quite common for our respondents to refer to the phone as the "other" in a comparative way. Here is a typical comment from a young parent (Diane):

Q.:[2]   Why do you use the e-mail over the phone?
Diane: Because it is sometimes easier and even quicker to e-mail her [referring to her mother] instead of pick up the phone.

In people's perceptions, there seems to be a strong interplay between the telephone and e-mail. Depending on the set of conditions of a scenario, a person will choose the appropriate method of initiating (or maintaining) communication by e-mail or telephone.

Strong Relationships through E-mail   E-mail may be initially perceived as a weaker form of socialization than the telephone, but not always because it is frequently used among spatially distant friends and family. After all, it makes economic sense to do so. Even in situations when it does not cost to make a telephone call, our respondents use e-mail instead of the telephone on a regular basis. So, an early conclusion is that e-mail is being used to bolster both weak and strong relationships, in distant and near locales.

However, we must keep in mind that keeping a relationship strong via e-mail does not necessarily mean that the e-mail communication entails a long or engaging conversation. A comment from a working female professional (Anita) that elucidates the point:

I email my mom sometimes a joke or some private information even though she lives close by. . . . But Matt's [husband] family is all back East, and they are in Florida and Virginia, and his brother is in Germany, so I just emailed them yesterday like a bunch of pictures like the house, or they wanted to see my stomach [she is pregnant] so I'm sending them pictures to see how it is getting really big.

The above scenario is a good example of how e-mail is being deployed. With her mother, Anita has a strong, local relationship, and she maintains it, in part, with a snappy e-mail and a joke. With the extended family, Anita has a strong, distant relationship and maintains it with nontext e-mails—just pictures.

Fast, Convenient, and Easy   For many of our respondents, e-mail is a faster and more convenient method of communication than the telephone. Although vocal communication is commonly regarded as *the* way to communicate, people justify their perspective by saying that with e-mail one does not have to spend so much time socializing. They can get "straight to the point" and "just send it out." Typing was not a major factor for a good number of our respondents. Most of them could type quickly and had no more qualms about tapping the keyboard than lifting the phone from its cradle. Here is Deborah, a working mother, explaining her preference for email because she can type fast:

Deborah: I can type really fast, so it's easier for me to send her a message just to say it's OK, "I can" and "what time do you want us to do [etc]. . . ." It's faster because I don't have to spend time socializing.

Q.: And you notice if there's any differences between e-mailing and calling? You do them for different reasons?

Deborah: Because in one I just go straight to the point and the other one I do just if I have more time. One is faster, e-mail is for me faster.

E-mail is more convenient for some because it does not exist as a real-time communicator. The need for another person to be at the other end is eliminated. It works better with people with conflicting schedules, or for people who are located in two different time zones, as Jane, an unmarried single female explains:

Q.:   For what do you prefer using the e-mail or the phone?

Jane: Phone for every day things, e-mail to send some messages if there's something interesting that I saw . . . I don't use my phone too much for my friends, I use phone mostly for my family than friends, if it's out of state I rarely call. And it is not because we can't afford it, it's just convenience, you know time difference. Otherwise you try to call and they are not there and then they try to call you back and it goes on . . .

E-mail as a Form of Collateral Exchange   With standard telephone calls, the exchange is audio in nature. E-mail is predominantly visual. One has to read the text and open and look at attachments. The active exchange of visual collateral differentiates e-mail from some other modes of communication. With the quick transmission of a picture, attachment, Web link, or a joke one can maintain a relationship with someone close or far away without writing a word.

E-mail as an Informational Medium   In addition to exchanging words, our respondents also use e-mail as a source for delivering and receiving informational objects. This is consistent with the theme of e-mail as being a snappy, low-socialization, and quick (but effective) networking medium. Short e-mail messages with appointments, or event scheduling information, are often sent by our respondents to members within their social network. A stay-at-home mother (Brenda) explains thus:

Probably family and . . . contact the school the kids go to, like their soccer team, baseball team, everything pretty much. People don't really call each others as much. We communicate via email when tournaments are going to be held or when games are going to be played. So just exchange information on the new schedule, when new practice is going to be and so on. Really I think most of my e-mail has to do more with kind of scheduling activities and that kind of thing.

E-mail also exhibits properties of physical mailboxes, in the sense that community and shopping information is delivered via e-mail. Here are some excerpts from interviews with some working mothers:

Anita: I contact like family and friends. I shop online a lot, so I get e-mail . . . confirmations of my orders, things like that. And like our neighborhood association here something that I contact through e-mail.

Diane: [O]ur local school has started the thing where you sign up to get on the distribution list and they'll send different things going on at the school, so there is a kind of information gathering via e-mail . . .

Bonnie: Because right now, I mean just going to the e-mail, like I was looking for tickets to go to visit my brother in law in Colorado; so, it was like I was waiting to get certain air fare, now I don't go to Travelocity every day to check if the fare is available. . . . I need to know once for all . . .

One of the drawbacks of e-mail, cited by a few, is typing. Here is a great opportunity for voice communications. Some see typing as another hassle, in addition to turning on and logging onto the computer. In two cases, the husbands seem to have a preference for picking up the phone, to enable quick communication.

Telephone

The telephone continues to be a powerful communication technology within the social network. In our study sample, the telephone is often used in strong network relationships (contacting significant others), in local contacts, and in urgent situations. Of the different modes of communication available to our respondents, the telephone is probably regarded as the most personal. Of course, the telephone can be found in other network relationships and is used as the situation warrants. Again, the telephone is often mentioned along with e-mail, as a basis of comparison.

Strong Local Relationships Unlike e-mail, the telephone is used to maintain strong relationships, to bolster a relationship in the social network that is most important to our respondents. Immediate family and close friends would fall under this category. They are called more often than e-mailed, although there were instances where friends were habitually e-mailed. Community institutions did not seem to be a strong part of the telephone network in this study, for e-mail seems to be the preferred mode of contact.

Immediacy/Urgency For all types of relationships, if there is an urgent message that needs to be transmitted, the telephone seems to be the obvious choice. Telephone calls are regarded as the "fastest" form of communication because of the potential for immediate feedback. However, when conversations get too involved, then phone conversations can actually be more tedious than a quick, simple e-mail—as explained by Carol, who is the mother of two pre-teenage sons:

Carol: Probably urgency. Like a visit being planned suddenly . . . Most things aren't as urgent. When she [my mother] is coming down to visit the next few days and I call her to make sure it's OK . . .

High Levels of Socialization   As previously explained, there is a higher level of socialization in telephone conversation than in e-mails. Certain respondents prefer e-mail in certain situations because they won't have to spend time socializing. The other side of the story is that if they *do* want to socialize, then they would pick up the phone.

Limitations of an Audio-Only Medium   Because telephones are audio-based communications, people cannot physically see the person they are talking to, and pictures and other sorts of collateral cannot be transmitted. So, in a sense, there is incentive to use e-mail (or a fax) if there is some visual information that needs to be sent.

High Vulnerability   Telephone calls, like face-to-face interactions, can make the actors highly vulnerable to each other. By this, we mean there is little "protection" for people when there is a topic that they want to avoid. Furthermore, people can be held accountable for what they say to each other on the telephone.

Paper Notes
Paper notes are low-tech ways to make a quick connection with family members, for example when the writer wants to remind them to do something. However, the nature of the task is *not urgent*. Paper notes are usually left when the writer of the note leaves the home. Their content is usually *informational*—notes are often used as reminders about tasks that need to be done. In rare cases, "I Love You" notes of affection are left for family members. And notes are almost always used for local, *in-house relationships* to communicate with other household members.

Visibility and Accessibility   Unlike other modes of communication, paper notes need to be found. There are areas in the home where paper notes are frequently placed, like the kitchen counter, but they are not fixed to that place in the same way a telephone is fixed to its cradle or e-mail is fixed into the computer. Thus, making a paper note visible and accessible is a key issue. Here is an excerpt of an interview with a husband and wife (John and Mary) regarding paper notes.

John: Sometimes, you know when we are going to be out we leave a note or something.

Mary: Once in a while but usually we call.

Q.: Where do you leave the note?

John: On the refrigerator or here [pointing to the kitchen counter] . . .

Mary: Probably on the calendar . . . I circle it.

Q.: Why the kitchen counter and why not the refrigerator?

John: Because it's just kind of a central point area and I usually sit next to the kids.

Wife: Because it is the main place in the house.

Cellular Phone (Family Context)

Types of Uses In some cases, our observations of cellular phone use within the family reveal that such calls tend to be short and are made for safety reasons or just to see how someone is doing (usually the spouse and kids) or check where someone is. Long calls over the cellular phone in the family context were not documented, as highlighted in a conversation with one of our respondents (Jane).

Jane: [Husband] probably will call me four times a day.

Q.: Just to check up?

Jane: Check up, what's going on. Yes, to let me know what he's doing or when he's coming home.

Q.: At certain times?

Jane: Usually, he always calls me in the afternoon, and then he'll call me on the way once he gets into his car, so I know exactly how long it's going to take to get home. And he calls me in the morning to see how I feel.

Strong Local Relationships Recipients of cellular phone calls tend to be involved in strong relationships, and tend to be local. There is no mention of calls to network actors on the periphery (weak relationships).

Cellular Phone (Children): A Prime Communication System

For some of our younger respondents, the only personal communication system besides e-mail is the cellular phone. Since the cellular phone is portable and very handy, it is becoming ubiquitous. It is also a means by which parents keep track of their children. Here is what a 14-year-old female (Samantha) said:

Samantha: The cell phone is my life line to the rest of the world. Sometimes, I forget to charge it and the whole world falls apart before my eyes.

One 17-year-old female (Christine) emphasized that she would like to see greater technological networking capabilities:

Christine: Wow, I don't know, if I could have my dream computer it would include, the video cam, so I can see the person I'm talking to but have to take the time, to make it, like, a real video projection sort of thing but I know it's hard. Um . . . I would also want—I know you can set up some timer ready, like, that would automatically turn on and off and also to check my e-mail. . . . Like it's connected to a pager that tells me you have new e-mail and maybe . . .

In the following excerpt, we see some communication strategies in the use of cell phones and complex networking scenarios. Ashley, 12-year-old female, commented in the following way:

Q.:   Use cell phone . . . ?
Ashley: Yes, I do.
Q.:   And do you use it regularly?
Ashley: Yes, I mean, I lost it a couple days ago. I don't know where it is, but without it, it's made me feel so helpless. But a pager is very helpful because, you know, if I'm going out and I want my friends to reach me they could page me and I can call them back at my convenience, like, if I'm in the middle of something . . . call them back right away, whereas if it's a cell phone if somebody's calling me and the cell phone is off . . . a call that might come at any time, and you have to answer it. Also my parents try to get hold of me here. . . . You feel more free to do what you need to do and then also be connected to people who they need to get a hold of you.
Q.:   So they tell you to carry it with you all the time, more or less?
Ashley: More or less. I don't—if I don't want to be bothered somewhere, like, I wouldn't carry it. I wouldn't go to the gym with it. . . . I'll go to a movie or play with it but I'll turn it off, like, I'll turn the sound off but leave it on . . .

Here is how another respondent, Kathy (16) feels about this:

Q.:   And a cell phone?

Kathy: Cell phone, um, I've only started using it in the last nine months.

Q.:   And you said you don't use it that much these days?

Kathy: Um, I use it a lot but I don't—I don't let people call me. I don't give out my number very much unless it's an emergency, and it's connected to my car so it's only on when I'm driving.

Q.:   And . . . ?

Kathy: Yeah, like if I'm coming home late or something, to say I'm on my way home. Um, or if I left the house and they paged me then I can call them right back.

In the following interview, Maria (12) explains why she thinks it is "cool" to have a cell phone. She also refers to her parents' role.

Q.:   Do you use a cell phone?

Maria: My mom does. I don't. All my friends—or some of my friends, like, have pagers and stuff but I don't really see what the big deal is about them anyway. I mean, some people have gotten into trouble at school, like, people stealing and stuff and that's been like a big deal so I don't really want to get into that—so I don't really want a pager or cell phone yet (laughing).

Q.:   Yeah, what do think they use it for?

Maria: I don't really know. I just think they think it's cool to have one.

Q.:   Uh-huh, just another sort of technology . . .

Maria: Yeah.

Q.:   Another gadget to have, another toy?

Maria: Yeah.

Another young female (Kim, 11) is similarly placed:

Q.:   And then cell phone, do you use your mom's cell phone?

Kim: Um, like I don't take it with me or anything. She doesn't even really use it, like, she just does it when she wants to call my dad, like, when he's home late or see to check where he is . . .

Q.:   Does she take it in the car with her?

Kim: Yeah, it's a safety thing.

Q.:   But she doesn't use it very much?

Kim: Yeah, she doesn't sit around the freeway and call . . .

Instant Messaging

The instant message experience seems to be something that is informal and long-term. Here is a young male (Don, 15) describing his experiences:

Q.:   What do you think about instant messaging?

Don: I think it's great. Specially . . . I think it's great for people that are kind of at a computer all the time. Not great for someone like me now because I never can use my computer for 45 minutes . . . If I know there's someone wanting to communicate with me I'm going to say pick up your phone and call me . . .

Q.:   Are there different things that you talk about? I mean, what are the things that you have talked about recently?

Don: Instant messaging is more like talking. . . . When I e-mail it's more like a letter, to say this is what is happening to us. Instant messaging is . . . I feel I'm almost conversing as opposed to some detailed stuff . . . more open I would say. I don't know . . . what's the word? Casual, natural . . . yes. Informal . . . perfect.

The Opportunities Provided Different Communication Technologies

This section has shown that different communication technologies provide different opportunities and structures of communication. No technology is better than the other, and each has some unique properties. Table 17.2 provides a comparison of the different technologies and their social character as revealed through our empirical analysis. Although one of the communication situations (face-to-face) was not specifically addressed in our empirical study, we have included it in the table for the sake of completeness.

All technologies have the potential to be transformative. We have provided some evidence on the transformative properties of some of the technologies. With so many new technologies coming into the marketplace, we are bound to see many more changes in the years to come.

Discussion and Conclusions

From the social dimensions and emerging trends we have analyzed in this chapter, certain patterns can be seen to be beginning to coalesce. For instance, as mentioned earlier, we are starting to see the significant role

Table 17.2
Summary of communication technologies and their social dimensions

	Relationships	Distant/local	In urgent situations	Social/informative	Intimacy	Engaged/terse dialogue	Personal/casual	Fast/slow	Convenience	Vulnerability
Telephone	Strong	Mainly local	High	Social	High	Engaged	Personal	Fast feedback, sometimes slow	Medium-high	High
E-mail	Strong and weak	Both	Low	Informative	Low and high	Terse	Both	Fast to send, slow feedback	High	Low
Cellular phone	Strong	Local		Both	Medium	Seemingly terse	Seemingly Personal	Seemingly fast	High	High
Instant messenger		Both			Low	Both		Slow	High	Low
Paper notes	Strong	Very local	Low	Informative	Low	Terse		Fast	Low	Very low
Face-to-face	Strong	Local		Social	High	Engaged		Slow	Low	Highest

of technology in communication in the form of network growth, network management, and in negotiation and empowerment.

Network Growth

We find that there is an inclination towards the expansion of social networks through the mutual interplay between human decision-making and technology. Technology and the human decision to employ it can expand the network in two ways: by reinforcing current relationships and by making it feasible for new relationships to enter the network. Before the era of e-mail, to maintain a relationship one had to call or write a letter. A rather intimate and potentially long process is characteristic of these two modes of communication. However, with e-mail one can just blast-off a note and maintain the relationship, without the time and emotional investment required with the other media. In other words, technology enables and creates new forms of relationships: ways having new properties that are not part of established forms of communication. One can now maintain a relationship with someone with whom he/she does not want to lose touch, be it a strong or weak relationship, through forms of communication that range from low to high in their level of intimacy.

It can also be argued that certain relationships are strengthened because of technology, especially distant ones. In relationships among actors separated by great distances, telephone contacts tend to be infrequent. However, the addition of e-mail to the communication arsenal changes this quite dramatically.

In certain cases, new relationships are forged and maintained because of the ease of communication. For instance, instant messaging or a buddy list is useful for a variety of contacts. If instant messenger were suddenly erased from the communication repertoire, a person accustomed to this would probably cease contact with others in the list. Buddy lists are another example of the role of technology in expanding and shaping social networks.

Network Management

Network management applies to social activities and priorities, rather than to identifying the players in the communication network (network growth). Another way of translating this concept is: How does access to communication technology impact the way people manage network activity and priorities? We argue that communication technology impacts network management in two ways: by improving the efficacy of scheduling events and by giving users more security measures.

Having a breadth of communication technology gives people more options for managing social and community events. For instance, the convenience of scheduling a get-together is displayed in e-mail scenarios where a mass of people can be contacted through a single message. Similarly, participating in community events (like team sports for children) is made more efficient via e-mail, where team members can mutually share game times and other notices. Without e-mail there would be a lack of such a convenient mode of communication.

Communication technology provides extra security as well as convenience for users. For example, when parents are away from the home, they can check up on the family with a quick cellular phone call.

Negotiation and Empowerment

We argue that people are empowered by the negotiation process made possible by communication technology. People now can choose from not just one or two, but five to six different ways to communicate with other network players. As described earlier, each mode of communication has its own set of unique characteristics and properties. The tech-enabled network allows its members to choose a mode of communication that suits their experiential needs.

Acknowledgments

This chapter was written as part of ongoing research under Project NOAH II and POINT Project at the Center for Research Technology (CRITO), University of California, Irvine. We acknowledge the financial support received from the National Science Foundation (NSF grant Nos. IRI 9619695 and SES-0121232) and CRITO Industry Consortium.

Notes

1. Note that all names from the sample used in this chapter are fictitious.
2. "Q." in interview transcripts indicates a question from the researcher to a respondent in the study.

References

Abel, Paul (2003) "How Will the Technology in the Home Evolve?" Paper presented at the HOIT 2003 Conference, University of California, Irvine, April.

Bell, Nancy, Pam Heath, and Wallace Henry (2003) "The Changing World of Home Technology: A Microsoft Perspective." Paper presented at the HOIT 2003 Conference, University of California, Irvine, April.

Bergman, Eric, ed. (2000) *Information Appliances and Beyond: Interaction Design for Consumer Products.* San Mateo, Calif.: Morgan Kauffman Publishers.

Bott, Elizabeth (1957) *Family and Social Network.* London: Tavistock.

Business Week Online (2001) *Still Waiting for the Networked Home.* February 5. Available online at http://businessweek.com/bwdaily/dnflash/feb2001.

Dutton, William, Peter Kovaric, and Charles Steinsfield (1985) "Computing in the Home: A Research Paradigm," *Computers and the Social Sciences* 1: 5–18.

Frye, Curtis (2003) *Faster Smarter Home Networking*, Redmond, Wash.: Microsoft Press.

Grieco, Margaret (1987) *Keeping It in Family: Social Networks and Employment Chance.* London: Tavistock Publications.

Hampton, Keith (2003) "ICT Mediated Place-Based Community." Paper presented at the HOIT 2003 Conference, University of California, Irvine, April.

Harper, Richard, ed. (2000) Special Issue on Domestic Computing. *Personnel Technologies* 4, no. 1: 1–69.

Kiesler, Sara (1997) *The Culture of the Internet.* Hillsdale, N.J.: Erlbaum.

Kraut, Robert (2003) "The Internet and Social Life: Details Make a Difference." Paper presented at the HOIT 2003 Conference, University of California, Irvine, April.

Kraut, Robert, William Sherlis, Tridas Mukhopadhaya, Jane Maning, and Sara Kiesler (1996) "The Homenet Field Trial of Residential Internet Services." *Communications of the ACM* 39, no. 12 (December): 55–61.

Lee, Wai On (2000) *Challenges and Issues in the Application of Living Space Model for Home-Based Information Technology.* Working Paper, Redmond, Wash.: Microsoft Corporation.

Magid, Lawrence (2000) "Home Networking Next Big Thing for Families With Multiple PCs." *Los Angeles Times*, June 26, section C3, p. 1.

Milardo, Robert M. (1988) *Families and Social Networks.* Thousand Oaks, Calif.:, Sage Publications.

Milardo, Robert M., and Graham Allan (1997) "Social Networks and Family Relationships." In *Handbook of Personal Relationships*, ed. S. W. Duck. New York: Wiley Publications, pp. 505–522.

Miyake, Kuriko (2002) *Networked Home Becomes Reality.* Available online at http://maccentral.macworld.com/news.

Neibauer, Alan (1999) *This Wired Home: The Microsoft Guide to Home Networking.* Redmond, Wash.: Microsoft Press.

Robinson, John (2003) "The Internet: Time and the Rest of Life." Paper presented at the HOIT 2003 Conference, University of California, Irvine, April.

Roschelle, Anne R. (1997) *No More Kin: Exploring Race, Class, Gender in Family Networks.* Thousand Oaks, Calif.: Sage Publications.

Ruhling, Nancy A. (2000) "Home Is Where the Office Is." *American Demographics.* June, pp. 54–60.

Salazar, Christine (2000) "Building Boundaries and Negotiating Work at Home." Working Paper, Seattle, University of Washington.

Scott, John (1991) *Social Network Analysis.* London: Sage.

Shih, Chuan-Fong, and Alladi Venkatesh (2004) "Beyond Adoption: Development and Application of a Use-Diffusion Model." *Journal of Marketing* 68, no. 1 (January).

Szinovacz, Maximiliane (1988) Series editor's foreword to *Families and Social Networks*, ed. Robert M. Milardo. Thousand Oaks, Calif.: Sage Publications, p. 7.

Turow, Joseph, and Andrea L. Kavanaugh (2003) *The Wired Homestead: An MIT Press Sourcebook on the Internet and the Family.* Cambridge, Mass.: MIT Press.

Venkatesh, Alladi (1996) "Computers and Other Interactive Technologies for the Home." *Communications of the ACM* 39, no. 12 (December): 47–55.

Venkatesh, Alladi, and Sanjoy Mazumdar (1999) "New Information Technologies in the Home: A Study of Uses, Impacts, and Design Strategies." In *The Power of Imagination*, ed. Thorbjoern Mann. Environmental Design Research Association (EDRA), pp. 216–220.

Venkatesh, Alladi, Erik Kruse, and Eric Shih (2003) "The Networked Home: An Analysis of Current Developments and Future Trends." *Cognition Technology and Work* 5, no. 1 (May): 23–32.

Wellman, Barry, ed. (1999) *Networks in the Global Village.* Boulder, Colo.: Westview Press.

Wellman, Barry, and Carolyn Haythornthwaite, eds. (2003) *The Internet in Everyday Life.* Oxford: Blackwell Publishing.

Wellman, Barry, and Barry Leighton (1979) "Networks, Neighborhoods and Communities." *Urban Affairs Quarterly* 14: 363–390.

Wellman, Barry, and David Tindall (1993) "Reach Out and Touch Some Bodies: How Telephone Networks Connect Social Networks." *Progress in Communication Sciences* 12: 63–93.

18

The Social Impact of the Internet: A 2003 Update

John P. Robinson and Anthony S. Alvarez

Reaching conclusions and generating ideas on the impact of new technologies may seem straightforward and simple, so speculation abounds about how daily life has been transformed by them. However, it is an extraordinarily complex task to demonstrate, document, and defend such speculation scientifically. First, these new technologies often arrive in a society that may be already anticipating their arrival, as in societies that place a premium on efficiency and saving time. In a related fashion, the new technology may simply accentuate or mimic social changes that are already taking place. Thirdly, the people who are the first users or "innovators" are known to be rather different in their social-psychological makeup from other members in society (Rogers 1962). Is it the technology making people different, or were the people different prior to the advent of the technology?

Television is perhaps the premier technology associated with societal changes in the twentieth century (Robinson and Godbey 1999; Robinson 1969). Like the communications media before it (radio, movies, comic books), TV is often blamed for deleterious effects on society, such as decreased literacy, indifference to violence, declines in social capital, and watered-down culture. Some of these allegations have been backed up with empirical scientific evidence (e.g., Putnam 2000; Gerbner 1978), while others haven't. What very detailed studies of time do show, however, is that unlike other technologies, such as the automobile and the washing machine, TV was associated with dramatic differences in how people spent time. Not surprisingly, they show that people took some of their viewing time from the movies, from radio, and from pulp fiction—content more readily transmitted on a home TV screen. But TV viewers around the world who filled out these time-diary forms in the 1960s also had lower socializing time with neighbors and relatives, while concentrating more of that social life around the screen with members of their nuclear families. Unrelated to these free-time communication activities, TV viewers also got less sleep and did less gardening, two activities for which it would be hard to argue that television provided the same functions for the viewers.

Thus, TV appears to have made inroads on both free-time and non-free-time activities to the extent that TV owners were spending 90 minutes a day on an activity that had not existed before TV. Since then, the time that TV viewers spent with their screens in the 1960s grew to between 120 and 150 minutes two to three decades later.

This carving out of new time for TV is unprecedented both in its clear time displacement and in the approximate two-hour magnitude of its displacement. To be sure, many of the activities it apparently displaced were ones that performed the same functions, such as radio and movies, but after the advent of TV personal communication activities such as conversations and parties saw declines as well. Eventually, TV would also make inroads on news communication via newspapers and on social capital activities such as volunteering and organizational participation, according to data analyzed by Putnam (2000). Technologies such as the automobile or the elevator might lead to profound spatial reorganizations of society, but they do not seem to affect the time spent traveling or in motion. TV seems unique in its ability to affect the temporal organization of social life.

These functions are made clearer in figure 18.1, which also conveniently shows the communications environment existing at the time of the Internet's arrival into society. Unlike TV, however, the Internet falls into three of the quadrants identified in figure 18.1. Like TV, it is a mass medium, shown in the lower left corner in figure 18.1. Unlike TV, it also can become a personal communications medium, like the telephone, through email, shown in the upper right cell of figure 18.1, and further unlike TV, it can combine personal and mass communication functions, shown in the lower right cell, as when bloggers and personal website users are able to "broadcast" their messages to people they have never met.

Figure 18.1 then suggests how the Internet might be expected to become a more potent force in society than TV, in its ability to both serve

	One-Way	Two-Way
Discriminate	Soliloquy Pager	Conversation Mail Telegraph Telephone
Indiscriminate	Books Journals Newspapers Movies Radio Television	Ham Radio Citizens Band Radio

Figure 18.1
Pre-Internet communication media typology.

and combine personal and mass communication functions. Since it can be individually tailored, as in allowing users to receive only news information or entertainment the user is interested in, one can expect users to be able to reduce their time on TV or newspapers. Since it allows users to reach each other at more convenient times during the day and provides an electronic record of previous conversations, it can reduce the time users otherwise communicate with each other. Moreover, in its ability to combine personal and mass messages, it provides an entirely new way of communicating and spending time.

Internet Data Collections

These are issues that can now be examined scientifically, because unlike TV, the changes in daily life associated with the Internet are now being tracked regularly by social scientists, government agencies, and commercial interests. Figure 18.2 outlines the many organizations that are doing this tracking by collecting systematic data and the years these studies (mainly surveys) were conducted. These surveys differ in several respects, as discussed in appendix A.

Thus, in contrast to the earlier case with the introduction of television, the Internet has been the subject of a plethora of rich data collections from which to draw conclusions about its initial societal impact. Moreover, the various data sets have offsetting strengths and weaknesses,

Organization	Sample size~	1995	1996	1997	1998	1999	2000	2001	2002	2003
Internet Diffusion?		(10%?)		(17-30%)	(33-37%)		(44-48%)	(54%)	(57%)	(60%)
EstimateData										
1. NTIA	50,000	X		X	X		X	X		(X)
2. GSS	2700						X	P	X	
3. a. PewPress	4000	X			X		X		X	
b. Pew Internet	2000						XXXX	XXX	XPX	(XXPX)
4. UCLA	2000						X	PX	PX	
5. Rutgers	1000	X		X	X		X			
Diary Data										
6. UMaryland	1000	X			X		(Parents)	X		
7. SIQSS	5000									
8. Canada	10,000				X					
9. UK	1000					X	P	P		(X)
10. Holland	3000	X					X			
Clickstream										
11. Nielsn/Net	50,000		(P) ─────────────────────────────────>							
12. ComScore			(P) ─────────────────────────────────>							
Volunteer										
13. Nat Geo	10,000				X		X			(X)
14. GaTech										
15. Harris Int.										

x – Cross Sectional
p – Panel

Figure 18.2
Years and sample sizes of national Internet use surveys. Source: Retrieved from www.webuse.umd.edu.

discussed in appendix A, that make it possible to use them in combination to triangulate differences in findings and implications. Their value becomes further magnified when the Internet analyst has the ability to locate and analyze them in a single source.

Moreover, it is now possible to access these diverse data together directly on a single website, www.webuse.umd.edu, the home page of which (figure 18.3) shows the diverse data sets available. Further, the data are organized using a new statistical software package called SDA that is designed to be highly user-friendly, as discussed in appendix B, along with other empirical data features at the website.

Before reviewing recent research from the above studies on such issues as reduced socialization and the digital divide, we first turn to what basic data from the figure 18.2 national surveys tell us about who and how many people are on the Internet and how much time they are spending on it.

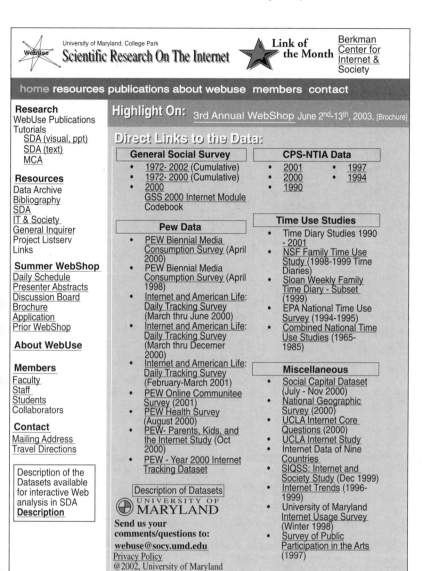

Figure 18.3
Home page of www.webuse.umd.edu.

How Many People and What Kind of People Are Online

As of August 2003, about 60 percent of the American public is online, a figure that figure 18.4 shows has not increased much over the previous year or two, according to the data regularly collected by the Pew Project. The Pew Internet access figures have been higher than those collected by the NTIA (National Telecommunications and Information Administration of the U.S. Department of Commerce), which are now almost two years out of date, in part because the NTIA data include non-telephone households. The Pew numbers in 2003 now seem very close to the 60 percent access rate reported in the in-home surveys conducted by the General Social Survey in the Spring of 2002.

A 70 percent penetration rate generally has been held out as the point at which the Internet will become a stable fixture in society. The lack of movement toward this 70 percent figure may be some cause for concern. It suggests that the diffusion of the Internet has "peaked" well short of innovations such as the VCR. At the same time, Lenhart and Horrigan (2003) further note that the 60 percent figure does not simply divide

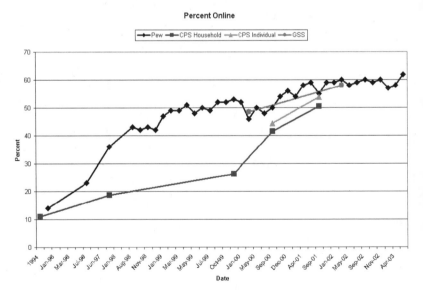

Percent Online

Figure 18.4
Growth in Internet diffusion from 1995 to 2003. Source: Retrieved from www.webuse.umd.edu.

those online from those never online. There are large numbers of "former users" and "proxy users," people who have gone offline because of lost access, lost appeal, or inability to pay monthly charges. On the other hand, the lack of recent growth could reflect fallout from a struggling economy, particularly one accustomed to rapid growth in IT. Additionally, Schement (2001) has shown that technologies that require subscription services, e.g., telephones, do not disseminate as fully as one-time purchase items, such as televisions. At the same time, the demographic predictors of who goes online are notably consistent across studies, across time periods, and across countries. Among "birth" factors, characteristics that one is born with, younger, white, and Asian people are notably more likely to be online than older, black, and Hispanic people. Males are slightly more likely to be online than females, but that overall "gender gap" has closed significantly since the early years of Internet adoption.

Among the biggest gaps are those by a person's social status, either by education, income, or occupational position; table 18.1 shows that of these, education is the most predictive, using a table derived by SDA. It is not the case that income (and occupational position) are unimportant but that of the three, education predominates, which is not surprising given the verbal and educational skills often needed to navigate the Web successfully. Among "role" factors that people take on during their lifetimes, taking on a job (being an employee) has more effect than getting married or having children (which may result in lower Internet access for women), but none is nearly as important as education or age as predictors. Finally, location factors (such as region of the country, urbanicity, home vs. apartment living) have little to do with access, once age, education, and other predictors are taken into account.

In addition to demographics, there are a number of more personal factors that are associated with being on the Internet, independent of these demographics. Robinson et al. (2002a) used General Social Survey (GSS) data to document a "diversity divide," separating users from nonusers in terms of users being more prodiversity, tolerant of nontraditional or less popular behaviors and attitudes. Thus, users were more tolerant of public speeches or library holdings of proponents of communism, racism, or antireligion; they were more tolerant of premarital sex or homosexuality, more approving of women working, and more

Table 18.1
Education and income as predictors of access

Main statistics

Cells contain
-Means
-N of cases

EDUC	INC					
	1 less than 20,000	2 20,000 to 34,999	3 35,000 to 49,999	4 50,000 to 74,999	5 more than 75,000	Row total
1: less than high school	.12	.20	.28	.38	.49	.22
	3,701	2,458	1,360	1,001	641	9,161
2: high school	.28	.41	.52	.61	.71	.50
	4,456	5,776	4,604	5,189	3,983	24,008
3: some college	.57	.63	.71	.77	.84	.73
	2,843	3,990	4,084	5,619	5,708	22,244
4: college graduate	.67	.79	.84	.88	.92	.87
	691	1,438	1,907	3,340	5,987	13,363
5: graduate school	.76	.80	.86	.91	.93	.90
	217	403	635	1,436	3,915	6,606
COL Total	.33	.48	.62	.73	.84	.63
	11,908	14,065	12,590	16,585	20,234	75,382

Source: From NTIA 2002 SDA table. Retrieved from www.webuse.umd.edu.

supportive of blacks improving their position in society. Users were also more upbeat, more satisfied with their incomes, and most of all, more trusting of their fellow citizens. These significant differences were not simply explained by Internet users being younger, better educated, or the other demographic factors above. Lenhart and Horrigan (2003) found additional (healthy) psychological factors in the Pew studies that were independently associated with being online—social contentment, feeling in control of one's life, belonging to social clubs, and having people one can turn to for support.

Digital Divides

Many social issues have surfaced as a result of the Internet's diffusion in society: its effects on social life, its relation to previous mass media, and its impact on politics and business, among others. However, probably none has attracted as much scholarly and policy attention as what is referred to as the "digital divide." The term has been widely interpreted to cover a variety of gaps in American society, as well as differences between the United States and other Western countries and the rest of the world (e.g., Norris 2001).

Overall Digital Divides

The NTIA's controversial 2002 report, which concluded that America's digital divide problem was essentially healing itself, is directly addressed by Martin (2003), particularly key methodological assumptions employed in the 2002 report. Martin found inherent problems in using "Gini coefficients" developed by economists to measure inequality, such as its inevitable increase as technologies diffuse and the report authors' failure to examine its inverse, that is, the changes in the probability of *not* being online (which is also decreasing across time). Instead, Martin employed an alternative measure of inequality, the odds ratio, to argue that Internet access is actually spreading less quickly among poor households. Moreover, while the poor may eventually reach universal access, that process could last as long as two decades.

Lenhart and Horrigan (2003) employed data from the Pew Internet and American Life Project to argue that the digital divide is better seen as a "digital spectrum." As in the case of panel research on the poor, people move in and out of the user category rather than being permanent users or permanent nonusers. The authors' digital spectrum thus distinguishes between intermittent users and longer term (or never) users, but it also differentiates the more intense broadband users from those who access via modems.

The digital divide phenomenon is not isolated to America, as de Haan (2003) documented in his study of Dutch Internet access, where again education and age are the main predictors of access and use, along with certain psychological factors. Innovators and early adopters also become more adept at understanding Internet features as well as using the Internet for making purchases and for meeting new people for social purposes.

Mason and Hacker (2003) argued that a major problem with digital divide research is that it has developed in much of a theoretical vacuum, often failing to tie in with the "increasing knowledge gap" hypothesis the diffusion of innovations literature and elements of various "structuration theories" of Giddens, Castells, Keane, and DeSanctis and Poole.

Cho and his colleagues (2003) used a "uses and gratifications" model in a secondary analysis of a Pew 2001 national survey to find that younger and higher status users use the Internet to satisfy their motivations strategically and to gain the desired gratifications.

Digital Divides in Specific Populations

Gender: While most observers seem to feel the "gender gap" in Internet and computer use has largely been overcome, Losh (2003) found consistent male–female differences in the context of gender studies dating back two decades. Kennedy, Wellman, and Klement (2003) also found important gender differences in IT use, with women in two large national surveys using the Internet more for social connections, while men used IT more for instrumental projects and for individual forms of recreation.

Race: Alvarez (2003) noted that both NTIA and GSS national surveys show that about half of the 20-point lower IT access by blacks still remains after status and other demographic predictors are taken into account, but that blacks and whites report rather similar levels of time spent online, of social support, of search strategies, and of knowledge of IT terms—and are equally as likely to visit capital-enhancing websites.

In the innovative HomeNetToo community project of Jackson et al. (2003), blacks and older participants used the Internet less than whites.

The infrequent use of email by participants, regardless of race or age, is attributed to their having few friends and relatives online. Gordo (2003) reported on the success of the "Plugged In" community project in the poorest part of Silicon Valley, a model of public policy intervention aimed at low-income populations at the grass roots, affecting children as well as adults.

Age: The age gap has remained an important and persistent finding from surveys that few analysts have taken the time or effort to explain, except Cho et al. (2003) found that younger (and higher status) users use the Internet to satisfy their motivations strategically and employ consumption strategies to attain connection gratifications.

Education: As noted in table 18.1, education emerges as the major predictor of Internet access when compared to income, but education is important not just in access but in the social processes and behaviors once access had been achieved. Robinson, DiMaggio, and Hargittai (2003) used GSS evidence to show education differences among Internet users in terms of five aspects of Internet sophistication, such as knowledge of Internet terms and more autonomous access, but particularly in terms of visiting sites that build on their human capital or enhance their life chances. Robinson and Neustadtl (2003) showed that greater trust in people expressed by Internet users was significantly higher among those who became or remained Internet users between 2000 and 2001. Mason and Hacker (2003) also documented several studies of the clearly positive social and psychological benefits of being online.

Social Activity and Internet Use

The studies reviewed on this issue are based on survey data collected from time-diary and time-estimate samples, almost all at the national level. The articles on this issue can be grouped under three headings, the first being the most comprehensive method of collecting behavioral data, namely, the time diary. In time diaries, respondents are expected to record all their daily activities, which allows the analyst to take advantage of the important "zero-sum" property of time, namely, that if time on one activity (such as the Internet) increases, time on some other activity (such as TV, social life, or meals) must decrease.

Time-Diary Studies

Nie and Hillygus (2002) used a new Internet-based diary to collect data from more than 6,000 adult respondents nationally, drawn from their carefully screened probability-sample panel; that panel is recruited to complete a 7–15 minute survey conducted every week in return for their being given a "Web TV," a device that allows them to gain access to the Web via their ordinary TV sets. Nie and Hillygus found that respondents who used the Internet on the diary day did engage in much less social activity and spent less time in contact with friends and family. They estimate that for every hour using the Internet, time with others declines by almost 40 percent.

Robinson et al. (2002a) also examined national time-diary data, but collected by the more conventional telephone method, from more than 1,700 respondents between 1998 and 2001. Respondents who reported any IT use in their diaries basically showed no difference from nonusers in their six main social activities, and when examined on a long-term basis, they reported slightly more time in these social activities. While IT users on diary day reported more time alone in their diaries and less time with work colleagues, the reverse was found on a long-term basis. IT users also reported slightly more conversation as a secondary activity.

Qiu, Pudrovska, and Bianchi (2002) examined complete 168-hour weekly paper-and-pencil diaries filled out personally by 800 respondents in middle-class dual-income families across the country. Once again, no difference was found in the social activities of those parents who reported use of the Internet or other IT during the diary week and those who didn't.

A 1998 diary study conducted by Statistics Canada was conducted by telephone with more than 5,000 respondents aged 18–64. As analyzed by Pronovost (2002), once again essentially no difference in social life was found between users and nonusers. In this way, the results are consistent with the Nie–Hillygus findings on the single-day basis but are more consistent with the American diary results on the longer term basis. Likewise, in Holland, de Haan and Huysmans (2002) found almost no difference in the socializing and visiting activity of users versus nonusers using diary data from 2000.

Gershuny (2002) used a sophisticated panel study design of more than 1,500 U.K. respondents to contrast those who began using the Internet later with those who were users in both Wave 1 and Wave 2 a year later, and with those who did not use the Internet in either wave. In none of the three groups is he able to find consistent evidence of declining social life associated with Internet use.

Behavior Estimate Studies

While faster and far less expensive than time diaries, estimates are subject to distortion, either because of social desirability or respondents' inability to make accurate estimates.

Using the most complete and targeted time estimates from GSS, Neustadtl and Robinson (2002) found no decrease in aggregate socializing since 1995, the initial year of significant Internet presence. While they find occasional areas of decreased socializing (e.g., with relatives and neighbors) among Internet users, these are offset by increases in other areas (e.g., friends) so that overall little difference can be found. Much the same conclusion emerges from analyses of new GSS questions on the extent of respondents' social networks by personal, telephone, postal mail, and organizational channels of communication. Indeed, Internet users report larger social networks than nonusers, and those who depend on email are able to keep in contact with more members of those networks without decreasing their contact by more traditional channels of communication.

Robinson and Shanks (2002) focused on two GSS social behaviors, religion and sex. Internet users report more attendance at religious services than nonusers. In contrast, after adjustment, lower sexual frequency is consistently associated with more Internet use.

Kiesler et al. (2002) first described the "Internet paradox," in which the new communications medium was associated with negative social effects. In their follow-up study, however, they concluded the opposite— especially for extroverted people and for those with more social support.

Horrigan and Rainie (2002) reviewed new national survey evidence from their Pew Center for the Internet in Everyday Life project that users did greatly increase their use of email for "serious" matters between 2000 and 2001.

Wellman, Boase, and Chen (2002) largely echoed these upbeat assessments with evidence from three of their recent projects that converged on the conclusion that the Internet largely adds to personal and telephone contact.

Perhaps the first studies to document the positive impact of the Internet were those from Project Syntopia. Katz and Rice (2002) described how from 1995 on, Internet users in their national samples consistently reported being more involved and socially active than nonusers.

World Internet Project
A number of articles are linked by their involvement in the World Internet Project, initiated at UCLA.

Coget, Yamauchi, and Suman (2002) reviewed the original UCLA data and found no evidence of a displacement of face-to-face relations with online relations and some evidence of decreased loneliness as well.

Like Coget et al., Cole and Robinson (2002) found no evidence of consistent declines in contact with household members, friends, or neighbors among users or heavier Internet users in 2001 as well as 2000. However, they did find a significant link between usage and lower feelings of loneliness, alienation, and other measures of negative outlook in both surveys.

Lee and Zhu (2002) reached much the same conclusion as Cole and Robinson in their analysis of common behavior estimate questions asked in random samples of Chinese. Liang and Wei (2002) reached the same conclusion in their analyses of data from a separate study of five different urban areas in mainland China. Mikami (2002) also reached similar conclusions from his multivariate analysis of national panel data in Japan, namely, no connection between Internet use and socializing with family and friends.

Mandrelli (2002) found more positive connections between Internet use and several indicators of sociability in her Italian sample, a finding she linked to theories and findings from the broad range of authors and literature she reviews.

Almost none of these results, then, support the most publicized article from Nie and Erbring (2000), which was the benchmark article for much of the controversy of the negative association between Internet use and sociability.

Overall

In the 20 articles reviewed here, one can find evidence that the Internet is associated with decreased social life, with enhanced social life, or with no difference in social life. Three studies suggest a strongly negative relation, one or two a strongly positive relation, and four or five others a mildly positive association. Most of the studies, however, are decidedly in the middle, as shown in table 18.2.

Other (Nonsocial) Activity

Time on the Internet may also impact other daily activities as well. New IT time could conceivably come from alternative free-time activities, such as hobbies, fitness, or games, simply because users prefer to spend time on the Internet. The time might come from time spent on travel, since telecommuting or online shopping could reduce the need to leave one's home. More broadly, it could come from work or education, perhaps as these activities are accomplished more quickly and easily with the new technology. Alternatively, time spent on house, child and family care could possibly be displaced, as people find an additional way to avoid facing these often unpleasant daily tasks. Finally, personal care activities, such as eating, sleeping, and grooming, could be affected by the allure of, or even pressure created by, the new technology (as seems to have been the case in the early days of television).

Time-Diary Studies

Based on a detailed diary approach with over 5,000 respondents aged 18 to 65, Nie and Hillygus (2002) found that the time spent on the Internet appears to displace many other daily activities. In terms of absolute time reductions, the activities that appear to be most affected by Internet use are sleep, TV, and work, but hobbies, socializing, and reading show the largest proportional reduction in time.

In the telephone-based diaries of Robinson et al. (2002b) IT users (defined both with "yesterday" diary measure and "long-term" follow-up measure) showed consistently lower times on sleep and TV. Among those using the Internet yesterday, significantly more time was spent on education and less time working (perhaps because the "yesterday" was a day off or a day at school) than nonusers. In contrast, the analysis of

Table 18.2
Studies on the Internet and sociability

	Strongly antisocial	Tending antisocial	Neither or both	Tending prosocial	Strongly prosocial
I. Time-diary data					
1. Nie and Hillygus (Stanford)	—				
2. Kestnbaum et al. (Maryland)			0	+	
3. Qiu et al. (Maryland)			0		
4. Gershuny (Essex)			0		
5. Pronovost (Quebec)			0		
II. Time-estimate data					
6. Neustadtl and Robinson (Maryland)			0	+	
7. Robinson and Shanks (Maryland)		—		+	
8. Kiesler et al. (CM)				+	
9. Horrigan and Rainie (Pew)				+	
10. Katz and Rice (Rutgers)				+	
11. Wellman, Boase, and Chen (Toronto)			0	+	
12. Cole and Robinson (UCLA)			0	+	
13. Coget, Yamauchi, and Suman (UCLA)			0		
14. Lee and Zhu (Hong Kong)			0		
15. Liang and Wei (China)			0		
16. Mikami (Japan)			0		
17. Mandrelli (Italy)				+	
18. Nie and Erbring (Stanford)	—				
III. Attitudes					
19. Robinson et al. (Maryland)				+	
20. Price and Capella (Pennsylvania)				+	++

those using the Internet in the last month ("long-term" users) found that Internet users report significantly *more* paid work than nonusers, while education differences are not significant. Rather than offering evidence than the Internet is displacing print media, these findings suggest that IT users are actually more active readers.

Fu, Wang, and Qiu (2003) examined 168-hour weekly diaries collected from all family members in approximately 450 dual-earner households, finding that IT users spent roughly 1 hour per week less on sleep and 5 hours less on paid work than nonusers. Among the children in the survey, the 5 hours given over to IT was offset by 1.5 hours less time on other hobbies/games, an hour less grooming, an hour less studying, and an hour less on other activities.

Pronovost's (2002) 1998 telephone diary study conducted in Canada with more than 10,000 respondents found that respondents reporting IT use in their day's diary reported less sleep than nonusers but no less time watching TV and more time reading, corresponding, and doing hobbies, after adjustment for several demographic predictors. In contrast, when the analysis focused on longer term users, users watched significantly less TV than nonusers but sleeping times were the same.

Gershuny (2002) found almost no significant displacement effects in a panel analysis of those who became new IT users over the one-year period of his study. They reported less time doing unpaid work (a trend found among nonusers and among previous users as well), studying, engaging in hobbies, and doing nothing. Gershuny concludes that these results show no evidence of a "privatization effect" of home IT use, since no notable changes in media use were found in any of the three groups across the span of the study.

In the Netherlands in the fall of 2000 with about 1,800 respondents aged 12 and above (de Haan and Huysmans 2002), IT use during the diary week has increased steadily from 18 percent in the 1985 sample to 60 percent in 2000, while usage per user has remained relatively stable— 3.5 hours in 1985 to 3.9 hours in 2000. While Internet users have about 3 hours more free time than nonusers (due to less paid work and family care), no specific free-time activities stand out as significantly different. While at the bivariate level users do sleep and watch TV less, these differences do not hold after adjustment for demographic factors, with reading (especially books) by IT users slightly above average. Van Rees and van Eijck (2002) examined this same Dutch TUS diary study from the perspective of media user types, taking advantage of the detailed media content incorporated into these Dutch diaries. Their analysis suggests that at the same time Internet use (1) is not a single media factor and depends on the purpose and audience for its use and (2) is independent of use of other media for other purposes. The authors conclude that these results indicate that media use is far more nuanced than suggested in Bourdieu's theoretically ideal types of media users.

Time-Estimate Questions

Neustadtl and Robinson (2002) examined traditional and new questions on TV and newspaper use in the 2000 GSS and found that Internet users report watching significantly less TV and reading newspapers more often than nonusers, but the differences did not vary systematically with extent of Internet use and did not hold after adjustment for demographic factors. New "benchmark" GSS questions on usage of the Internet compared to other media for health and politics showed heavy Internet users were much more likely than nonusers to consult other media sources, while lighter Internet users used these sources less than nonusers. Overall, the authors conclude that there is little evidence for a displacement effect of Internet use, since Internet users appear to spend more time on traditional media, especially print media.

Much the same conclusion emerged from Cole and Robinson's (2002) analysis of the media and other time-estimate questions in the UCLA national surveys. Internet users report more reading of print media, more playing of video games, and more listening to music than nonusers, both before and after multiple classification analysis (MCA) adjustment for demographic factors. While users reported 5 hours less weekly TV viewing than nonusers, that difference was mainly accounted for by these demographic factors.

In their analysis of 1999 Stanford Institute for the Quantitative Study of Society questions asked of 4,133 national respondents, Nie and Erbring (2000) found that the more time users estimated using the Internet, the less time they estimated spending watching TV, reading newspapers, shopping, and commuting in traffic. Only for the TV question, however, did almost as many respondents say that after being on the Internet, their time on an activity had decreased (46 percent) than remain unchanged (50 percent).

Finally, Horrigan and Schement (2002) examined the alleged reductions in purchasing of recordings as a result of accessing that content free from Napster and relevant Internet sites. In a 2001 Pew Internet and Daily Life Project survey of more than 2,000 national respondents. The authors found that people who use the Internet for information seeking are 2.5 times more likely to be online purchasers of products (including music recordings) than those who do not, and they argue that the act of downloading represents a step toward ultimate consumer purchase, a step widely misunderstood by the music industry.

Overall

As with most evidence in the social sciences that attempts to integrate the results from several sources, methods, and authors, there is far less than unanimity of results and conclusions about time displacement and the Internet. An overview of the findings on this issue is summarized in tables 18.3 and 18.4. The designation "+" in these tables refers to positive relations (more Internet use is associated with increases in that activity), the designation "–," with negative relations or time displacement, and "0," with no significant relation. In each cell of tables 18.3 and 18.4, the bivariate or unadjusted relation is shown first and the multivariate relation after the slash (/) sign. Results are also shown separately for short-term comparisons (yesterday) and long-term (weekly, monthly, or longer run IT use). The term "NA" means that the study contains no applicable data to test the hypothesis in question.

Concerning television, most studies in table 18.3 found no relation with Internet use, especially after control for other factors. Four of the present studies (Sloan, TUS, GSS, and UCLA), in contrast, did find a significant bivariate difference initially, but one that is largely explained by Internet users having higher education, lower age, and the like. The U.K. panel study similarly found an overall lower TV viewing figure for IT users, but one that did not hold when new users, prior users, and nonusers were examined separately. The main exceptions are the SIQSS Internet diaries and the University of Maryland telephone and the

Table 18.3
Summary of findings regarding TV and other media

Article	TV		Reading		Radio/stereo	
	Long term	Short term	Long term	Short term	Long term	Short term
SIQSS	NA	–/–	NA	–/–	NA	0/0
Univ. of Maryland	–/–	–/–	+/+	0/0	0/0	0/0
Sloan family	–/0	NA	0/0	NA	0/0	NA
Canada	–/–	0/0	+/+	+/+	0/0	0/0
United Kingdom	0/0	NA	0/0	NA	0/0	NA
Netherlands A	–/0	NA	0/0	NA	0/0	NA
Netherlands B	NA		NA		NA	
Pew 1998–2000	0/0	0/0	+/0	+/+	NA	
GSS	–/0	NA	+/0	NA	NA	
UCLA	–/0	NA	+/+	NA	+/+	
SIQSS	–/NA	NA	–/NA	NA	NA	
Pew	NA		NA		NA	

Source: Introduction to IT & Society, Vol. 2.

Table 18.4
Summary of findings regarding other daily activities

Article	Sleep		Work/education		Family care	
	Long term	Short term	Long term	Short term	Long term	Short term
SIQSS	NA	–/–	NA	–/–	NA	0/0
Univ. of Maryland	–/–	0/0	+/+	–/–	–/–	–/0
Sloan family	–/–	NA	–/–	NA	0/0	NA
Canada	0/0	–/–			–/0	–/0
United Kingdom	0/0	NA	–/0	NA	0/0	NA
The Netherlands	–/0	NA	–/–	NA	–/–	NA

Source: Introduction to IT & Society, Vol. 2.

Canadian telephone diary studies, and these do provide the most persuasive evidence of a displacement effect of TV by IT.

The more interesting, and counterintuitive (given the zero-sum nature of time as a variable), findings concern higher reading times among IT users. This shows up after multivariate adjustment in almost half of the studies, and the relation usually holds after these multivariate adjustments for various background predictors are taken into account. Notably, only the SIQSS study shows significantly lower reading times among IT users.

Outside of the UCLA study showing *higher* music listening the more one uses the Internet, none of the other studies found any significantly higher or lower radio or music listening by IT users—as shown in the third column of table 18.3.

Table 18.4 covers nonmedia activity, and it provides some support for Internet users getting less sleep, but mainly in the SIQSS and University of Maryland telephone diary studies, and to a lesser extent in the Sloan family and Canadian diaries. Using an estimate question, no such sleep difference was found in the UCLA estimate question (not shown in table 18.4).

There is also some evidence of lower paid work time and higher education among users, a finding that needs further study to understand whether this means Internet users are working at home, are interviewed on a day off from work, or just work fewer hours for other reasons. Almost no studies provide consistent or persuasive support that more IT use is associated with less housework or child or family care.

Summary and Conclusions

Unlike the earlier technology of television, table 18.1 documents the impressive body of national social survey data now available to track the changes in daily life associated with the introduction of the Internet and other new information technology into society. Moreover, these data are now immediately accessible on a statistically interactive website for more intensive and detailed analysis, in order to verify whether results in one study are replicated in others.

As of mid-2003, the Internet had not progressed much beyond the 60 percent diffusion rate achieved two years earlier—partly for economic reasons, but also partly because of Internet dropouts and proxy users. Among the demographic predictors of access, education was a stronger predictor than income or occupation; age and race/ethnic factors were also important predictors, even after taking these status factors into account. While women had largely overcome the "gender gap" in terms of access, it was clear that they used it more for social rather than "instrumental" functions, which is consistent with traditional gender roles in society.

Contrary to the upbeat 2002 Bush administration report of a closing of economic and racial "digital divides," more sophisticated and appropriate reanalyses show that the divides may be increasing across time, and that less advantaged potential users faced up to a 20-year lag in access to IT. Moreover, these divides showed up not only in terms of access, but in terms of the more sophisticated and "capital-enhancing" websites visited, awareness of the features available, and autonomy of usage. Other studies also documented important advantages accruing to those who had Web access.

Time and Activity

Most of the studies described in table 18.1 included behavioral questions about how Internet users spent their daily time differently than nonusers, including detailed time-diary surveys examining all behaviors over the 24 hours of the day (or 168 hours per week). In both of these diary studies, as well as those respondent estimates of time spent, little evidence was found to replicate earlier findings that Internet use was associated with declines in prosocial activity; indeed, evidence of a "diversity divide" suggested that users were more tolerant and trusting of other people.

Across all of these behavioral studies, there is some evidence that Internet use is associated with less time at work, in TV viewing, and in sleep. However, like the findings about social life above, the differences are often insignificant and sometimes explained by simple background factors. To offer the more historical perspective afforded by figure 18.1, this collection of articles suggests that IT has not affected the communications environment to the same degree that TV did nearly more than half a century ago.

Television still continues to consume 3 to 10 times as much of people's time as IT. Historically, and unlike other modern technologies, TV truly revolutionized how people spent time (Robinson and Godbey 1999). It seems doubtful that IT will ever have a similar temporal or consistent impact on our daily lives.

Appendix A: Methodological Differences in Internet Surveys

Figure 18.2 outlines certain differences in the various national surveys related to Internet access and use, but to describe these in further detail, the following factors are of importance:

Type of Organization

The NTIA data are collected by the U.S. Bureau of the Census, both government agencies, as are the time-diary data collected by Statistics Canada and the Dutch Social Planning Bureau. The GSS data are collected by a university-based survey firm, the National Opinion Research Center at the University of Chicago; the time-diary studies conducted at the University of Maryland were collected by its Survey Research Center. The UCLA, Rutgers, and Stanford (SIQSS) University surveys are also conducted for university organizations, but the data are collected by commercial survey firms. The Pew studies are also collected by a commercial survey firm, as are the "clickstream" data collected by Nielsen and by ComScore.

When Studies Began

The NTIA, Pew (Press), and Rutgers studies were all conducted in 1995, when Internet penetration in the public was 10 percent or less, and these were repeated in 1997 and 1998, as penetration reached the 30 percent level. The GSS and UCLA longitudinal studies were first undertaken in the spring of 2000 at a time of 40–50 percent penetration. The national time-diary studies in the United States (University of Maryland), Canada, England, and Holland were all part of general time-use tracking studies (not connected to the role of the Internet) begun prior to 1995, while the new SIQSS Internet diary was begun in 2001 specifically to follow how and where the Internet and other IT were being used. The commercial clickstream data began in the later 1990s, when many more firms than now began collecting such data. The first online *National Geographic* survey was conducted in 1998 and was partly replicated in 2001.

Sample Size

By far the largest data collection is the government NTIA effort, which is tied to the Census Bureau's Current Population Survey (CPS) and collects data from 50,000 households each month, representing almost 150,000 individuals above the age of 3. The GSS, Pew, and UCLA surveys interview 2,000–4,000 respondents per survey, while the Rutgers, University of Maryland, and U.K. diary surveys interviewed closer to 1,000 respondents. The SIQSS, Canada, and Dutch time-diary studies include 3,000 to 10,000 respondents, while the Nielsen and ComScore clickstream samples include 50,000 respondents; the *National Geographic* sample involved more than 45,000 users around the world, nearly 90 percent of them Americans, who volunteered to participate.

Sample Response Rate

The NTIA surveys, being part of the U.S. government's attempt to track unemployment in the country, can achieve response rates of higher than 80 percent. The GSS personal in-home surveys typically achieve response rates of 70 percent and higher, while the University of Maryland, U.K., and Statistics Canada diary studies average about 60 percent. The other organizations usually have response rates below 50 percent, although most of them are based on random sample methods, in which all adults in the country have an equal chance of being included in the sample. That is not true for the volunteer samples from ComScore and *National Geographic*, for which no response rate can be calculated.

Sample Frequency

The NTIA surveys are conducted at irregular intervals, now about every two years. The GSS survey is conducted every two years, the UCLA study every year, and the diary studies at irregular intervals, except in Holland where they are done every 5 years and in the United States, where they are now conducted continuously with 20,000 respondents per year by the Census Bureau and the Bureau of Labor Statistics (although with no data on Internet access). The most frequently collected data are the Pew studies (which conducted a study each month in 2001 but has cut back to 4–6 surveys per year) and the Nielsen and ComScore continuous data collections basically done every day.

Amount of Information Collected

The NTIA and GSS in-home studies collect about 20 minutes of information as part of a larger data collection, while the telephone surveys of Pew, UCLA, and Rutgers averaged closer to 30 minutes, similar to the telephone diary studies of the University of Maryland, Canada, etc., except here only 1–3 minutes of Internet questions can be asked because of the lengthy diary. The *National Geographic* survey was able to include closer to 40 minutes of information. The clickstream studies, which collect data on all strokes on the computer keyboard, are the most detailed and complicated in terms of information collected; indeed, probably not even 1 percent of the data collected ever finds its way into use. The value of the NTIA data is severely limited by the fact that only dichotomous (yes-no) questions were asked, rather than extent of usage for particular purposes.

Panel Component

One of the most valuable aspects of survey data is the ability to follow the same respondents across time in order to make dynamic or causal inferences about the impact of the Internet. This is one of the great strengths of the UCLA data collection and the special U.K. diary study. In 2001, respondents in GSS 2000 were reinterviewed, as was true for a Pew survey. The clickstream data are of course exemplary in their tracking of behavior across time. Otherwise, other data collections are single time only, limiting conclusions that can be drawn.

Appendix B: Features of WebUse Website

Developed at the University of California at Berkeley, SDA was designed to be extremely easy to use, requiring little statistical experience beforehand, and it is remarkably fast, providing almost instantaneous results even when sample sizes exceed 100,000 entries. It comes with an easy-to-use online codebook that can be easily accessed with a window command on a back-and-forth basis as the analysis proceeds. It shows significant results with the use of colors to help the less experienced analyst gain some intuitive understanding of the statistical significance involved in a particular data table or regression analysis. Because of these user-friendly features, SDA has been awarded special merit as a social science innovation both by the American Political Science Association and the American Association of Public Opinion Research. The home page of the University of Maryland WebUse website is shown in figure 18.3 in the text, with the various publicly available data collections from GSS, Pew, NTIA, etc., of figure 18.2 listed on the right-hand side of the page. Clicking on these data links allows data analysts instant access to the Internet variables included in each study. To assist first-time users of these analyses in following the outputs possible from the data and their interpretation, links to simplified tutorials are provided in the upper left-hand corner of the page.

However, the data sources and analysis methods available to webuse.umd.edu users are not confined to reanalysis of survey data. Users can conduct online content analysis of documents and text using the General Inquirer program. They can access the extensive bibliography of existing literature on the impact of the Internet that has been assembled. They can access the rich "observational" data on user profiles collected from small random samples of users interviewed in depth by experienced interviewers, which use extensive sets of open-ended questions. They can conduct experimental studies, such as that by Kiesler et al. (2002).

There is also a link to the new online journal *IT & Society*, which is dedicated to providing readers with the most up-to-date integrated research on important Internet phenomena, such as the digital divide and user frustration; many of its articles are based on data in the archive, allowing interested readers to check the authors' conclusions with their own. The site further can be used by students wishing to apply to the summer "WebShop" at which more than 50 prominent Internet researchers share their findings and insights with a select group of graduate students across the country and the world. Proceedings of these WebShops are also available on the website.

References

Alvarez, A. S. 2003. "Behavioral and Environmental Correlates of Digital Inequality." *IT & Society* 1(5):97–140.

Cho, J., H. Gil De Zuniga, H. Rojas, and D. Shah. 2003. "Beyond Access: The Digital Divide and Internet Uses and Gratifications." *IT & Society* 1(4): 46–72.

Coget, J. F., Y. Yamauchi, and M. Suman. 2002. "The Internet, Social Networks, and Loneliness." *IT & Society* 1(1): 180–201.

Cole, J., and J. P. Robinson. 2002. "Internet Use and Sociability in the UCLA Data: A Simplified MCA Analysis." *IT & Society* 1(1): 202–218.

De Haan, J. 2003. "IT and Social Inequality in The Netherlands." *IT & Society* 1(4): 27–45.

De Haan, J., and F. Huysmans. 2002. "Differences in Time Use Between Internet Users and Nonusers in the Netherlands." *IT & Society* 1(2): 67–85.

Fu, Shu-jen, Rong Wang, and Yeu Qiu. 2002. "Daily Activity and Internet Use in Dual-Earner Families: A Weekly Time-Diary Approach." *IT & Society* 1(2): 37–43.

Gerbner, G. 1978. "Cultural Indicators." *Journal of Communication* 28, 178–207.

Gershuny, J. 2002. "Social Leisure and Home IT: A Time Diary Approach." *IT & Society* 1(1): 54–72.

Gordo, Blanca. 2003. "Overcoming Digital Deprivation." *IT & Society* 1(5): 166–180.

Horrigan, John, and Lee Rainie. 2002. "Emails That Matter: Changing Patterns of Internet Use Over a Year's Time." *IT & Society* 1(2): 135–150.

Horrigan, J., and J. Schement. 2002. "Dancing with Napster: Predictable Consumer Behavior in the New Digital Economy." *IT & Society* 1(2): 142–160.

Jackson, L. A., G. Barbatsis, A. Von Eye, F. Biocca, Y. Zhao, and H. Fitzgerald. 2003. "Internet Use in Low-Income Families: Implications for the Digital Divide." *IT & Society* 1(5): 141–165.

Katz, J., and R. Rice. 2002. "Project Syntopia: Social Consequences of Internet Use." *IT & Society* 1(1): 166–179.

Kennedy, T., B. Wellman, and K. Klement. 2003. "Gendering the Digital Divide." *IT & Society* 1(5): 72–96.

Kiesler, S., R. Kraut, J. Cummings, B. Boneva, V. Helgson, and A. Crawford. 2002. "Internet Evolution and Social Impact." *IT & Society* 1(1): 120–134.

Lee, B., and J. Zhu. 2002. "Internet Use and Sociability in Mainland China and Hong Kong." *IT & Society* 1(1): 219–237.

Lenhart, A., and J. Horrigan. 2003. "Re-Visualizing the Digital Divide as a Digital Spectrum." *IT & Society* 1(5): 23–39.

Liang, G., and B. Wei. 2002. "Internet Use and Social Life/Attitudes in Urban Mainland China." *IT & Society* 1(1): 238–241.

Losh, S. 2003. "Gender and Educational Gaps: 1983–2000." *IT & Society* 1(5): 56–71.

Mandrelli, A. 2002. "Bounded Sociability, Relationship Costs and Intangible Resources in Complex Digital Networks." *IT & Society* 1(1): 251–247.

Martin, S. 2003. "Is the Digital Divide Really Closing? A Critique of Inequality Measurement in a Nation Online." *IT & Society* 1(4): 1–13.

Mason, S., and K. Hacker. 2003. "Applying Communication Theory to Digital Divide Research." *IT & Society* 1(5): 40–55.

Mikami, S. 2002. "Internet Use and Sociability in Japan." *IT & Society* 1(1): 242–250.

National Telecommunication and Information Administration (NTIA). 1997. *Falling Through the Net II: New Data on the Digital Divide.* Washington, D.C.: U.S. Department of Commerce.

National Telecommunication and Information Administration (NTIA). 2002. *A Nation Online: How Americans are Expanding Their Use of the Internet.* Washington, D.C.: U.S. Department of Commerce.

Neustadtl, A., and J. Robinson. 2002. "Media Use Differences Between Internet Users and Nonusers in the General Social Survey." *IT & Society* 1(2): 100–120.

Nie, Norman H., and Lutz Erbring. 2000. "Internet and Society: A Preliminary Report." Stanford Institute for the Quantitative Study of Society.

Nie, N., and D. Hillygus. 2002. "The Impact of Internet Use on Sociability: Time-Diary Findings." *IT & Society* 1(1): 1–20.

Price, V., and J. Capella. 2002. "Online Deliberation and Its Influence: The Electronic Dialogue Project in Campaign 2000." *IT & Society* 1(1): 303–329.

Pronovost, G. 2002. "The Internet and Time Displacement: A Canadian Perspective." *IT & Society* 1(1): 44–53.

Putnam, R. D. 2000. *Bowling Alone: The Collapse and Revival of American Community.* New York: Simon and Schuster.

Qiu, Y., T. Pudrovska, and S. Bianchi. 2002. "Social Activity and Internet Use in Dual-Earner Families: A Weekly Time Diary Approach." *IT & Society* 1(1): 38–43.

Robinson, John P. 1969. "Television and Leisure Time: Yesterday, Today, and (Maybe) Tomorrow." *Public Opinion Quarterly* 33(2): 210–222.

Robinson, John P., and Geoffrey Godbey. 1999. *Time for Life: The Surprising Ways Americans Use Time.* University Park: Penn State Press.

Robinson, J., and A. Neustadtl. 2003. "An Expanding Digital Divide? Panel Dynamics in the General Social Survey." *IT & Society* 1(4): 14–26.

Robinson, J. P., and M. Shanks. 2002. "Sex, Church, and the Internet." *IT & Society* 1(1): 103–119.

Robinson, J. P., P. DiMaggio, and E. Hargittai. 2003. "New Social Survey Perspectives on the Digital Divide." *IT & Society* 1(5): 1–22.

Robinson, J. P., M. Kestnbaum, A. Neustadtl, and A. Alvarez. 2002a. "Information Technology and Functional Time Displacement." *IT & Society* 1(2): 21–36.

Robinson, J. P., A. Neustadtl, and M. Kestnbaum. 2002b. "The Online 'Diversity Divide': Public Opinion Differences Among Internet Users and Nonusers." *IT & Society* 1(1): 284–302.

Rogers, Everett. 1962. *The Diffusion of Innovations*. New York: The Free Press.

Schement, Jorge Reina. 2001. "Of Gaps by Which Democracy We Measure." In *The Digital Divide: Facing a Crisis or Creating a Myth?*, ed. B. Compaine. Cambridge, Mass.: MIT Press.

Van Rees, K., and K. van Eijck. 2002. "The Internet and Dutch Media Repertoires." *IT & Society* 1(2): 86–99.

Wellman, B., J. Boase, and W. Chen. 2002. "The Networked Nature of Community Online and Offline." *IT & Society* 1(1): 151–165.

19

Charting Digital Divides: Comparing Socioeconomic, Gender, Life Stage, and Rural-Urban Internet Access and Use in Five Countries

Wenhong Chen and Barry Wellman

The Multifaceted and Persistent Digital Divide

The chapters in *Transforming Enterprise* often assume ubiquity: that almost all of those who are truly modern and transforming will be connected. They portray a world system transformed by the Internet and other new information and communication media, from relations among people within organizations to relations between large economic sectors. They see computer-supported communication and knowledge to be crucial for the innovation and dissemination of products and services. Indeed, the very foundation of this book rests on the belief that the informed use of the Internet is already widely available and growing swiftly.

In short, the authors assume that the digital divide—the numbers of people who are not connected to the Internet—is small, shrinking, and becoming irrelevant. However, it is not. The digital divide is here for some time to come. It is large, multifaceted, and in some ways, it is not shrinking. Moreover, the divide is socially patterned, so that there are systematic variations in the kinds of people who are on and off the Internet. These patterns vary between nations and over time. Country A's divide does not necessarily resemble Country B's, and last year's divide often does not resemble this year's. There is no one digital divide; there are many divides. Indeed, it is more accurate to use the plural—digital *divides*—because the digital divide is multifaceted and varies within and between countries, both developed and developing.

To be sure, the Internet has grown rapidly in the past decade, so much so that many people in the developed world assume that almost everyone is connected to it. Although there are no reliable data on the size of the world's online population, educated estimates show that use of the Internet has diffused at an unprecedented speed. The number of Internet users around the globe has surged from 900,000 in 1993 (AC Nielsen, 2001), 25 million in 1995 (Pew, 1995), 83 million in 1999 (Intelli-Quest, 1999, as cited by DiMaggio et al., 2001), 513 million in 2001, to more than 600 million by the end of 2002 (NUA, 2003). More recently, other new media, such as web-enabled mobile phones, have further fostered computer-mediated technology (Ito, 2004).

Yet, widespread diffusion does not equal ubiquity, even within developed countries. In this chapter, we demonstrate that the uneven diffusion of the Internet is still persistent. The first digital divide appeared at the very start of the Internet. Early users were disproportionately affluent, male, white, better educated, and from developed countries, especially the United States.

Rather than shrinking with expanding Internet use, the global digital divide between developed countries and developing nations continues to be huge. Affluent residents of economically developed countries sometimes forget what a small percentage of the world's population is online. After all, the majority of their country-mates are on the Internet, as are the economically advanced segments of developing countries.

Yet, only 10 percent of the world's population was on the Internet in 2002, and 88 percent of these Internet users resided in industrialized countries (World Economic Forum, 2002). Within countries, the uneven diffusion of the Internet appears along familiar lines of social inequality such as socioeconomic status, gender, age, geographic location, and ethnicity. Moreover, having access to computers and the Internet and possessing the ability to use them effectively are two different issues (Warschauer, 2003). While marketers, media, and governments often report only the number of people who have access to the Internet, the digital divide is not a binary yes/no question of having access to the Internet. The question is not whether people have ever glanced at a monitor or put their hands on a keyboard but the extent to which they regularly use a computer and the Internet for meaningful purposes.

Although the digital divide has been a continuing buzzword in public discourse and international agendas, analyses have largely been confined to the boundaries of national states. Moreover, research on Internet diffusion has followed the evolution of the Internet itself. With the Internet born and raised in the United States, most research has been American. With Internet use increasing in other developed countries, research about their situations has been on the rise. Accordingly, most studies have taken place in OECD (Organization for Economic Cooperation and Development) countries, especially the United States, Canada, Japan, and Western European countries. Even though there is a deeper and wider global digital divide, little research has been paid to how Internet access and use fit into everyday life in developing countries.

The Technological and Social Aspects of the Divide

The digital divide has both technological and social aspects. Table 19.1 shows an integrative framework developed to scrutinize Internet access and use within and between countries. We examine the digital divide from four perspectives:

1. *Technological access.* "The Internet is not a single innovation but is a cluster of related technologies that must be present together to support adoption decision by end users" (Wolcott et al., 2001, p. 5). Around the world, people, groups, and countries use different levels and combinations of technologies (e.g., hardware, software, and bandwidth) to access computers and the Internet. These differences can affect the efficiency, volume, and diversity of Internet use in important ways (DiMaggio and Hargittai, 2001). Being able to hook up to the Internet via a 56K dial-up modem is markedly different from accessing the Internet via

Table 19.1
An integrative framework for the digital divide

Access	Use
Technological Access	Technological Literacy
ICT infrastructure	Technological skills
Hardware, software, bandwidth	Social and cognitive skills
Social Access	Social Use
Affordability	Information seeking
Awareness	Resource mobilization
Language	Social movements
Content/usability	Civic engagement
Location	Social inclusion

high-speed broadband connections such as telephone-based DSL or cable modems.

2. *Technological literacy.* Having access to the Internet and having the ability to use the Internet effectively are two distinct aspects of the digital divide. Using a computer and the Internet is more complicated than changing channels on a television or dialing a telephone. Meaningful and productive Internet use requires computer, social, and cognitive skills for such things as seeking information, developing community networks, accumulating social capital, or participating in political activities (Wellman, 2001; Hargittai, 2003).

3. *Social access.* Economic, organizational, and cultural factors affect equal access to the Internet.

a. Income is the most important factor that affects Internet diffusion. The International Telecommunications Union estimates that within countries, inequalities in Internet access and use are likely to be twice as high as inequality in income (ITU, 2003a). Among OECD countries, the income divide varies from country to country, ranging from a gap of more than 60 percentage points in the United Kingdom to less than 20 percentage points in Denmark (OECD, 2002).

b. Barriers to Internet access begin with the lack of awareness and interest. For instance, in Japan, 47 percent of nonusers found that high cost and 36 percent felt that uneasiness or lack of trust with the Internet deterred them from using it. However, 73 percent of nonusers said they had "no interest" or felt the Internet was "incomprehensible" (Digital Opportunity Site, 2001). Research in the United States and Canada indicates that the elderly, women, and nonwhites are less aware of the Internet than youth, men, and whites (Reddick and Boucher, 2002; Katz and Rice, 2002).

c. Language, content, and location barriers become salient after access to the Internet becomes available. What users bring with them online has a profound impact on what they can gain from the Internet. The diffusion of computers and the Internet is contingent on their affordability, user-friendliness, and relevance in people's everyday lives.

d. Inequality in Internet access is not solely contingent on individuals' capabilities, resources, and attitudes. Interpersonal and institutional contexts that foster Internet use also are important. For example, informal training through social support plays a crucial role in equipping people with the necessary computer and navigation skills to use the Internet.

4. *Social use.* The digital divide is a social as well as a technological divide. Ultimately, the digital divide is a matter of who uses the Internet, for what purposes, under what circumstances, and how this use affects

socioeconomic cohesion, inclusion, alienation, and prosperity. Interpersonal emailing and web searching for information are the two most widely used Internet activities worldwide that people carry out on the Internet. Other important uses include civic organizing and accessing online services such as online banking, job searching, and interacting with public authorities (OECD, 2002). However, users in developing countries tend to consume rather than to produce information on the Internet. This is because of the higher cost of Internet access, censorship, the lack of Internet culture in organizations, and the small percentage of peers online (Bazar and Boalch, 1997).

Internet Access and Use in Five Countries

To our knowledge, this chapter is the first to systematically compare and synthesize research on the digital divide over time and in a global range of developed and developing countries. The time span we cover is from the advent of the commercialized Internet in the early 1990s to 2002. We examine five countries[1]: the United States, Germany, Japan, the Republic of Korea, and China. We first present a framework of Internet diffusion and its social impact on people and societies in the global context. Using this framework, we then describe trends and patterns of access and use of computers and the Internet in each country.

The five countries we examine were selected for several reasons. First, they make up a majority of the world's Internet population. Together, they account for about 56 percent of all Internet users in the world in 2002. Second, the five countries present diverse patterns of Internet access and use. For example, as the birthplace of the Internet, the United States has been the leader in both computers and Internet use. As recently as 1997, 54 percent of the world's Internet users were American-based. Even though the Internet has been spreading quickly outside the United States, users in America still accounted for 29 percent of world online populations in 2002 (Nielsen/NetRatings, 2003a).

To demonstrate alternative patterns of Internet diffusion, we also include one developed European country (Germany) and two East Asian countries (Japan and South Korea), where wireless computing and mobile phones have proliferated more than in North America. We also include one developing nation—China—to broaden understanding of how the developing world is both increasing Internet use and being left behind. China is the world's most populous country, with rapidly expanding populations of computer and Internet users. With 58 million Internet users in 2002, China had the world's second largest online population after the United States.

We caution that these data are necessarily rough approximations. Getting a perspective on the Internet is like tracking a perpetually moving and mutating target. The lack of internationally comparable data has led researchers studying the global diffusion of the Internet to rely on statistics gathered country by country that often employ different measurements (for detailed accounts see Jordan, 2001, and Norris, 2001). Meaningful comparison and knowledge accumulation are hindered by the lack of measurement comparability of data from different countries.

1. *The definition of the online population* often differs from country to country. While some countries focus on adult users, other countries include children and teenagers in the online population.
2. There is a lack of a *standard definition of who is an Internet user.* The frequency of Internet use and length of Internet experience are two important criteria for defining who are Internet users. Again, different studies in different countries use different measures. Some generously embrace everyone who has ever accessed the Internet as a user while some more strictly count as users only those who use the Internet at least once a week. Similarly, the definition of "heavy users" varies widely between studies and nations.
3. To make things more complicated, some studies use households, and not individuals, as *the units of analysis.* In addition to hindering comparability, this masks how individual members within the household use the Internet.

To increase reliability and comparability, we primarily draw data from national representative surveys conducted by government agencies and scholarly researchers and from policy reports issued by international organizations such as the ITU (International Telecommunication Union, 2001a, b, 2002a, b, 2003a, b), UNDP (The United Nations Development Programme, 2001), and OECD (2001, 2002; see table 19.2). In each country, we review the trend of Internet diffusion and the extent of the digital divide in terms of socioeconomic status, gender, life stage, ethnicity, geographic location, place of accessing the Internet, and the social use of the Internet.

Table 19.2
Sources of data for each country

Country	Data source
United States	NTIA 1995, 1998, 2000, 2002
	UCLA Internet Report, 2003
	Pew Internet and American Life Studies
Germany	ARD/ZDF online surveys, 1999–2002
Japan	Japan Statistics Bureau, MPHPA
	(Ministry of Public Management, Home Affairs) 2002, 2003
	World Internet Project Japan, 2002
South Korea	Korea National Computerization Agency, 2002
	Korea National Statistical Office, 2003
	Supplemented by scholarly research
China	CNNIC, 1997–2003

United States

The "Falling through the Net" series by the National Telecommunications and Information Administration (U.S. NTIA, 1995, 1998, 2000, 2002) provides a long-term view of how computers and the Internet have entered the everyday lives of Americans (15 years and older) between 1995 and 2002. In addition, the UCLA Internet Report Year Three provides recent data on the Internet access and use of Americans (12 years of age and over; UCLA, 2003). Furthermore, the Pew Internet and American Life surveys (www.pewinternet.org) offer more fine-grained understanding of how Americans (18 years and older) use Internet-connected computers and the impact of these technologies on their lives.

In 2002, 169 million Americans were online, accounting for about 60 percent of the country's total population and 29 percent of the world's Internet population (Nielsen/NetRatings, 2003a). The online population is becoming more like the general population as many Americans have moved from being newcomers ("newbies") to the Internet to being veterans. However, the steady growth of Internet users has lost its momentum since late 2001, with the penetration rate hovering at above 60 percent. One possible reason might be that Internet dropouts offset newcomers to cyberspace (Lenhart et al., 2003). With the majority of the population online, the population of potential dropouts now exceeds the population of potential new adopters.

The digital divide is narrowing in the United States in terms of gender, age, and geographic location. However, the sociodemographic divide is still wide. The April 2003 Pew report (Lenhart et al., 2003) shows that younger, well-to-do, white, well-educated, urban and suburban Americans are still more likely to be on the Internet than older, less well off, black and Hispanic, less educated, and rural Americans. Moreover, Internet users are more socially connected than nonusers, have a stronger sense of efficacy (perceived control over one's life), and consume more media (including newspapers, TV, and mobile phone usage, etc.) than nonusers (Lenhart et al., 2003; Hampton and Wellman, 2003).

Internet Access
Socioeconomic status: Income has been the most important factor determining Internet access. For example, more than 60 percent of Americans with a household income of $35,000 or higher were online in 2000, whereas only 42 percent of those with a household income of less than $15,000 were online (U.S. NTIA, 2000). In 2002, the share of American Internet users with a household income of less than $30,000 (18 percent) continued to be lower than its share in the general American population (28 percent). Moreover, those with a high school education or less made up merely 5 percent of the American online population but one-quarter of nonusers (Lenhart et al., 2003).

Gender: The gender divide has been decreasing in the United States. Although just 34 percent of American women were using the Internet by the end of 1998, 44 percent of them had become Internet users by August 2000 (ITU, 2003b). In 2002, 73 percent of American men and 69 percent of American women were Internet users (UCLA, 2003).

Life stage: The "gray gap" remains. Younger Americans have the highest level of Internet access and use. More than 80 percent of Americans aged between 12 and 35 were using the Internet. By contrast, only 34 percent of Americans over 65 were online in 2002 (UCLA, 2003).

Ethnicity: The racial/ethnic digital divide is pronounced in the United States. Although 63 percent of Asian-Americans and 55 percent of white Americans were online in 2000, only 30 percent of blacks and 28 percent of Hispanic-Americans were online then (Fong et al., 2001). In 2003, only 8 percent of African-Americans were online despite being 11 percent of the population. The divide was much narrower for Hispanic-Americans: 9 percent were online, despite being 10 percent of the American population (Lenhart et al., 2003).

Geographic location: The geographic divide persists to some extent. Americans living in rural areas (42 percent) and central cities (44 percent) had substantially lower levels of Internet access than suburbanites (55 percent) in 2000, regardless of their socioeconomic statuses (Fong et al., 2001). In 2002, 63 percent of suburbanites were using the Internet as compared to less than half of rural inhabitants (Lenhart et al., 2003).

When the digital divide occurs at the intersection of class, race/ethnicity, gender, and geographic location, it can come in mutually reinforcing ways. For example, the average Internet penetration rate among high-income Americans (annual household income of $75,000 or higher) reached 78 percent in 2000. Yet, within this affluent group, there was a 31-percentage point gap in Internet access in 2000 between those with a college education (82 percent) and those with less than a high school education (51 percent; U.S. NTIA, 2000). Race/ethnicity further complicates the picture. Even at the same income levels, African-Americans are less likely than other racial/ethnic groups to be connected to the Internet (Lenhart et al., 2003). Although Asian-Americans as a racial/ethnic group have had a high Internet penetration rate, there has been a wide gender gap among them (58 percent male vs. 42 percent female users; Spooner, 2001).

Internet Use   American Internet users spent an average of 11 hours online every week in 2002. About 60 percent of American users used the Internet from home, up from one-fifth in 1995. Sixty-one percent of users considered the Internet as an important source of information (UCLA, 2003).

The Internet has become an important tool at many American workplaces, where emailing has become common. In 2002, 62 percent of all American employees had Internet access at work and 98 percent of those with such access used email on the job. About two-thirds of American workers with Internet access reported that email was the most effective way of arranging meetings, making appointments, and editing or reviewing documents (Fallows, 2002).

There are gender, age, and ethnic gaps in Internet use as well as in access. The ways in which Americans communicate online are associated with gendered styles of maintaining relationships offline (Boneva et al.,

2001; Kennedy et al., 2004). Women are more enthusiastic in emailing family members and friends than men, reflecting women's domestic role as the ones responsible for keeping contact with family and friends (Horrigan and Rainie, 2002). Furthermore, teenagers spend more time on the Internet than adults (Kraut et al., 1998).

Asian-Americans have been the heaviest Internet users of any racial/ethnic group and have the longest experience. They have been more likely than users in other ethnic groups to use the Internet for work-related reasons or school research (Spooner, 2001). By contrast, African-American Internet users have spent less time on the Internet, initiated fewer sessions, and browsed fewer web pages (Nielsen/NetRatings, 2003b).

Language is a barrier that becomes salient after basic Internet access becomes available since people feel more comfortable surfing websites in their first language. For instance, Hispanic-American Internet users have been more likely to spend their online time in Spanish than in English (Greenspan, 2002).

Germany

We use data from the ARD/ZDF-Online annual surveys from 1999 to 2002 (ARD/ZDF, 1999, van Eimeren amd Gerhard, 2000, van Eimeren et al., 2001, and van Eimeren et al., 2002). ARD and ZDF are the country's major public television broadcasters and have conducted nationally representative surveys since 1997, targeting German Internet users aged 14 or older.

The Internet penetration rate has risen generally since the mid-1990s in Germany. Among the German population aged 14 and older, 7 percent used the Internet in 1997, 10 percent in 1998, and 18 percent in 1999 (van Eimeren et al., 2002). Unlike North America, there was a substantial gap in Germany between computer ownership and Internet use as late as 1999, when 45 percent of households in Germany owned a computer but only about one-quarter of these households (11 percent of all households) were connected to the Internet (ITU, 2002a; Welling and Kubicek, 2000). Internet diffusion has accelerated since then. Twenty-nine percent of the German population was wired in 2000, 39 percent in 2001, and 44 percent in 2002 (van Eimeren et al., 2002).

The Internet is widely accessed in German public places. About 98 percent of schools in Germany were connected to the Internet in 2001, a dramatic increase from 15 percent in 1998. By 2000, all of the 1,270 public libraries were connected and provided Internet access to the public. The number of college students majoring in computer science during 2001 (27,000) was more than twice as high as in 1997 (German Federal Government, 2001). The personal computer is the primary means of going online in Germany. For example, less than 1 percent of Internet users went online via Web TVs or PDAs (van Eimeren et al., 2002).

Internet Access

Socioeconomic status: German Internet users in the mid-1990s were predominately male, young, employed, well off, and well educated. Early Internet adopters often have higher income and better education. Education is the deepest fault line of the digital divide in Germany. In 2000, 86 percent of Germans with a college or higher degree had Internet access, while only 8 percent of those with high school qualifications were online. Furthermore, the percentage of better-educated Internet users has been increasing faster than that of high school graduates. On average, users with a postsecondary education adopt the Internet 19 months earlier than those with high school or grammar school education (van Eimeren et al., 2001, 2002).

Gender: There is a sizeable gender divide in Germany, with men more likely to be Internet users. Moreover, even though more women have been gaining access to the Internet, the gender divide widened from 7 percentage points in 1997 (10 percent of men vs. 3 percent of women were online) to 18 percentage points in 2001 (48 percent of men vs. 30 percent of women online). The growing gender divide was the result of significantly higher Internet diffusion among German men. In 2002, 53 percent of men and 36 percent of women were using the Internet, a continuing gap of 17 percentage points. Thus, the gender divide does not seem to be narrowing (van Eimeren et al., 2002).

Life stage: The rate of Internet access declines sharply with older age. In 1997, 73 percent of all Internet users in Germany were younger than 40. Even though the percentage of such younger adult Internet users dropped to 65 percent in 2002, the age divide has remained large. For example, only 5 percent of people aged 60 years or older accessed the Internet in 2002 (van Eimeren et al., 2002).

Geographic location: On average, users from the more developed former West Germany have five months more experience with the Internet than those living in the former East Germany (van Eimeren et al., 2002).

Internet Use

For the great majority of Germans, the Internet is primarily a means of communicating and seeking information. In 2002, 81 percent of German Internet users sent and received emails, and 55 percent searched for information on the Web. In addition, 32 percent of wired Germans used the Internet for online banking, 23 percent for chatting and newsgroups, 15 percent for online games, and 13 percent for online auctions (van Eimeren et al., 2002).

Place of Internet access and use: In the mid- and late 1990s, Germans were more likely to use the Internet outside their homes rather than from their homes. Even in early 2000, the number of Internet users who had access from outside their homes was almost twice as high as those accessing the Internet from inside their homes. However, more German users have gained home access since then. In 2002, 50 percent of Internet users accessed the Internet exclusively from home, while 34 percent had access at both home and at work (van Eimeren et al., 2002).

Length of Internet experience: The average German Internet user has been online for three years (van Eimeren et al., 2002).

Japan

The percentage of Japanese households owning PCs increased 2.6 times in six years, from 1996 to 2002. About 22 percent of Japanese households owned a PC in 1996, 29 percent in 1997, 33 percent in 1998, 38 percent in 1999, 51 percent in 2000, and 58 percent in 2001 (Japan MPHPT, 2002).

However, there has been a gap between PC access and Internet access. The diffusion of the Internet, especially the PC-based Internet, started relatively late in Japan. For instance, while 40 percent of American households were online in 1999, only 12 percent of Japanese households were online that year (Dewey Ballantine and Cyberworks Japan, 2001). The number of Internet users (6 years and older) was 12 million in 1997, 17 million in 1998, 27 million in 1999, 47 million in 2000, 56 million in 2001, and 69.4 million in 2002 (Japan MPHPT, 2002, 2003).

The high rate of mobile phone use in Japan may explain the gap between the high rate of PC use and the relatively low rate of PC-based

Internet use. Japan is a world leader in the use of mobile phones to access the Internet (Miyata et al., 2004). In 1996, 25 percent of Japanese households had a mobile phone. The percentage tripled in four years, increasing to 46 percent in 1997, 58 percent in 1998, and 64 percent in 1999. It reached 75 percent in 2000 and remained at this level in 2001 (Japan MPHPT, 2002). When NTT DoCoMo launched its mobile Internet service in February 1999, the number of subscribers skyrocketed from zero to 20 million in just 18 months (Dewey Ballantine and Cyberworks Japan, 2001). The ownership of Internet-capable mobile phones among Japanese households jumped from 9 percent in 1999 to 27 percent in 2000 and reached 30 percent in 2001 (Japan MPHPT, 2002).

Among Japanese who are 12 years or older and connected to the Internet, 85 percent access the Internet through PCs, 63 percent through mobile phones, and 5 percent through other technologies (World Internet Project Japan, 2002). Although the share of users accessing the Internet via PCs remained the same from 2000 to 2001, the share of users accessing the Internet through mobile phones soared during this same period. In 2002, 83 percent of Japanese mobile phone users (62 million people) were mobile Internet subscribers (Japan MPHPT, 2003).

Internet Access
Socioeconomic status: The higher the household income, the greater the likelihood of Japanese residents accessing the Internet through PCs or mobile phones. In 2001, about 50 percent of households with an income of 8 million yen (approximately US $70,000) or higher had access to the Internet, while about one-quarter of those households with an income under 2 million yen (approximately US $17,500) were connected, indicating an income gap of 25 percentage points. However, the income divide shrank rapidly to less than 20 percentage points in 2002 (Japan MPHPT, 2003). In all income groups, the PC-based digital divide was bigger than the gap in mobile Internet use. Internet use is also positively associated with education, although the educational divide has narrowed slightly over the years. For instance, the gap between university graduates and college graduates was diminishing in 2002. Yet, Japanese with high school or lower education continue to have much less Internet access (World Internet Project Japan, 2002).

Gender: Overall, 68 percent of Japanese men and 56 percent of Japanese women were Internet users in 2002. There was a slight decrease in the gender gap of Internet access between 2001 and 2002 (Japan

MPHPT, 2003). Although men are ahead of women in terms of access to both PC-based and mobile Internet, the gender gap for mobile Internet users is slightly smaller than the gap for PC Internet users (FY 2001 White Paper on Telecommunications in Japan, cited by Digital Opportunity Site, 2001). The gender gap is reversed among young Japanese aged between 19 and 24. There are more female users than male users, because many young women are using mobile phones to access the Internet (World Internet Project Japan, 2002).

Life stage: Japan has a sizeable generational divide in Internet use. Young Japanese in their twenties were 30 times more likely than people in their seventies to be connected to the Internet in 2001. Japanese in their twenties had the highest rate of Internet access in 2001: About 80 percent had access to the Internet with 48 percent through PCs and 53 percent via mobile phones. By contrast, only 15 percent of people in their sixties were using the Internet, with the majority accessing the Internet exclusively via PCs. At the other end of the age spectrum, the mobile Internet was embraced by 53 percent of Japanese teenagers, while only 30 percent of them accessed the Internet through PCs (World Internet Project Japan, 2002).

Geographic location: In Japan, the percentage of home access via PCs declines with the size of the city. Major cities have higher Internet penetration rates than smaller cities, followed by towns and villages. However, the digital divide in terms of city size dropped 2 percentage points between 2001 and 2002 (Japan MPHPT, 2003). The proliferation of mobile Internet use may also be shrinking the city-size divide. For example, in 2000, residents in smaller cities were slightly more likely to access the Internet via mobile phones than those in major cities (17.3 percent and 16.8 percent, respectively) (FY 2001 White Paper on Telecommunications in Japan, cited by Digital Opportunity Site, 2001).

Internet Use   A random sample survey in Yamanashi prefecture in 2002 found that the use of PC-based and mobile Internet varied by age and gender (Miyata et al., 2004). Compared to users connecting to the Internet via personal computers, those who connect via mobile phones tend to come from a different group with different sociodemographic profiles. PC-based Internet users are more likely to be male, older, and better educated, while mobile Internet users are more likely to be female, younger, and less educated (World Internet Project Japan, 2002).

By 2002, 77 percent of all Japanese Internet users were accessing the Internet through mobile phones, either as their only access point or as complements to PC-based access. Mobile Internet use is not so very mobile. People most use their mobile phones from home to access the Internet: 63 percent, with an average of 88 minutes per week. The amount of email exchanged by mobile phone (26 emails per week) is 6 times higher than by PC. Mobile Internet is not only used for emailing but also for accessing websites, including search sites, weather forecasts, and "transportation/travel course/maps" (World Internet Project Japan, 2002; see also Ito and Daisuke, 2003; Ito, 2004; Miyata et al., 2004).

Japanese differ in the ways in which they use the PCs and mobile phones to exchange email and surf the web. Twenty-nine percent of Japanese users email through PC-based Internet. The use of email through PC-based Internet is the highest among those with at least a university education. However, the average number of emails sent on PC-based Internet is less than 4 per week. The most frequent email correspondents are "friends who seldom meet" (37 percent) or coworkers (16 percent). By contrast to PC-based Internet, email through mobile Internet was most frequently exchanged among "friends who one often meets" or family members. Hence, the World Internet Project Japan (2002) has concluded that "email is not a replacement of face-to-face communication, but has a strong supplementary role to communication" (p. 50; see also Miyata et al., 2004).

Korea

From 1998 to 1999 the number of Internet users in Korea (defined as users over 7 years old using the Internet at least once a month) increased threefold, jumping from 3 million to 11 million. By 2001, 57 percent of Koreans over 7 years old (24 million) were online (Soe, 2002). The number of Internet users continued to grow to 26 million by June 2002, nearly 9 times greater than five years ago. There were also 27 million mobile Internet subscribers in Korea in June 2002, although many of them presumably also had PC-based access.

The penetration rate of PCs has been increasing markedly since the mid-1990s, from 21 percent in 1994 to 52 percent in 1999. Two-thirds of Korean households owned a PC in 2000 and four-fifths in 2001 (Park, 2001).

It is not just use that has skyrocketed. Korea recently became the world leader in broadband Internet access, with 14 broadband subscribers per 100 inhabitants in June 2001 (Yun et al., 2002). While only 14,000 Korean households had a broadband connection in 1998, 8.7 million of them were enjoying broadband connections by 2002 (Korea National Computerization Agency, 2002).

Nevertheless, the lack of computer skills has been a barrier to Internet diffusion in Korea. According to the National Statistical Office in Korea, 60 percent of all Koreans were computer illiterate in 1997. Another national survey in 2000 reported that 46 percent of Koreans lacked the necessary computer skills to surf the web. In particular, a disproportionate share of women, older people, people with low education, and blue-collar workers had insufficient or no computer skills (Park, 2001). The high rate of current Internet use suggests that this skills gap has markedly lessened in recent years. This implies that computer skills can be taught quickly and informally.

Internet Access
Socioeconomic status: Despite the proliferation of Internet use, the income divide in Korea has been widening since 1999. While 70 percent of Koreans with a monthly income higher than 2.5 million won (approximately US $2,000) were connected in 2001, only 37 percent of those with a monthly income less than 1.5 million won (approximately US $1,250) were connected. Moreover, there has been a growing educational divide in Internet access and use. The divide between those with a college degree or higher and those with a high school degree has grown from 28 percentage points in 1999 to 40 percentage points in 2001. The gap between college graduates and upper middle school graduates increased even more during the same period, from 37 to 65 percentage points (Soe, 2002).

Gender: Korean men are more likely than women to use the Internet. While half of Korean men had already been online by the end of 2000, only 39 percent of women were online at that time. Korean women have been making progress, since more than half of them were connected to the Internet by the end of 2002. However, since more Korean men have also been going online, the gender gap has lingered at 12 to 14 percentage points since 2000 (Korea National Computerization Agency, 2002).

Life stage: There is a clear and growing age divide between younger and older Koreans because young Koreans have embraced the Internet at a high rate. The already large divide between the age group of 7 to 19 and the age group of over 50 has been widening rapidly: from 40 percentage points in 1999 to 70 in 2000 and 84 in 2001 (Soe, 2002).

Geographic location: Internet diffusion is uneven among different regions. The capital city of Seoul, the most wired area in the country, has an Internet penetration rate that is at least 10 percentage points higher than any other region in Korea. However, the gap between metropolitan and nonmetropolitan areas has narrowed as the government has fostered broadband connectivity in most of the country (Soe, 2002).

Internet Use

Place of Internet access and use: Home is the primary place where Koreans access the Internet, followed by workplaces, cybercafes, and schools. Internet cafes are popular access points. Cafes are more attractive than schools to young Koreans under 30 because they are places where many users sit side-by-side playing online games at high speeds (Soe, 2002).

The most frequent activities Koreans carry out online are information seeking, emailing, gaming, browsing newspapers and magazines, shopping, and making reservations. Young Koreans, aged 7 to 19, use the Internet for playing games more than for any other purpose. By contrast, users over 50 years old are more likely to use the Internet for online banking, or browsing newspapers and magazines (Soe, 2002).

The expansion of broadband connections has profoundly affected the pattern of Internet use in Korea. Koreans are the heaviest Internet users worldwide, spending more than twice as much time online as American users (Yun et al., 2002). Korean time spent online is correlated with education, gender, and age. In 2001, well-educated Korean Internet users (with a college degree or higher) spent 5 more hours per week than those with an elementary school education or less. Men spent 3 hours more per week than women on the Web. Moreover, users in their twenties are the heaviest users of the Internet, spending an average of 16 hours per week on the Internet. Yet, once online, older Koreans aged 60 or older also spent a considerable amount of time (9 hours per week) on the Internet (Korea National Statistical Office, 2002).

The high penetration of broadband connections enabled 71 percent of Korean users to enjoy streaming audio and 54 percent to play online games in 2001 (Yun et al., 2002). The National Computerization Agency recently reported that half of Korean Internet users "are actively involved in cyber community activities, proving the existence of a newly formed networked culture" (Yun et al., 2002, p. 6).

China

The China Internet Network Information Center (CNNIC) has conducted semiannual surveys on Chinese computer and Internet users since 1997. Our review is primarily based on the CNNIC's semiannual report published in January of 2003 that defines Internet users as Chinese citizens who use the Internet an average of at least one hour per week. Survey respondents included both adults and those below 18 years old (CNNIC, 2003).

China is a relatively late starter in the Internet race but has been catching up quickly. Because of the large population of China, the low penetration rate of less than 5 percent provides both a great many users and much room for growth. There has been a dramatic increase in Internet users, from 620,000 in 1997 to 22.5 million in 2001 and about 60 million in 2003 (CNNIC, 2003). The number of Internet-connected computers has increased from about 0.3 million in 1997 to 12 million in 2002. China's Internet population probably ranks second in the world and is growing rapidly. "It's going to be the largest Internet market in the world," says Safa Rashtchy, analyst at US Bancorp (quoted in Waters, 2003, p. 17). However, another analyst warns, "a lot of people are wondering if it's a bubble" (Dickie, 2003, p. 18).

Internet Access

Socioeconomic status: Chinese Internet users are much better educated than the general population. In 2003, 57 percent of Internet users have at least a college-level education, while an additional 31 percent of users have a high school degree. The digital divide in terms of education is striking since 14 percent of the Chinese population aged 15 and above have a high school degree, while less than 5 percent have an education of college or higher (China National Statistics Bureau, 2002). Forty-four percent of Internet users have a monthly income of less than 1000 Yuan (about US $125) and an additional 17 percent have no income. Yet, income has an impact on the Chinese digital divide, although its significance is distorted by the 28 percent of Chinese Internet users who are (often lower-income) students going online at universities and schools.

Gender: There is somewhat of a gender gap in China, although the percentage of female users has steadily climbed from 13 percent in 1997 to 41 percent in 2003. The predominance of male users is also reflected in the gender diffusion rate in the general population: 5.3 percent of the male population is online, as compared to 3.9 percent of the female population. However, the digital divide in terms of gender has closed rapidly. The online gender gap has decreased from a highly male ratio of 7 : 1 in 1997 to 1.6 : 1 in 2002.

Life stage: Younger Chinese form the majority of Internet users in the country. In 2003, people aged 50 or older made up only 3.7 percent of all Internet users, while about two-thirds of users were younger than 35. However, the age divide is narrowing, since young Chinese under 35 years old accounted for an even higher percentage of all Internet users, about four-fifths, just one year prior in 2002.

Geographic location: As a result of the different levels of socioeconomic development between the richer east coast region and the poorer hinterland, the regional distribution of Internet access is uneven. The regional divide continues to be large but is shrinking. The share of Internet users from the more developed areas of Beijing, Shanghai, and the province of Guangdong decreased from 30 percent in 2002 to 23 percent in 2003 (CNNIC, 2003). There is an enormous divide between urban and rural areas. Peasants, accounting for approximately 80 percent of the Chinese population, make up only 1 percent of all Internet users in the country (CNNIC, 2002).

Internet Use

On average, Chinese users spent about 10 hours a week on the Web in 2003, although many users do not go online every day. Chinese use the Internet for a variety of activities, including seeking information (53 percent of users), entertainment (25 percent), romance/friendship (7 percent), and study/research (5 percent). The most used online services are email (93 percent), search engines (68 percent), chat (45 percent), downloading and uploading documents (45 percent), and newsgroups (21 percent). Eighty-one percent of the information browsed on the web is in the Chinese language. Most Chinese Internet users report that the Internet plays a positive role in their everyday lives: Seventy-three percent of users feel that the Internet is helpful or very helpful for study, 67 percent for work, and 61 percent in daily life.

Place of Internet access and use: Chinese Internet users are more likely to access the Internet from public places than are users from developed countries. A sizeable minority of users goes online at public places: 20 percent from schools and 19 percent from cybercafes. In addition, 63 percent of Chinese Internet users access the Internet from home while 43 percent access it from work (CNNIC, 2003). We caution that there is some double counting in these statistics for people accessing the Internet in more than one place.

The Multifaceted Nature of the Digital Divide

Fundamentally, the digital divide is about the gap between individuals and societies *that have the resources* to participate in the information era and those that do not. This digital divide remains real worldwide. While we caution once again that the lack of standardized measurements and definitions weakens the precision and comparability of all statistics, our international comparative study clearly suggests that the uneven diffusion and use of the Internet are shaped by—and are shaping—social inequalities.

Digital divides occur at the intersection of international and intranational socioeconomic, technological, and linguistic differences. Telecommunications policies, infrastructures, and education are prerequisites for marginalized communities to participate in the information age. High costs, English language dominance, the lack of relevant content, and the lack of technological support are barriers for disadvantaged communities using computers and the Internet. For instance, while about one-half of the world's Internet users are native English speakers, about three-quarters of all websites are in English (World Economic Forum, 2002).

The diffusion of Internet use in developed countries may be slowing and even stalling. Currently, Internet penetration rates are not climbing in several of the developed countries with the most penetration. Compared to the explosive growth of Internet access and use in the past decade, this is a new phenomenon. It is too soon to tell if this is a true leveling off of the penetration rate or a short-term fluctuation as Internet use continues its climb to triumphant ubiquity.

The digital divide between first-movers and latecomers among developed countries is narrowing. Countries such as Germany, Korea, and Japan have caught up to the level of Internet connectivity in the United States (see table 19.3, columns 2 and 4). In some ways, the Internet is expanding in developed countries in similar ways to its expansion in the United States, although with a time lag (Bazar and Boalch, 1997). For example, the demographic profile of users in developed countries looks roughly similar to that of American Internet users a half-decade earlier: young, well-educated men.

The nature of the digital divide varies between countries. Although all sorts of digital divides are narrowing in the United States, this is not true

Table 19.3
Number and percentage of population online in the five countries

Country	Number of Population Online in 2002 (millions, month)[a]	Percent of Population Online in 2002[a]	Number of Population Online in 2001 (millions)[b]	Percent of Population Online in 2001[b]	Number of PC Users in 2001 (millions)[b]	Percent of Female Users in Population Online 2002[b]
United States	166 (Apr)	59	143	50	178	51 (2001)
Republic of Korea	26 (July)	54	24	52	12	45
Japan	56 (June)	44	56	44	44	41 (2001)
Germany	32 (Aug)	39	31	37	32	37 (2001)
China	58 (Dec)	4.8	34	2.5	25	39

a. NUA, 2003, http://www.nua.ie/surveys/how_many_online/.
b. ITU, 2002b.

for other developed (or developing) countries. Different sorts of social structures, demographics, political policies, and technological dynamics influence the use and the impact of the Internet in different countries. For instance, Japan is leading the development of mobile Internet, and Korea is the world leader of broadband connections. Although most countries lag behind the United States in PC-based Internet use, they are quickly adopting mobile phones.

The digital divide remains substantial between developed and developing countries. The digital divide reflects the broader context of international social and economic relations: a center–periphery order marked by American predominance. There are large disparities of Internet access between affluent nations at the core of the Internet-based global network and poor countries at the periphery that lack the skills, resources, and infrastructure to log on the information era. For instance, the average Internet penetration rate in developed countries was 30 percent in 2001, 10 times as high as in developing nations (ITU, 2003b; International Labor Office, 2001).

The digital divide can widen even as the number and percentage of Internet users increases. The widening of the divide can happen when the newcomers to the Internet are demographically similar to those already on. For example, while poorer and less-educated people are accessing the Internet, the rate of increased access is higher among the more affluent and better-educated segments of society in developing countries. To take another example, if men come to the Internet at a higher rate, the gender divide will grow.

The digital divide is wide and deep in developing countries. It is wide in the sense that only a small percentage of the population uses the Internet, and deep in the sense that the consequences for not being online may be greater when moving beyond a subsistence level. There are stark contrasts in the developing world between those connected to the Internet and those who are even more on the periphery. Those with Internet access are much more likely to be living in major urban centers and to have more education and income. They are better connected to developed countries culturally and economically, and of course, the Internet increases their connectivity.

To be sure, the sizeable populations of some developing countries— such as China and India—mean that there are different impacts of the percentage and the absolute number of people accessing the Internet. China already has the second largest population of Internet users, despite its low percentage of users. This sizeable population of users will be catered to—witness the proliferation of Mandarin-language websites—and can have an impact on the future development of the Internet.

The digital divide has profound impacts on the continuation of social inequality. People, social groups, and nations on the wrong side of the digital divide can be excluded from the knowledge economy. If preexisting inequalities deter people from using computers and the Internet, these inequalities may increase as the Internet becomes more consequential for getting jobs, seeking information, and engaging in civic and entrepreneurial activities.

There are multiple digital divides, not just a single digital divide. Moreover, some of these divides are widening in some countries. Across the five countries studied, socioeconomic status, gender, life stage, and geographic location significantly affect people's access to and use of the Internet (as summarized in tables 19.4 and 19.5).

Socioeconomic status: Internet users are more likely to be better off and better educated than nonusers in all five countries surveyed. In general, the lower the Internet penetration rate in a country, the more elite the online population (see also Chen, Boase, and Wellman, 2002). Although the socioeconomic divide is narrowing in the United States and Japan, the digital divide elsewhere seems to be widening along the lines of income and education (table 19.4, column 2).

Gender: Men are more likely than women to access and use the Internet. With the exception of the United States, the share of female Internet users is lower than their share in the general population in each of the countries surveyed. Yet, this gender divide is narrowing, except in Germany, and sometimes the divide is small (table 19.4, column 3).

Life stage: In both developed and developing countries, the Internet penetration rate among younger people is substantially higher than that among older people. In general, the life stage divide is declining in most countries, except for Korea. Students who go online via school connections make up a large share of Internet users in developing countries (table 19.4, column 4).

Geographic location: Geographic location affects access to and use of the Internet. Richer regions have higher Internet penetration rates than poorer ones. The overall trend across the five countries shows a narrowing, yet persistent, digital divide in terms of geographic location (table 19.4, column 5).

Table 19.4
Summary of Internet access in the five countries

Country	Socioeconomic status	Gender	Life stage	Geographic location
United States	Declining yet persistent	51% of Internet users are female. However, the Internet penetration rate among men is still higher than that among women	Declining yet persistent	Declining yet persistent
Germany	Increasing	Increasing	Declining yet persistent	Declining
Japan	Declining yet persistent	Declining yet persistent (reversed digital divide in mobile Internet)	There is a generational divide	Major cities have higher Internet diffusion than smaller cities
Republic of Korea	Increasing	Persistent	Increasing	Declining. However, Seoul is still the most wired area in the country
China	Huge yet slightly declining	Declining yet persistent	Slightly declining	Huge yet slightly declining

Table 19.5
Trends of the digital divide by country

Country	Socioeconomic status	Gender	Life stage	Geographic location
United States	↓	↓	↓	↓
Germany	↑	↑	↓	↓
Japan	↓	↓ reversed in mobile Internet, women higher	→	↓
Korea	↑	→	↑	↓
China	↓	↓	↓	↓

As shown in table 19.5, the digital divide has diverse manifestations along these fault lines. For instance, the gender divide is especially wide in Germany: The percentage of female Internet users is lower than in all other countries reviewed in this chapter—not only in the developed countries but also in China (table 19.3, last column). Moreover, the intersection of socioeconomic status, gender, age, language, and geographic location tends to increase the digital divide in mutually reinforcing ways within, and between, countries. The largest gap is between (a) better-educated, affluent, younger, English-speaking men in developed cities and (b) less-educated, poor, older, non-English-speaking women in underdeveloped rural areas.

Bridging the Digital Divide

Since the diffusion of the Internet is global, ongoing, and consequential, understanding the causes and impacts of the multiple digital divides has substantial policy implications. We need nuanced understanding and action. Understanding the causes of the uneven diffusion of the Internet and other telecommunication technologies across countries is the first step in narrowing the digital divide. Such understanding can provide practical information for decision making: both for targeting market segments with different social and economic backgrounds in different parts of the world and for building public–nongovernmental organization (NGO)–private partnerships for narrowing the digital divide.

Governments, private sectors, and NGOs have initiated and sponsored numerous programs to narrow the digital divide through Internet connectivity in public places such as schools, community centers, and public libraries. To do this properly, there is a need to evaluate systematically the impacts of such programs, going beyond merely documenting that X percent of group Y in country Z are now connected.

Given the multiple causes and manifestations of the digital divide, narrowing the divide is more complicated than merely providing computers and Internet connections. Bridging the divide has to promote both broader access to and effective use of the Internet. It requires cooperation between governments, the private sectors, and nongovernmental organizations. The question is how disadvantaged individuals and groups could be enabled to obtain the necessary resources and afford the leapfrog into a digital future. A simple commitment to "close the digital divide" is simplistic because there is no one digital divide.

Acknowledgments

We appreciate the advice and assistance of Soonhoon Bae, Vicky Hung, Ken McEldowney, Carlos Scheel, and Hideyo Waki of the AMD Global Consumer Advisory Board; Brenda Forsythe, Sarah Russ, Patrick Moorhead, and Timothy Martin of AMD; Monica Prijatelj, Phuoc Tran, and Uyen Quach of our NetLab; and Professors Leopoldina Fortunati and Kakuko Miyata. We appreciate the financial support provided by AMD and the Social Sciences and Humanities Research Council of Canada. The opinions expressed here are solely those of the authors.

Note

1. A longer version of this chapter is available at: http://www.amd.com/usen/assets/content_type/DownloadableAssets/FINAL_REPORT_CHARTING_DIGI_DIVIDES.pdf. The longer version contains additional analysis of the United Kingdom, Italy, and Mexico and a fuller discussion of digital divide issues.

References

AC Neilson (2001). "429 Million People Worldwide Have Internet Access, According to Nielsen/NetRatings." http://www.eratings.com/news/20010611.htm.

ARD/ZDF Arbeitsgruppe Multimedia (1999). "ARD/ZDF-Online-Studie 1999: Wird Online Alltagsmedium?" *Media Perspektiven* 8: 401–414.

Bazar, B., and G. Boalch (1997). "A Preliminary Model of Internet Diffusion within Developing Countries." Paper presented at AusWeb97: Third Australian World Wide Web Conference, Lismore.

Boneva, B., R. Kraut, and D. Frohlich (2001). "Using E-Mail for Personal Relationships: The Difference Gender Makes." *American Behavioral Scientist* 45(3): 530–549.

Chen, W., J. Boase, and B. Wellman (2002). "The Global Villagers: Comparing Internet Users and Uses around the World." In *The Internet in Everyday Life*, ed. B. Wellman and C. Haythornthwaite. Oxford: Blackwell, pp. 74–113.

China National Statistics Bureau (2002). *Statistic Book of China—2002.* Beijing. http://www.gse.pku.edu.cn/dataset/yearbook/yearbook02/indexC.htm.

CNNIC (China Internet Network Information Center) (2002). "Semiannual Survey Report on the Development of China's Internet." http://www.cnnic.org.cn/.

CNNIC (China Internet Network Information Center) (2003). "Semiannual Survey on the Development of China's Internet." http://www.cnnic.org.cn/.

Dewey Ballantine and Cyberworks Japan (2001). "The Internet in Japan: Catalyst for Change?" http://www.dbtrade.com/ecommerce/the_Internet_in_japan.htm.

Dickie, M. (2003). "Déja Vu as Leading Internet Stocks Keep Rising." *Financial Times,* June 27.

Digital Opportunity Site (2001). "Differences in the Use of the Internet." http://www.dosite.jp/e/do/j-state_net.html.

DiMaggio, P., and E. Hargittai (2001). "From the 'Digital Divide' to 'Digital Inequality': Studying Internet Use as Penetration Increases." Princeton, N.J.: Princeton University Center for Arts and Cultural Policy Studies, Working Paper Series no. 15.

DiMaggio, P., E. Hargittai, W. R. Neuman, and J. Robinson (2001). "Social Implications of the Internet." *Annual Review of Sociology* 27: 287–305.

Fallows, D. (2002). "Email at Work: Few Feel Overwhelmed and Most Are Pleased with the Way Email Helps Them Do Their Jobs." Washington: Pew Internet and American Life Project. http://www.pewinternet.org/reports/pdfs/PIP_Work_Email_Report.pdf.

Fong, E., B. Wellman, R. Wilkes, and M. Kew (2001). *The Double Digital Divide.* Ottawa, Canada: Office of Learning Technologies, Human Resources Development Canada.

German Federal Government (2001). "'Internet for All'—Taking Stock after a Year." http://eng.bundesregierung.de/top/dokumente/Background_Information/ The_Information_Society/Internet_for_Everyone/ix4365_33889.htm? template=single&id=33889&ixepf=4365_33889&script=0.

Greenspan, R. (2002). "In Any Language, Hispanics Enjoy Surfing." Cyber Atlas. http://cyberatlas.internet.com/big_picture/demographics/article/ 0,,5901_1369131,00.html.

Hampton, K., and B. Wellman (2003). "Neighboring in Netville." *City and Community* 2 (Fall): 277–311.

Hargittai, E. (2003). "How Wide a Web? Inequalities in Accessing Information Online." Doctoral dissertation, Department of Sociology, Princeton University.

Horrigan, J. B., and L. Rainie (2002). "Getting Serious Online." Washington: Pew Internet and American Life Project.

International Labor Office (2001). "World Employment Report 2001: Life at Work in the Information Economy." Geneva.

International Telecommunication Union (2001a). "Regulatory Implications of Broadband Workshop." ITU New Initiatives Programme, Geneva.

International Telecommunication Union (2001b). "Telecommunication Indicators in the Eurostate Area, 2001." Geneva.

International Telecommunication Union (2002a). "Basic Indicators: Population, GDP, Total Telephone Subscribers and Total Telephone Subscribers per 100 People." Geneva. http://www.itu.int/ITU-D/ict/statistics/at_glance/basic01.pdf.

International Telecommunication Union (2002b). "Mobile Cellular, Subscribers per 100 People." Geneva. http://www.itu.int/ITU-D/ict/statistics/at_glance/ cellular01.pdf.

International Telecommunication Union (2003a). "Final Report—DRAFT. World Telecommunication/ICT Indicators Meeting." Geneva.

International Telecommunication Union (2003b). "Female Internet Users as Percent of Total Internet Users, 2002." Geneva. http://www.itu.int/ITU-D/ict/ statistics/at_glance/f_inet.html.

Ito, M., ed. (2004). *Portable, Personal, Intimate: Mobile Phones in Japanese Life.* Cambridge, Mass.: MIT Press.

Ito, M., and O. Daisuke (2003). "Mobile Phones, Japanese Youth, and the Replacement of Social Contact." Paper presented at the international workshop Front Stage—Back Stage: Mobile Communication and the Renegotiation of the Social Sphere, Grimstad, Norway.

Japan Ministry of Public Management, Home Affairs, Posts and Telecommunications, Japan (2002). "Information and Communications in Japan White Paper 2002: Stirring of the IT-Prevalent Society." Tokyo.

Japan Ministry of Public Management, Home Affairs, Posts and Telecommunications, Japan (2003). "Information and Communications in Japan White Paper 2003: Building a 'New, Japan-Inspired IT Society.'" Tokyo.

Jordan, T. (2001). "Measuring the Internet: Host Counts versus Business Plans." *Information, Communication and Society* 4(1): 34–53.

Katz, J. E., and R. E. Rice (2002). *Social Consequences of Internet Use: Access, Involvement, and Interaction.* Cambridge, Mass.: MIT Press.

Kennedy, T., B. Wellman, and K. Klement (2004). "Gendering the Digital Divide." *IT & Society* 1(5): 49–72.

Korea National Computerization Agency (2002). "Informatization White Paper 2002: Global Leader e-Korea." Seoul.

Korea National Statistical Office (2003). "Hours Used on Internet." Seoul. http://www.nso.go.kr/cgi-bin/html_out.cgi?F=Xe6b0_le6b0.html.

Kraut, R., M. Patterson, V. Lundmark, S. Kiesler, T. Mukopadhyay, and W. Scherlis (1998). "Internet Paradox: A Social Technology that Reduces Social Involvement and Psychological Well-Being?" *American Psychologist* 53(9): 1017–1031.

Lenhart, A., J. Horrigan, L. Rainie, K. Allen, A. Boyce, M. Madden, and E. O'Grady (2003). "The Ever-Shifting Internet Population: A New Look at Internet Access and the Digital Divide." Washington: Pew Internet and American Life Project.

Miyata, K., J. Boase, B. Wellman, and K. Ikeda (2004). "The Mobile-izing Japanese: Connecting to the Internet by PC and Webphone in Yamanashi." In *Portable, Personal, Intimate: Mobile Phones in Japanese Life,* ed. M. Ito. Cambridge, Mass.: MIT Press.

Nielsen/NetRatings (2003a). "Global Internet Population Grows an Average of Four Percent Year-Over-Year." Nielsen/NetRatings. http://www.nielsen-netratings.com/pr/pr_030220.pdf.

Nielsen/NetRatings (2003b). "More Than 10 Million African-Americans are Online, According to Nielsen/NetRatings." Nielsen/NetRatings. www.nielsen-netratings.com/pr/pr_030226.pdf.

Norris, P. (2001). *Digital Divide? Civic Engagement, Information Poverty and the Internet in Democratic Societies.* Cambridge: Cambridge University Press.

NUA (2003). "How Many Online?" http://www.nua.com/surveys/how_many_online/index.html.

OECD (Organization for Economic Co-operation and Development) (2001). *Understanding the Digital Divide.* Paris: OECD Publications.

OECD (Organization for Economic Co-operation and Development) (2002). *Measuring the Information Economy.* Paris: OECD Publications.

Park, H. W. (2001). "Digital Divide in Korea: Closing and Widening Divide in 1990s." Paper presented at the Symposium on the Digital Divide, the International Association of Mass Communication Research and the International Communication Association, Austin, Texas.

Pew Center for the People and the Press (1995). "Americans Going Online . . . Explosive Growth, Uncertain Destinations—Technology in the American Household." http://people-press.org/reports/display.php3?ReportID=136.

Reddick, A., and C. Boucher (2002). *Tracking the Dual Digital Divide.* Ekos Research Associates.

Soe, Y. (2002). "The Digital Divide: An Analysis of Korea's Internet Diffusion." Washington: Georgetown University, Master's Thesis. Unpublished.

Spooner, T. (2001). "Asian-Americans and the Internet: The Young and the Connected." Washington: Pew Internet and American Life Project. http://www.pewinternet.org/.

UCLA Center for Communication Policy (2003). "The UCLA Internet Report: Surveying the Digital Future Year Three." Los Angeles. http://www.ccp.ucla.edu.

United Nations Development Programme (2001). "Human Development Report, 2001: Making New Technologies Work for Human Development." New York: Oxford University Press.

U.S. National Telecommunications and Information Administration (1995). *Falling through the Net: A Survey of the "Have Nots" in Rural and Urban America.* Washington: U.S. Department of Commerce.

U.S. National Telecommunications and Information Administration (1998). *Falling through the Net II: New Data on the Digital Divide.* Washington: U.S. Department of Commerce.

U.S. National Telecommunications and Information Administration (2000). *Falling through the Net III: Toward Digital Inclusion.* Washington: U.S. Department of Commerce.

U.S. National Telecommunications and Information Administration (2002). *A Nation Online—How Americans Are Expanding Their Use of the Internet.* Washington: U.S. Department of Commerce.

van Eimeren, B., and H. Gerhard (2000). "ARD/ZDF-Online-Studie 2000: Gebrauchswert Entscheidet über Internetnutzung." *Media Perspektiven* 8: 338–349.

van Eimeren, B., H. Gerhard, and B. Frees (2001). "ARD/ZDF-Online-Studie 2001: Internetnutzung Stark Zweckgebunden." *Media Perspektiven, 8,* 382–397.

van Eimeren, B., H. Gerhard, and B. Frees (2002). "ARD/ZDF-Online-Studie 2002 Entwicklung der Online-Nutzung in Deutschland: Mehr Routine, Wengier Entdeckerfreude." *Media Perspektiven 8,* 346–362.

Warschauer, M. (2003). *Technology and Social Inclusion: Rethinking the Digital Divide.* Cambridge, Mass.: MIT Press.

Waters, R. (2003). "Bears Gather as China's Portal Prices Soar." *Financial Times,* June 25.

Welling, S., and H. Kubicek (2000). "Measuring and Bridging the Digital Divide in Germany." Paper presented at the Stepping-Stones Into the Digital World Conference, Bremen, Germany.

Wellman, B. (2001). "Physical Place and Cyberspace: The Rise of Personalized Networks." *International Urban and Regional Research 25(2),* 227–252.

Wolcott, P., L. Press, W. McHenry, S. Goodman, and W. Foster (2001). "A Framework for Accessing the Global Diffusion of the Internet." *Journal of the Association for Information Systems* 2: 1–50.

World Economic Forum (2002). "Annual Report of the Global Digital Divide Initiative." Geneva.

World Internet Project Japan (2002). "Internet Usage Trends in Japan Survey Report." Tokyo: Institute of Socio-Information and Communication Studies, University of Tokyo.

Yun, K., H. Lee, and S.-H. Lim (2002). "The Growth of Broadband Internet Connections in South Korea: Contributing Factors." Stanford, Calif.: Asia/Pacific Research Center.

Appendix

Introductory Address

Rita Colwell

We are very pleased to be able to help bring together a wide range of people to share insights on how IT is/will be transforming almost all enterprises in our society.

This conference is an excellent example of the convergence of disciplines and of NSF's involvement in the research that is transforming all types of enterprises.

This conference is also a good example of cooperation—interagency cooperation, international cooperation, and interdisciplinary cooperation.

In terms of interagency cooperation, we at NSF are very happy to cooperate with the Department of Commerce to put together a conference like this as we have complementary goals and missions. NSF funds the basic research on e-commerce, e-business, e-work and the IT workforce while the Department of Commerce is very interested to know the latest research in these areas so as to apply it for policy purposes.

In terms of international cooperation, we are very fortunate to have representation from the European Commission-Information Society Technologies Program, including one of their senior scientific officers and a number of researchers. We intend to continue this cooperation in 2004 with a similar conference hosted by the Europeans.

In terms of interdisciplinary cooperation, we know now that to understand the implications of IT, we need to bring together social and economic scientists with IT researchers. This conference includes a vast array of disciplinary scientists as well as interdisciplinary researchers all grappling with the various issues related to IT and economic and societal transformation.

NSF was and is involved in generating many of the fundamental ideas underlying information technology—basic computational techniques, data base techniques, software systems to support applications, data mining, etc. Sometimes this research becomes very practical quite quickly—e.g. the Internet, web browsers (NCSA Mosaic), Google (which started on an NSF-funded project).

NSF is also involved in basic support for the social and behavioral sciences, but not just in the Directorate for Social, Behavioral and Economic (SBE) Sciences; this conference is supported out of the Directorate for Computer and Information Science and Engineering (CISE), for example. This is a good example of integration. Some of the most striking and farthest-reaching transformations are taking place in science and engineering now and will continue to over the next decades.

Prof. Dan Atkins, who will be speaking later this morning, has just finished an extensive report to NSF that illustrates this and suggests the unique opportunity facing all of Science and Engineering research and education.

As you begin your conference, I would suggest that the same transformative opportunities exist or will soon exist in all types of enterprises and that it behooves us all to try to learn from each other how to best exploit these opportunities.

Keynote Address

John Marburger

I am delighted to have this opportunity to speak at a conference which in other circumstances might have seemed too academic to excite much interest among practical folk. The very fact that this meeting is sponsored by the U.S. Department of Commerce, and that outstanding representatives from industry, government, and academia are participating, speaks to the widespread recognition that the explosive growth in information technology is a truly revolutionary phenomenon that is transforming our way of life.

Last weekend my wife and I visited Harpers Ferry just up the Potomac River in West Virginia. Even in the bitter cold, Civil War buffs were seeking out the famous battle sites, or John Brown's fort, or the remains of Lock 33 on the C&O Canal across the river. For me, however, the most important thing that happened at Harpers Ferry was John Hall's early 19th century introduction of methods that later came to be called "the American system of manufacture." His factory along the Shenandoah River was the first place in the world where machinery was employed to produce rifles whose parts were so identical they could be interchanged freely among different units. It led to an international market for American manufactured products and precision machinery early in the industrial revolution.

Harpers Ferry manufacturing was destroyed during the Civil War and never recovered. But the ideas it generated influenced industry throughout the world. What has always impressed me about this story is that no special scientific breakthrough made it possible. The machinery was made of wood and iron and was powered by water, all ancient technologies. What made the difference were ideas. As the ideas spread, they transformed how things were made, and at the same time they transformed the things themselves, and the behavior of the people who used the things. John Hall knew at the time that he was doing something important, but he could not have foreseen the impact his ideas would have upon the future. Nor can we foresee how the ideas driving today's information technology will affect the future, but we know something important is happening, and that it will make enormous and permanent changes in our way of life.

Like the use of machinery to increase the precision of manufacturing, the use of electronic circuits to store and process information simply allows us "to find better ways to do the same things" as the author Michael Lewis put it. At some point, however, the improvements make a qualitative change in how we do things, and the economy shifts massively, and the world changes. Having Google on your desktop, as Secretary of State Colin Powell does, changes how you learn about the world, and cuts the time needed to make informed decisions. Being able to search large databases quickly reduces the risk of credit card transactions, the expense of airline reservations, and the time to recognize trends in retailing, crime, or public health. Computing power replaces wind tunnels with workstations, model shops with monitors, experimental trials with simulations. Much of business, especially in a service economy, offers customers a way to cut costs of processes that are intermediate steps on the way to a final product. Advances in computing and communication have removed some of these steps entirely, and redefined the economics of business transactions. Today the information technology industry, which represents only eight percent of all enterprises in our economy, produces 29 percent of U.S. exports.

John Hall's rifle factory at Harpers Ferry was located around the corner from the Federal Armory on the Potomac River side of town. One was a private enterprise, funded through government contracts. The other was publicly owned and operated. It was John Hall, the private entrepreneur, who introduced the novel machinery and transformed manufacturing. The Federal Armory, according to MIT historian Merritt Roe Smith, lagged in implementing the new technology. Today too, the private sector forges ahead of government agencies in using the Internet and related information technologies, despite the fact that government funds supported the research upon which the new technology is based, either directly or through procurements.

President Bush would like to close that gap. He wants to "expand the use of the Internet to empower citizens, allowing them to request customized information from Washington when they need it, not just when Washington wants to give it to them." OMB's e-government czar Mark Forman will speak to this effort in one of tomorrow's panels. And OSTP Associate Director for Technology Richard Russell will speak tomorrow on this administration's policy of investment to keep up the momentum of change in information technology.

Of course the growing complexity of the systems we use in everyday life presents unprecedented challenges. New technologies bring new problems and create new vulnerabilities. Last weekend's "SQL worm" virus demonstrated once again the vulnerability of the Internet to deliberate disruption. The means to prevent and remediate such disruption were readily available in this case, but had not been implemented uniformly by system administrators for the Internet servers that spread the virus. This situation dramatizes how difficult it is to maintain the integrity of systems whose very ownership is distributed, and not only the physical system itself. Many of the challenges that Homeland Security Secretary Tom Ridge must address share this feature of wide distribution and fragmented ownership. Whatever the solution, most of us believe that information technology will play a key role.

Congress has already mandated improved technology for increasing awareness at our borders. The Enhanced Border Security Act requires the establishment of a border entry-exit system with biometric authentication of all aliens entering the United States at any of the 422 Ports of Entry by October 2004. OSTP is working closely with the Office of Homeland Security and the transition team for the new Department to identify and evaluate information technology responsive to this mandate. Air travelers have seen a rapid increase in the use of electronic ticketing and more sophisticated information technology associated with post 9/11 security measures. Geo-positioning devices linked to geographical databases are being used by fleet managers to increase security and improve management of commercial vehicles and rental cars. Nearly all plans to improve first responder capabilities include upgrades in speed and capability of data communication and information processing, and new tools that permit sharing among diverse databases.

Speakers at this conference will consider a wide spectrum of implications that the revolution in information technology holds for our society. On balance, these implications are positive, but they do entail change. Perhaps the greatest changes will be a shrinking of distances and times, so global transactions are accomplished as easily and as inexpensively as local conversations. Globalization, so closely tied with telecommunications and bandwidth improvements, brings unprecedented challenges. It is here that the social sciences have much to offer, and I am delighted that one of today's panels focuses on social transformation.

In meeting these challenges, whether of national or homeland security or maintaining economic competitiveness in a globalizing marketplace, there is a clear role for federal government. Federal funds have supported much of the basic research upon which information technology rests. In a deregulated environment, the private sector is unlikely to invest heavily in long lead time, high risk research. Even John Hall's experiments with high precision machinery were included in the costs of his contracts for supplying breech-loading rifles to the federal government, and it is clear from the records that the procurement officials were aware of this aspect of his work.

President Bush's budget request for Fiscal Year 2003, not yet implemented by Congress, advances federal R&D funding beyond $100 billion for the first time in history. And last year the president signed into law funding increases for science and technology at the National Institutes of Health, the National Science Foundation, the Environmental Protection Agency, and the U.S. Departments of Energy, Agriculture, Interior, Commerce, and Transportation.

The multi-agency Networking and Information Technology R&D program (called NITRD) is one of the nation's priority R&D initiatives. This effort coordinates activities within federal agencies engaged in fundamental research and development in all aspects of large scale and broadband networking, advanced computing, software, and information management technologies. NITRD's aim is to provide the base technologies necessary for the U.S. to maintain its leadership position in the application of information technology to critical national defense and national security needs, as well as scientific research, education, and economic innovation. The program includes a broad range of interdisciplinary technical activities cutting across twelve agencies. In the president's

2003 budget, NITRD increased by 3%, bringing the overall investment to $1.9 billion in this mature, but still critically important area.

Our nation played a significant role in the industrial revolution, and in the subsequent evolutions that have led to the modern information-oriented society. Now we are poised again at the threshold of another vista, barely discernible, but rich with promise. I look forward to working with you and your colleagues on behalf of this administration, not only to exploit, but to lead the transformations that will be necessary to reap the rewards of this revolutionary technology.

Thank you.

Keynote Address

Donald L. Evans

Thank you. Welcome to the Commerce Department—and the second day of our Transforming Enterprise Conference. As Phil [Bond] mentioned, I'm the second engineer to serve as Secretary of Commerce. The first was Herbert Hoover. He went on to be president. I've already told President Bush not to worry! I love the job I have.

It is a real privilege to be able to serve this president—I've known him for a long time . . . he's an extraordinary person and an extraordinary leader. And it's a real privilege to serve this great nation, to be a public servant.

We are all called to serve something greater than ourselves. We also have a responsibility to ensure that each and every man and woman has the opportunity to achieve the American dream. It's a universal dream, really, the dream of peace and prosperity for our families, our communities, our countries.

I want to take a moment here to thank all of you for your contributions to this effort, here in America . . . and around the globe.

Technology is transforming lives. We've gone from a world that used to be connected by airport terminals to a world that is connected by computer terminals. And with the explosion of information technology has come a fundamental shift in productivity. IT-intensive industries contributed to the positive productivity growth during 2001 (1.1%). And for the first three quarters of 2002, productivity growth climbed at an unusually robust pace of about 5%. In addition, IT-intensive industries continue to greatly contribute to slow price growth, tempering overall inflation.

Everywhere I go, I see the difference that technology is making: helping window manufacturers in Paducah be more productive and efficient . . .

giving graduate students at Notre Dame access to all the information on the World Wide Web.

As policymakers, it's important for us to understand the transformations taking place in our society so we can craft information-age policies nationally and internationally. To me, it boils down to government's fundamental role . . . creating the conditions for growth.

When we do this, a spirit of competition takes hold, leading to more innovation . . . which leads to greater productivity . . . which leads to more economic growth . . . which leads to a better quality of life . . . which leads to a world that lives in peace and prosperity.

Clearly, it begins with education. That is the first condition for growth. You cannot have a successful information economy without an educated workforce. And every American child is capable of being part of an information workforce. That's why this president's very first act was the Leave No Child Behind education legislation. An educated workforce is necessary to compete in this global economy.

The second condition is a vibrant entrepreneurial sector . . . with people ready to innovate, take risks and compete. Toward that end, this administration has proposed record amounts of R&D funding to support continued innovation and scientific discovery at our universities and federal labs.

But it's the private sector, entrepreneurs, who move discoveries from the lab to the marketplace. And when they do that they take a risk. That's why the president's growth and jobs package proposes to make equity financing more available to entrepreneurs. The president's plan would eliminate penalties on some productive investments . . . such as the double tax on dividends. And it will triple the tax write-off for equipment purchases from $25,000 to $75,000 . . . an incentive for small businesses run by those entrepreneurs to expand and hire. This also should increase opportunities for industries that produce IT hardware and software.

The president has offered a bold plan to deal with a big challenge. The plan would provide $674 billion in tax relief over 10 years. Nearly $56 billion would come in 2003. Those are pretty modest amounts for a $10-trillion-a-year economy . . . over $100 trillion for ten years. The key is that the money goes where it will do the most good.

We need to keep this economy moving ahead. And we urgently need more job creation. We won't be satisfied until every American who wants a job has a job.

Having just returned from the World Economic Forum in Switzerland, I can tell you the entire global community is looking to America to increase our growth rate. And I can also tell you this: this president is focused on growth . . . and has been from day one . . . from the education bill . . . to the tax cut . . . to the stimulus bill of last year . . . to the growth and jobs package of this year.

And we know that technology is key to a prosperous and productive future.

I think Congress understands this and I hope members in both parties work together to quickly move the president's growth and jobs package forward.

I'm confident that with strong presidential leadership, with spending discipline, with pro-growth policies . . . and with a vibrant and innovative private sector . . . we will expand the economy and help create more jobs. And I'm equally confident that the people in this room and throughout the tech industry will continue to play a critical role.

So, let's get on with it.

Thank you and God bless you.

Contributors

Anthony S. Alvarez
University of Maryland

Uday M. Apte
Southern Methodist University

Daniel E. Atkins
University of Michigan

Erik Brynjolfsson
Massachusetts Institute of Technology

Wenhong Chen
University of Toronto, Canada

Rita Colwell
Director, National Science Foundation

Panagiotis Damaskopoulos
European Institute of Interdisciplinary Research, France

Jason Dedrick
University of California, Irvine

William H. Dutton
Oxford Internet Institute, UK

Donald L. Evans
Secretary of the U.S. Department of Commerce

Dominique Foray
Centre National de la Recherche Scientifique, France

Lorin M. Hitt
University of Pennsylvania

Mun S. Ho
Resources for the Future, Washington, D.C.

Dale W. Jorgenson
Harvard University

Brian Kahin
University of Michigan

Sara Kiesler
Carnegie Mellon University

John Leslie King
University of Michigan

Kenneth L. Kraemer
University of California, Irvine

Boy Lüthje
University of Frankfurt, Germany

Kalle Lyytinen
Case Western Reserve University

John Marburger
Director of the Office of Science and Technology Policy, Executive Office of the President

Richard O. Mason
Southern Methodist University

Anna Nagurney
University of Massachusetts

Ramon O'Callaghan
Tilburg University, The Netherlands

Dirk Pilat
Organisation for Economic Co-operation and Development, France

Paul Resnick
University of Michigan

John P. Robinson
University of Maryland

Kevin J. Stiroh
Federal Reserve Bank of New York

Lee Sproull
New York University

Alladi Venkatesh
University of California, Irvine

Barry Wellman
University of Toronto, Canada

Andrew W. Wyckoff
Organisation for Economic Co-operation and Development, France

John Zysman
University of California, Berkeley

Index